FOLLOWING THE WATER

ENVIRONMENTAL HISTORY AND THE
HYDROLOGICAL CYCLE IN COLONIAL
GIPPSLAND, AUSTRALIA, 1838–1900

FOLLOWING THE WATER

ENVIRONMENTAL HISTORY AND THE
HYDROLOGICAL CYCLE IN COLONIAL
GIPPSLAND, AUSTRALIA, 1838–1900

KYLIE CARMAN-BROWN

In memory of
Mum and Dad

Published by ANU Press
The Australian National University
Acton ACT 2601, Australia
Email: anupress@anu.edu.au

Available to download for free at press.anu.edu.au

ISBN (print): 9781760462840
ISBN (online): 9781760462857

WorldCat (print): 1122806616
WorldCat (online): 1122806567

DOI: 10.22459/FW.2019

This title is published under a Creative Commons Attribution-NonCommercial-NoDerivatives 4.0 International (CC BY-NC-ND 4.0).

The full licence terms are available at creativecommons.org/licenses/by-nc-nd/4.0/legalcode

Cover design and layout by ANU Press

This edition © 2019 ANU Press

Contents

List of maps, figures and tables ix

Acknowledgements xiii

Maps xv

1. Introduction 1

2. Making the circle round: Perceptions of hydrology through time 21

3. The earth's thoughtful lords? Nineteenth-century views of water and nature 39

4. 'Notwithstanding the inclemency of the weather': The role of precipitation in the catchment 85

5. 'Fair streams were palsied in their onward course': The desirability of flowing waters 127

6. 'A useless weight of water': Responding to stagnancy, mud and morasses 167

7. Between 'the water famine and the fire demon': Drying up the catchment 213

8. Mirror, mirror? The reflective catchment 255

Bibliography 263

Index 291

List of maps, figures and tables

Maps

Map 1: Gippsland Lakes catchment area....................................xv
Map 2: East Gippsland locations xvi
Map 3: West Gippsland locations...xvii
Map 4: Topography of central Gippsland xviii
Map 5: Squatting runs, 1842.. xix
Map 6: Squatting runs, 1857.. xx
Map 7: Landmarks of early Sale.. xxi
Map 8: Ecological vegetation classes, 1750 xxii
Map 9: Ecological vegetation classes, 2005 xxiii
Map 10: Punt Lane and Swing Bridge.................................... xxiv

Figures

Figure 1.1: Catchment signage between Cann River and Lakes Entrance .. 3
Figure 1.2: 'Tarrago Creek', near Neerim, Latrobe River catchment,
    Thomas Henry Armstrong Bishop, c. 1901........................... 3
Figure 1.3: Nicholas Caire's image of the Reeves River, actually part
    of the Lakes, c. 1900... 4
Figure 1.4: Cleared hillsides surrounding Walhalla, Thomas Henry
    Armstrong Bishop, c. 1901... 8
Figure 1.5: Flora Gregson's painting of wildflowers from around
    Lakes Entrance, c. 1870... 10
Figure 2.1: Leonardo da Vinci's view of the hydrological cycle.............28

Figure 2.2: Schematic representation of the hydrological cycle 33

Figure 3.1: Church at Sth Yinnar. A typical example of the small community-built churches that dotted the landscape. 44

Figure 3.2: Tambo Valley, illustrating the gentle slopes that were so sought after 44

Figure 3.3: A bark hut in the Neerim area, Archibald J. Campbell, 1877 53

Figure 3.4: An example of trained hydrology, channels constructed in Walhalla 56

Figure 3.5: The now thickly overgrown Mosquito Creek in the Moe Drainage Scheme. 63

Figure 3.6: A fragment of grassy woodland, photographed along the road between Glengarry and Cowarr. 68

Figure 3.7: Depictions of Maffra from 1882. These montage scenes showing progress were common. 69

Figure 3.8: Eugene von Guerard's painting of Angus McMillan's Bushy Park, 1861, depicting a small wetland in the foreground with cattle drinking 71

Figure 3.9: Gregory's Cottage, Cooper's Creek, Walhalla Road, c. 1885 74

Figure 3.10: The A1 Mine at Gaffney's Creek, Thomas Henry Armstrong Bishop, c. 1901. 76

Figure 3.11: New Public Hall, Sale, FA Sleap engraver, 1879 76

Figure 4.3: Flood debris piled up against the railway bridge to Walhalla across the Thomson River, 2008. 105

Figure 4.4: Flats at the mouth of the Latrobe River, Thomas Henry Armstrong Bishop, n.d. (c. 1894–1909). 105

Figure 4.5: A Gipps Land track after rain, *The Australasian Sketcher with Pen and Pencil*, 1882. 106

Figure 4.6: Boggy Creek, J Williamson engraver, 1865 109

Figure 4.7: The floodplain between the Swing Bridge and Longford, November 2011. 115

Figure 5.3: Detail of *Acacia mearnsii*. 141

Figure 5.4: Livingstone Creek, below Omeo, illustrating the steepness of banks 144

Figure 5.5: Nicholas Caire's image of the Dargo River, n.d. 145

LIST OF MAPS, FIGURES AND TABLES

Figure 5.6: A bend in the Avon River 145

Figure 5.7: Crossing place at Wombat Creek, Thomas Henry
    Armstrong Bishop, n.d. Buckley recorded a log crossing
    on 25 February 1849 146

Figure 5.8: The Sale Swing Bridge, showing the central piers around
    which the deck pivots..153

Figure 5.9: Looking upstream from the Sale Swing Bridge at the new,
    and substantially higher, Latrobe River bridge.....................153

Figure 5.10: Oriental claims area, near Omeo 155

Figure 5.11: Bank erosion on the Avon River, 1872 157

Figure 5.12: Washaway at bridge over Avon River near Bushy Park,
    Victorian Railways, 1893...157

Figure 5.13: Flora Gregson's watercolour of the *Lady of the Lake* trying
    to pull a stranded vessel off the bar at Lakes Entrance, 1878......159

Figure 5.14: The permanent entrance from Jemmy's Point...................163

Figure 6.1: Sale Common, now a wildlife refuge...........................174

Figure 6.2: Shooting swans by moonlight, first published in
    The Australasian Sketcher with Pen and Pencil, 22 May 1880 176

Figure 6.3: Macalister Swamp, on the outskirts of Maffra.................177

Figure 6.4: A part of MacLeod's Morass, Bairnsdale. The low
    featureless expanse of reeds made perfect hiding grounds,
    but would exacerbate the troubles of the genuinely lost............183

Figure 6.5: 'Cleared' land near Mirboo, Thomas Henry Armstrong
    Bishop, n.d. (c. 1894 – c. 1909) 184

Figure 6.6: Selections in the parishes of Warragul, Moe, Yarragon
    and Narracan in 1883. Moe Swamp is not delineated except
    on the south by the main road.....................................200

Figure 6.7: Parish map of Yarragon, 1908 201

Figure 6.8: The former Moe Swamp, now the Moe Flats 203

Figure 6.9: Township of Warragul, by Nicholas Caire, 1886 209

Figure 7.1: AW Howitt's hop kiln, Frederick Cornell, 1872. Hops were
    dried with the use of specially constructed hop kilns 222

Figure 7.2: Bed of the Tambo River, Thomas Henry Armstrong
    Bishop, n.d. 229

Figure 7.3: The homestead saved, An incident of the Great Gippsland Fire of 1898, JA Turner, published 1908 234

Figure 7.4: Fire near Holey Plains State Park, close to Traralgon in the Latrobe Valley, July 2006..................................... 235

Figure 7.5: Cowarr Butter Factory 239

Figure 7.6: Senescent pine windbreak, Churchill area, November 2011... 244

Figure 7.7: Carl Walter's image of a prospector on Freestone Creek, 1867 252

Figure 8.1: Swans feeding at dusk, Lakes Entrance, November 2011...... 261

Tables

Table 4.1: Mean monthly and annual rainfall in mm across measuring stations in the GLC 91

Table 4.2: Extracted from *Blackie's Literary and Commercial Almanac*, 1848 99

Table 5.1: Historic uses of rivers in Europe................................ 130

Table 5.2: Table of water supply activities 137

Acknowledgements

This book is based on my PhD thesis. A thesis, and then turning that thesis into a book, is a lengthy and collaborative process, and many people have assisted me along me the way.

My supervisory panel, John Dargavel, Nicholas Brown, Barry Higman and Sara Beavis, deserve the academic version of a sainthood for getting me through when my life was falling apart. The strengths of this work reflect their patient and loving guidance, and the errors reflect my inability to translate their good advice into words on a page. The university was also kind enough to grant me many extensions so that I could go home to Perth to be with my dying mother, and then support my father as he adjusted to life without Mum. Sadly, neither of them lived long enough to know I had passed.

I would also like to thank Tom Griffiths and Libby Robin who accepted my application to participate in the 2006 Environmental History Postgraduate Workshop, as well as all the staff at the Fenner School and the School of History who were always willing to help. The College of Arts and Social Sciences were kind enough to award me their publishing prize, a true boon. I thank Geoff Hunt and Morgan Burgess for their assistance in editing the revised manuscript and also Christine Huber for her kindness and additional support during the publication process. ANU Press was a dream to work with and I thank Emily Hazlewood and her team for putting together a beautiful-looking book.

I'm grateful to Meredith Fletcher for her assistance in navigating the collection at the Centre for Gippsland Studies. Theses cannot get written without all the behind-the-scenes librarians, database builders, indexers and stack retrievers at all the libraries I consulted, and I thank them for their anonymous work.

Nor can theses get written without a substantial support crew in the rest of life. Seven deaths in five years would test anyone, and it is a testament to the cheering, cajoling and hornswoggling abilities of my support crew that I have managed to complete this and finally get published! I am totally in your karmic debt. You know who you are.

Maps

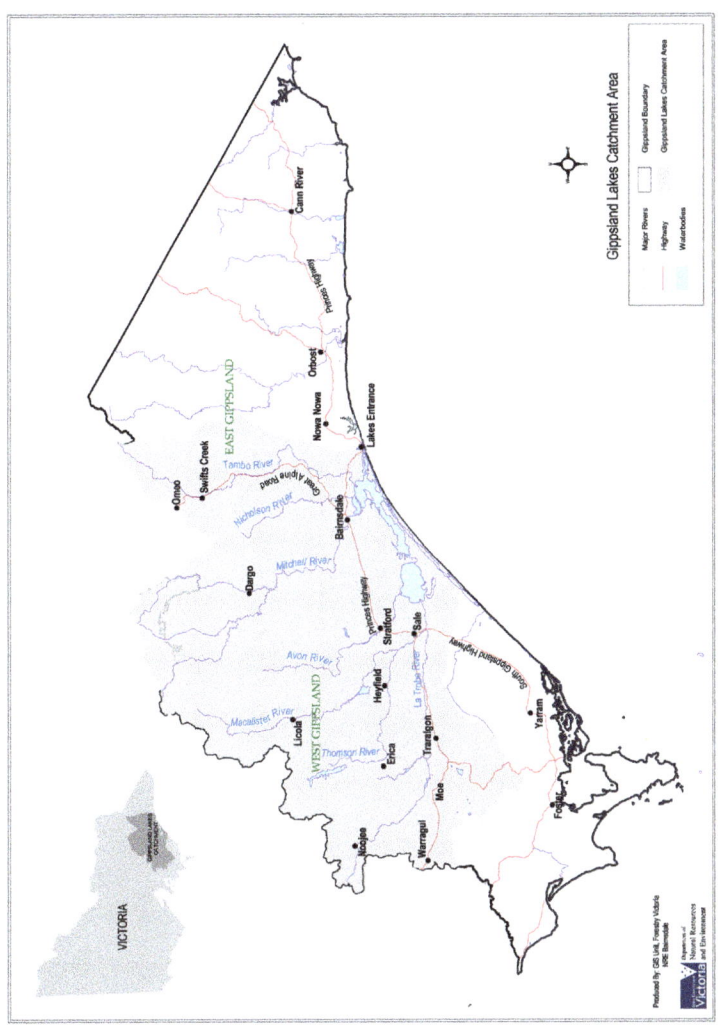

Map 1: Gippsland Lakes catchment area.
Source: State of Victoria, Department of Natural Resources and Environment.

FOLLOWING THE WATER

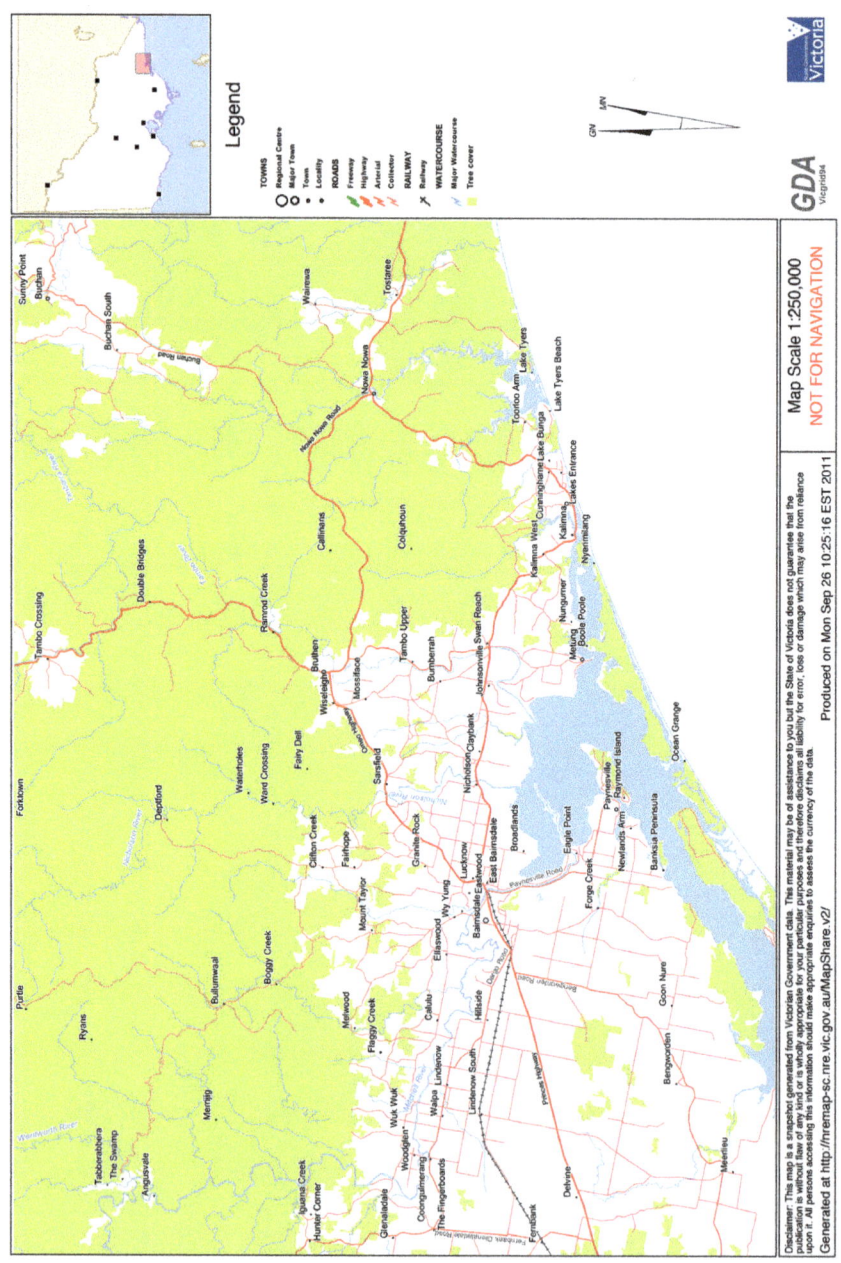

Map 2: East Gippsland locations.
Source: State of Victoria, Department of Sustainability and Environment.

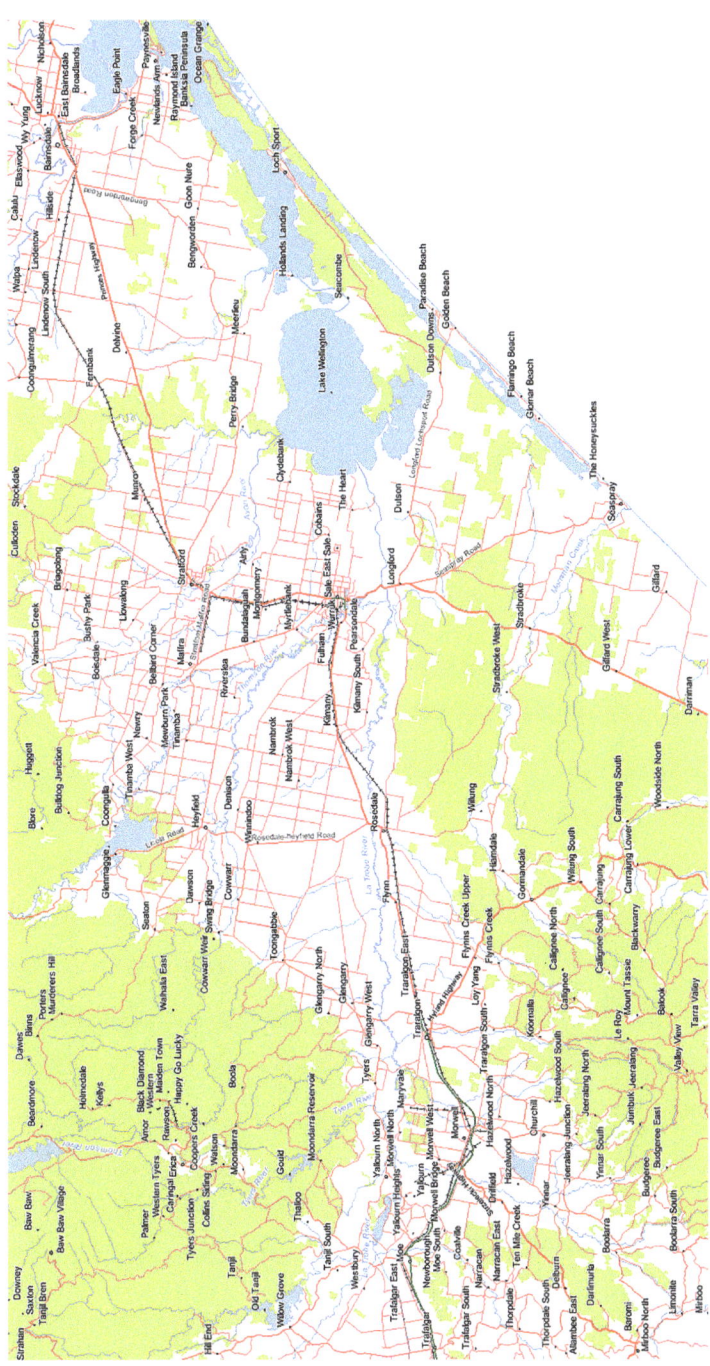

Map 3: West Gippsland locations.
Source: State of Victoria, Department of Sustainability and Environment.

FOLLOWING THE WATER

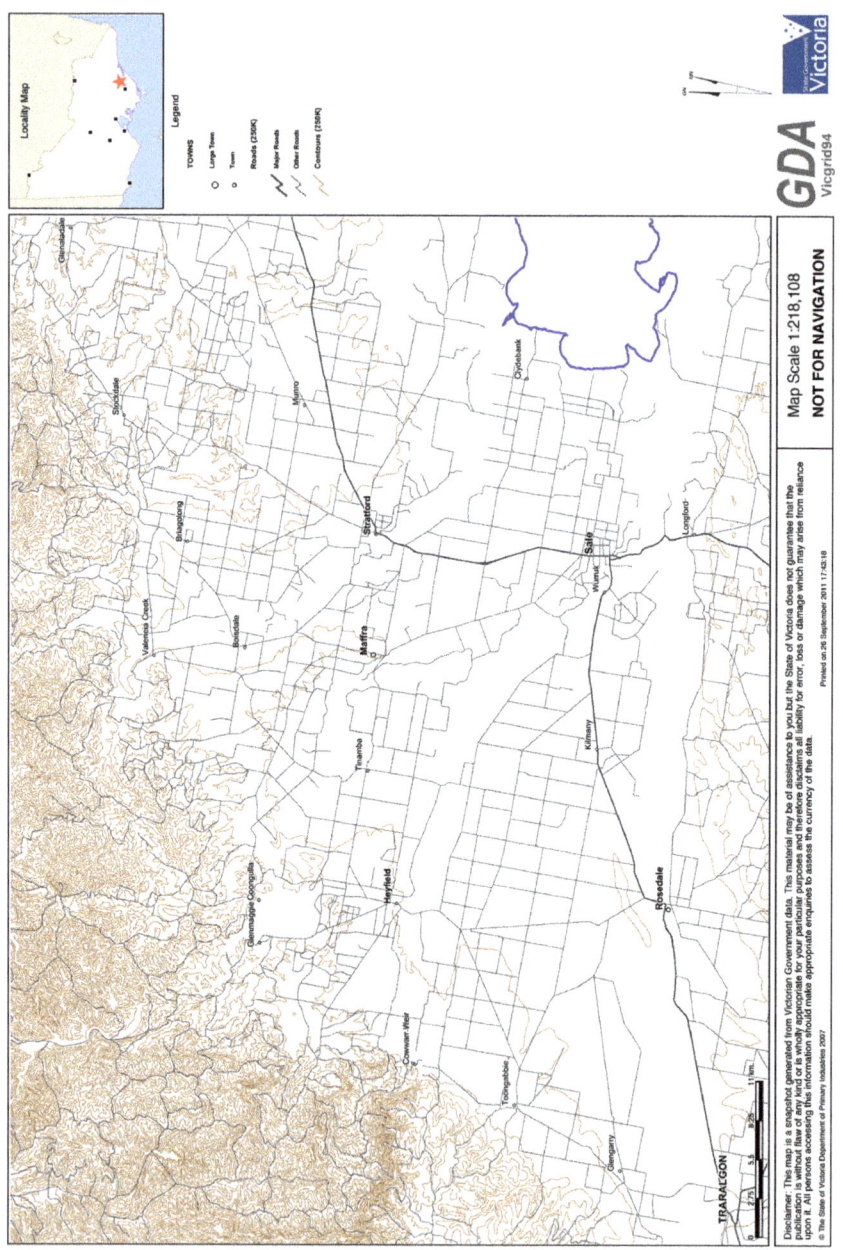

Map 4: **Topography of central Gippsland.**
Source: State of Victoria, Department of Primary Industries.

MAPS

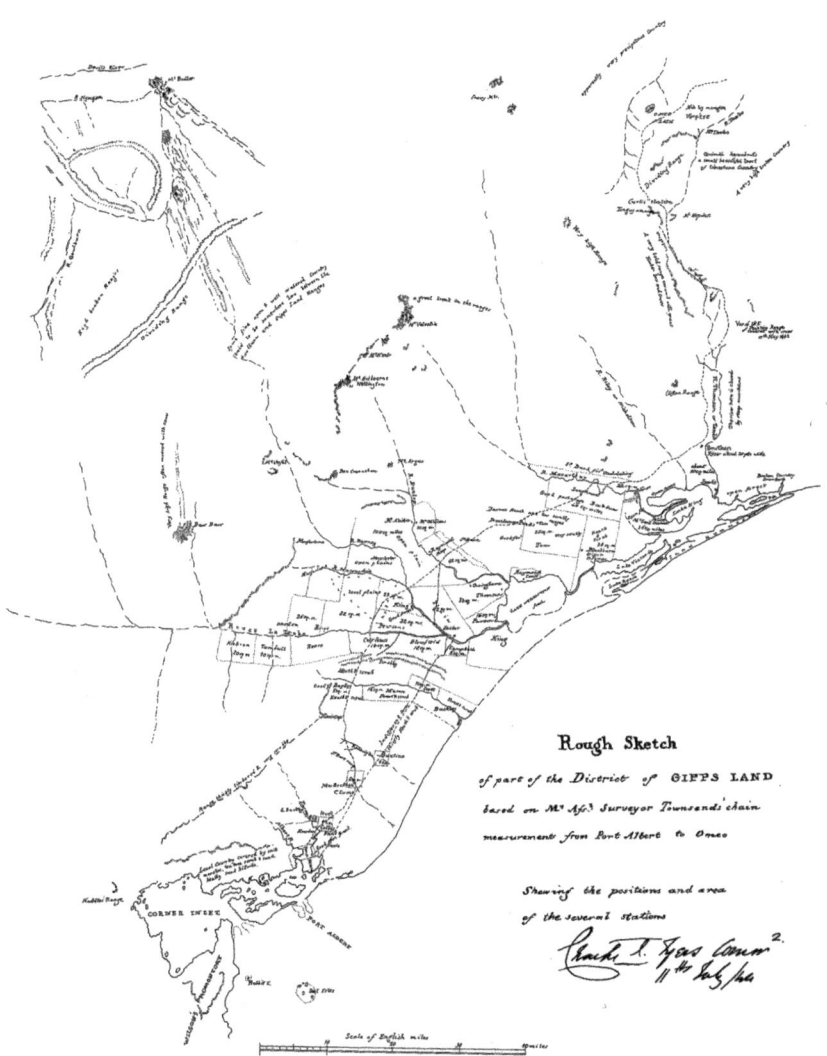

Map 5: Squatting runs, 1842.
Source: *Gippsland Heritage Journal*, vol. 2, no. 2, 1987.

FOLLOWING THE WATER

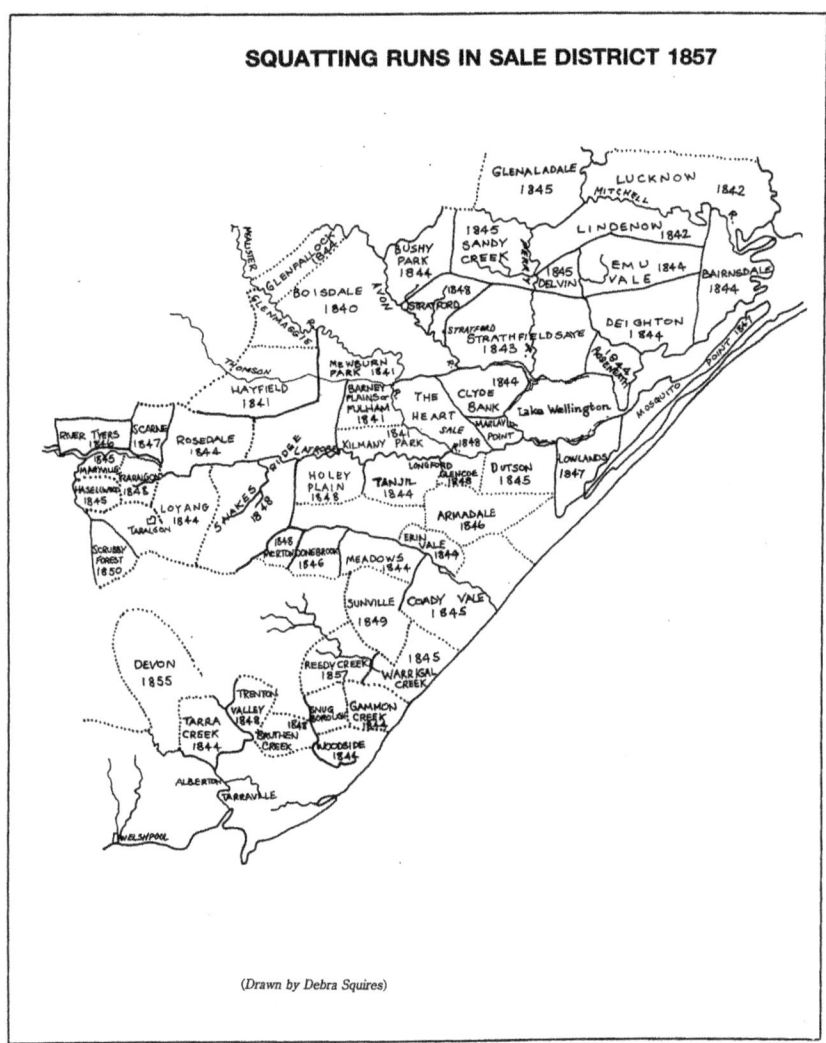

Map 6: Squatting runs, 1857.
Source: Peter Synan, *Gippsland's lucky city: A history of Sale*, City of Sale, Sale, 1994.

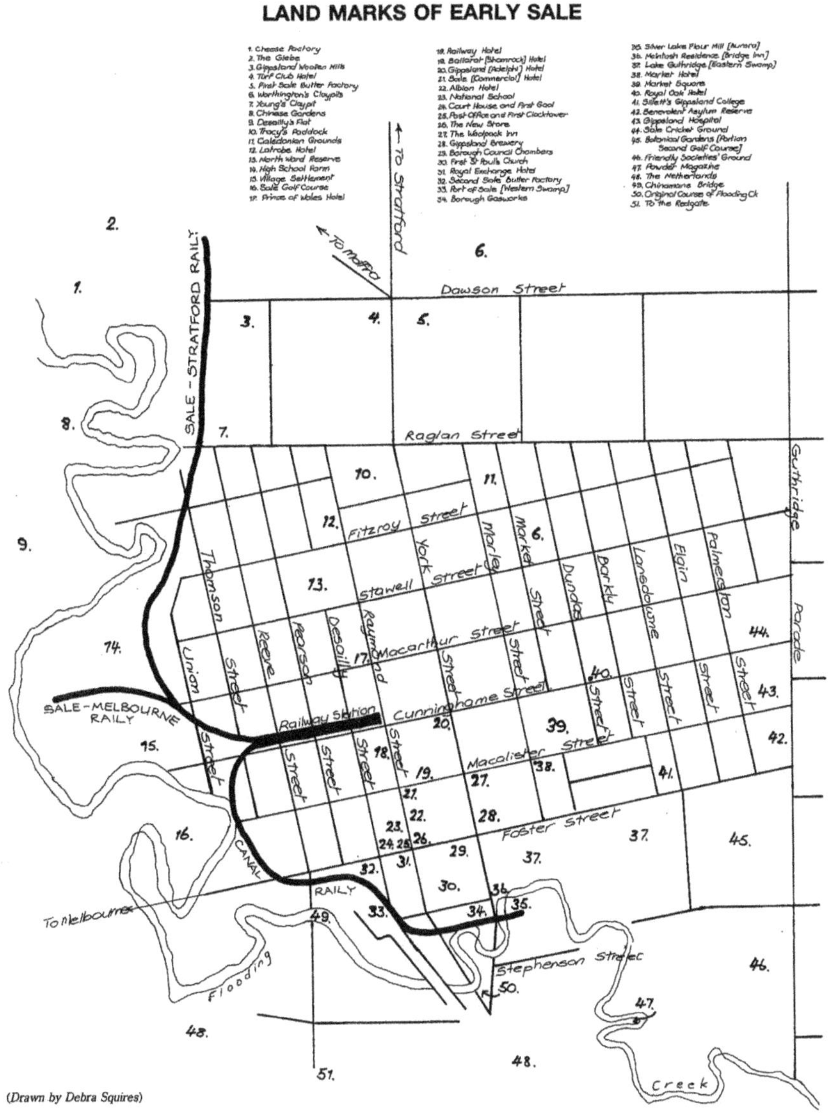
LAND MARKS OF EARLY SALE
1. Cheese Factory
2. The Glebe
3. Gippsland Woollen Mills
4. Turf Club Hotel
5. First Sale Butter Factory
6. Worthington's Claypits
7. Young's Claypit
8. Chinese Gardens
9. Desailly's Flat
10. Tracy's Paddock
11. Caledonian Grounds
12. Latrobe Hotel
13. North Ward Reserve
14. High School Farm
15. Village Settlement
16. Sale Golf Course
17. Prince of Wales Hotel
18. Railway Hotel
19. Ballarat [Shamrock] Hotel
20. Gippsland [Adelphi] Hotel
21. Sale [Commercial] Hotel
22. Albion Hotel
23. National School
24. Court House and First Gaol
25. Post Office and First Clocktower
26. The New Store
27. The Woolpack Inn
28. Gippsland Brewery
29. Borough Council Chambers
30. First St Paul's Church
31. Royal Exchange Hotel
32. Second Sale Butter Factory
33. Port of Sale [Western Swamp]
34. Borough Gasworks
35. Silver Lake Flour Mill [Aurora]
36. McIntosh Residence [Bridge Inn]
37. Lake Guthridge [Eastern Swamp]
38. Market Hotel
39. Market Square
40. Royal Oak Hotel
41. Sillett's Gippsland College
42. Benevolent Asylum Reserve
43. Gippsland Hospital
44. Sale Cricket Ground
45. Botanical Gardens [Portion Second Golf Course]
46. Friendly Societies' Ground
47. Powder Magazine
48. The Netherlands
49. Chinamans Bridge
50. Original Course of Flooding Ck
51. To the Redgate
To Stratford
To Maffra
SALE - STRATFORD RAILY
Dawson Street
Raglan Street
Fitzroy Street
Stawell Street
Macarthur Street
Cunninghame Street
Macalister Street
Foster Street
Stephenson Street
Guthridge Parade
Thomson Street
Union Street
Reeve Street
Pearson Street
Desailly Street
Raymond Street
York Street
Marley Street
Market Street
Dundas Street
Barkly Street
Lansdowne Street
Elgin Street
Palmerston Street
Railway Station
SALE-MELBOURNE RAILY
CANAL
RAILY
To Melbourne
Flooding Creek
(Drawn by Debra Squires)

Map 7: Landmarks of early Sale.

Source: Peter Synan, *Gippsland's lucky city: A history of Sale*, City of Sale, Sale, 1994.

FOLLOWING THE WATER

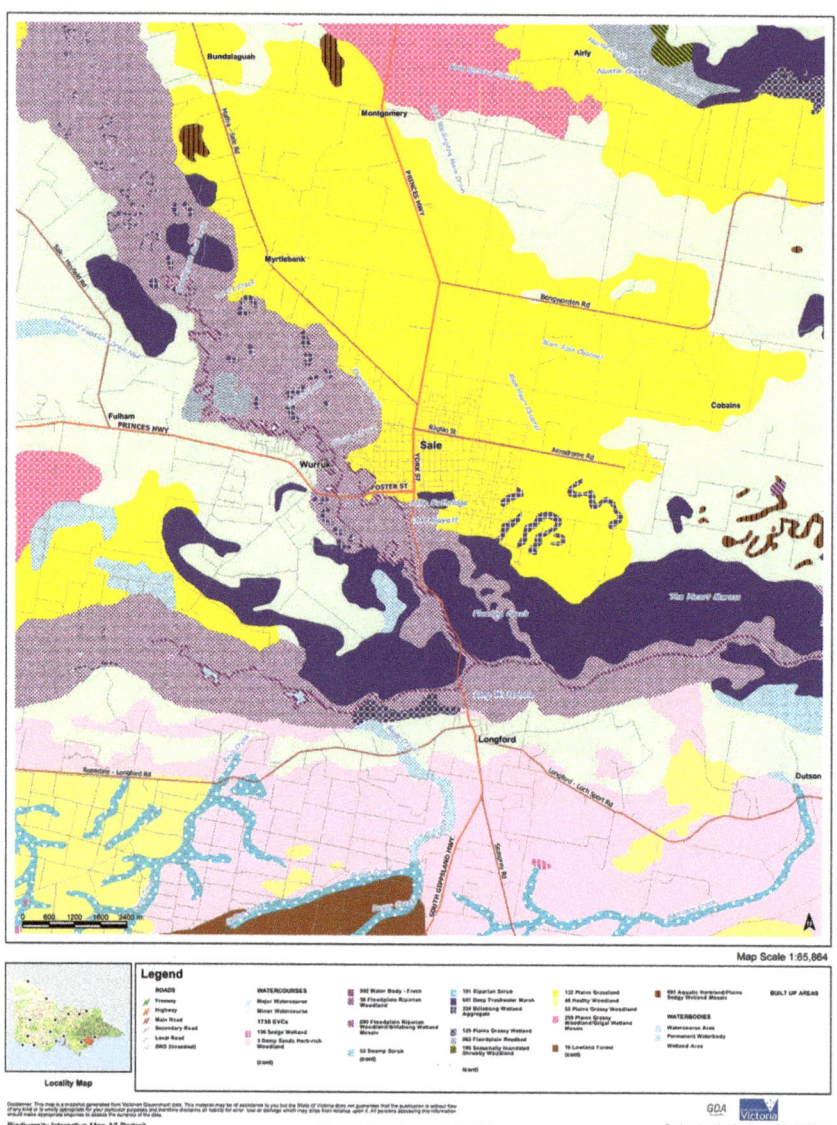

Map 8: Ecological vegetation classes, 1750.
Source: State of Victoria, Department of Sustainability and Environment.

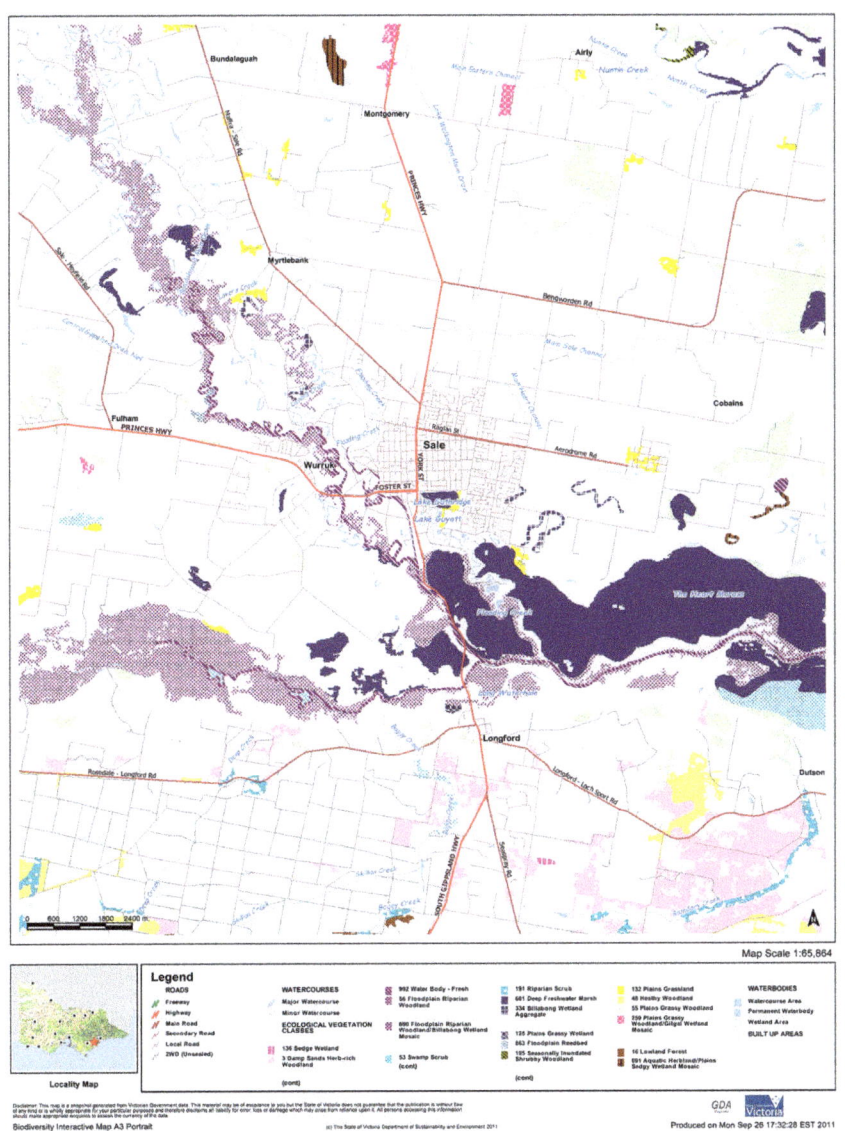

Map 9: Ecological vegetation classes, 2005.
Source: State of Victoria, Department of Sustainability and Environment.

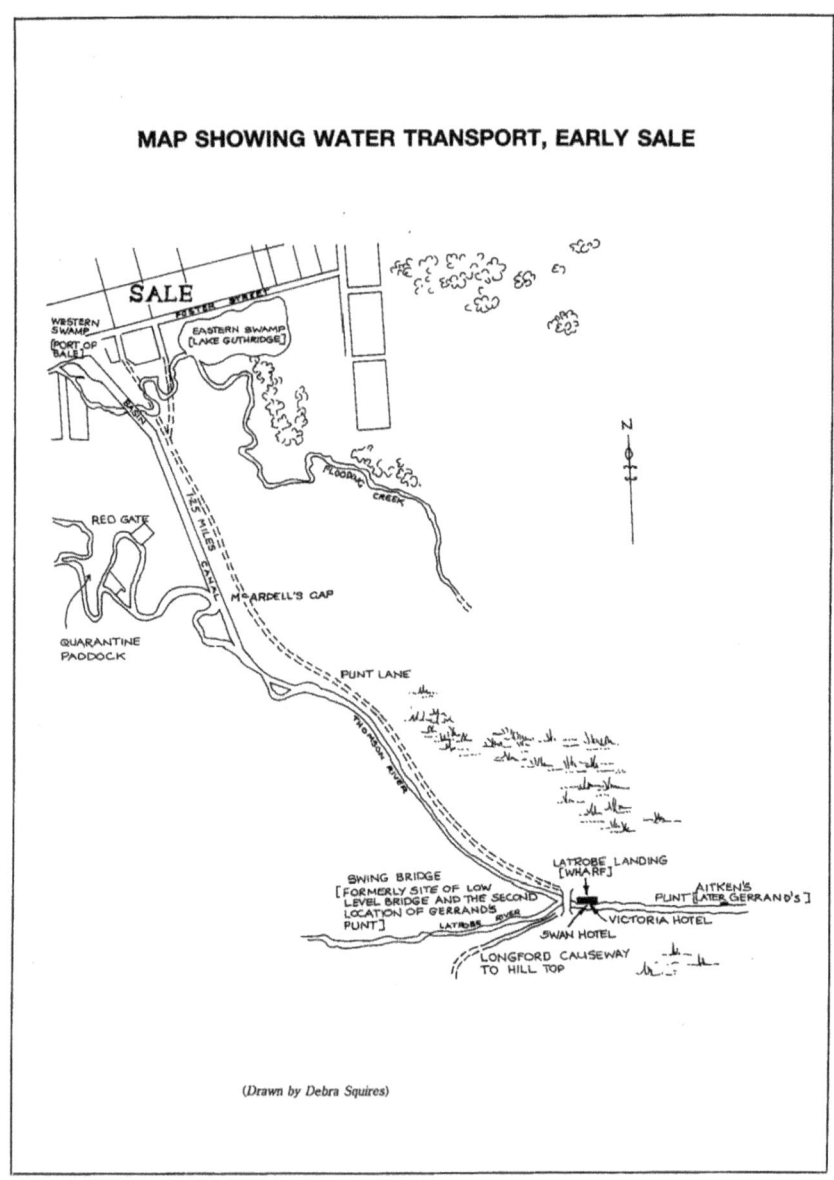

Map 10: Punt Lane and Swing Bridge.
Source: Peter Synan, *Gippsland's lucky city: A history of Sale*, City of Sale, Sale, 1994.

CHAPTER 1
Introduction

> I think that perhaps the ultimate mystery is not that there are no clear impenetrable boundaries in the universe but that we live as if there are.
>
> Barbara Hurd[1]

In the early 1990s, the Gippsland Lakes were in the worst ecological condition they had ever experienced, with massive algal blooms and ongoing fish deaths. At the same time, the $200 million Biosphere II project, in the Santa Catalina mountains near Tucson, folded in ignominy. It was the best known and largest experiment in artificial life support ever undertaken in human history.[2] In 1991 eight men and women entered a sealed dome, which included an ocean, savannah, rainforest, marsh and desert, and livestock such as pygmy goats. They intended to try and live independently of the surrounding biosphere for two years. The survival of the so-called Biospherians was contingent on the fresh oxygen pumped into the sealed dome.

What the failure of Biosphere II conclusively demonstrated was human inability to replicate the sophistication and interconnectedness of ecological life support processes. It is not entirely fanciful to begin a book about the colonial settlement of Australia with the image of asphyxiating Biospherians, because both are lessons in how little modern, white Europeans and their offspring have understood about the multitude of ecological processes that sustain our daily life.

---

1   B Hurd, *Stirring the mud: On swamps, bogs and human imagination*, Beacon Press, Boston, 2001, p. 77.
2   D Day, 'Beyond the biodome', The Space Review, 17 January 2005, www.thespacereview.com/article/305/1, accessed 3 January 2011.

This book is about the importance of ecological processes to sustaining life on a daily basis. It is a case study of how one group of people in a particular place perceived one process, the hydrological cycle, in the nineteenth century. It examines the relationship between the hydrological cycle and the nineteenth-century white emigrant colonisers of the Gippsland Lakes catchment in south-eastern Australia.

The case study area is bounded to the north by the Victorian Alps and to the south by Bass Strait (see Maps 1, 2 and 3). Its eastern boundary is the edge of the Tambo River catchment, which includes Lakes Entrance, and Omeo to the north. To the west, the catchment takes in the Latrobe Valley, past Warragul and Noojee up to the Dandenong Ranges. This topographical isolation from Melbourne was a defining characteristic of the area's history.

The catchment is made up of a number of smaller catchments, which fall into three large connected lakes. The catchment is drained by seven rivers, although, globally speaking, Gippsland's rivers are mere streamlets. At the eastern edge of the catchment, the Tambo River rises above Omeo and discharges into Lake King. The next and smallest river is the Nicholson, which falls out between the Tambo and Bairnsdale. The Mitchell River, with its internationally significant silt jetties, discharges at Bairnsdale, a major town. The Avon is next westward, and the major town associated with it is Stratford. The most westerly part of the catchment is drained by a combination of three rivers: the Macalister, the Thomson and the Latrobe. The former two join the latter just above Sale before their combined waters flow into Lake Wellington. The outfall of the whole system is at Lakes Entrance, a town that grew up once the entrance was made permanent. These lakes make up the largest navigable inland waterbody in Australia.

Sale and Bairnsdale were the major urban settlements in the nineteenth century and intense rivalry characterised their relationship. Each of the rivers supported settlements, with a number of mining towns being established in the Alps – for example, Walhalla and Dargo.

Figure 1.1: Catchment signage between Cann River and Lakes Entrance.
Source: Author.

Figure 1.2: 'Tarrago Creek', near Neerim, Latrobe River catchment, Thomas Henry Armstrong Bishop, c. 1901.
Source: Pictures Collection, State Library Victoria, Accession no. H36688.

Concern for the health of the lakes was expressed in the 1970s, but reached a critical point in the 1990s because of a series of major algal blooms. Algal blooms are caused by an excessive amount of nutrients that allow bacteria to multiply at the expense of other species. They are frequently toxic. The 1998 Gippsland Lakes Environmental Audit concluded that 'despite sporadic but generally high quality research and numerous management plans, no real improvement in health has been achieved in the lakes'.[3] In addition to the algal blooms, species loss, shoreline erosion and sedimentation were all beginning to claim public attention, indicative of a greater contemporary appreciation of the role of ecological processes.[4] This book situates the perception and actions in regard to the hydrological cycle of the region's settlers within their own terms, and simultaneously offers a reflection on their legacy.

Figure 1.3: Nicholas Caire's image of the Reeves River, actually part of the Lakes, c. 1900.
Source: National Library of Australia, Accession no. 2593652.

---

3   G Harris, G Batley, I Webster, R Molloy & D Fox, *Gippsland Lakes environmental audit: Review of water quality and status of aquatic ecosystems of the Gippsland Lakes*, prepared for the Gippsland Coastal Board by CSIRO Environmental Projects Office, Melbourne, October 1998, p. 2.
4   P Synan, *Highways of water: How shipping on the Lakes shaped Gippsland*, Landmark Press, Drouin, 1989, p. 195.

1. INTRODUCTION

The *Encarta Concise English Dictionary* provides five definitions of the word 'process', of which two are the most relevant. The first definition is 'a series of actions directed towards a particular aim', while the second is 'a series of natural occurrences that produce change or development'. The first implies human agency, while the second encompasses humans as part of nature and includes ecological processes. Aren't ecological processes what environmental historians write about anyway? The answer is a qualified no. My reading of the field suggests that ecological processes remain subordinate to better known concepts such as landscape, space and place.[5] This tends to reflect a preference for what is easy to perceive, such as a river, lake or a mountain range. The immediacy of the seen, the felt and the observed helps to explain the preponderance of river and wetland histories in the discipline.[6] There has also been considerable research on problems

---

5   For example, P Hubbard, R Kitchin & G Valentin, *Key thinkers in space and place*, Sage Publications, London, 2004.

6   A selection of surface water histories large and small: for the US see DJ Pisani, 'Beyond the hundredth meridian: Nationalising the history of water in United States', *Environmental History*, vol. 5, no. 4, October 2000, pp. 466–82. doi.org/10.2307/3985582; J Sellye, *Beautiful machine: Rivers and the republican plan*, Oxford University Press, New York, 1991; M Reisner, *Cadillac desert: The American West and its disappearing water*, Viking, New York, 1986; C Sheriff, *The artificial river: The Erie Canal and the paradox of progress, 1817–1862*, Hill and Wang, New York, 1996; B Black, 'Oil Creek as industrial apparatus: Re-creating the industrial process through the landscape of Pennsylvania's oil boom', *Environmental History*, vol. 3, no. 2, April 1998, pp. 210–28. doi.org/10.2307/3985380; CF Meindl, 'Past perceptions of America's great wetland, Florida's Everglades in the early twentieth century', *Environmental History*, vol. 5, no. 3, July 2000, p. 378. doi.org/10.2307/3985482; WD Solecki, J Long & CC Harwell, 'Human–environment interactions in South's Florida's Everglades region: Systems of ecological degradation and restoration', *Urban Ecosystems*, vol. 3, nos 3–4, 1999, pp. 305–43. doi.org/10.1023/A:1009560702266; R White, *The organic machine: The remaking of the Columbia River*, Hill and Wang, New York, 1995; D Worster, *Rivers of empire: Water, aridity, and the growth of the American West*, Pantheon Books, New York, 1985; A Vilesis, *Discovering the unknown landscape: A history of America's wetlands*, Island Press, Washington DC, 1997; Hugh Prince, *Wetlands of the American Midwest: A historical geography of changing attitudes*, University of Chicago Press, Chicago, 1997. doi.org/10.7208/chicago/9780226682808.001.0001.

For Europe, see P van dam, 'Sinking peat bogs: Environmental change in Holland 1350–1550', *Environmental History*, vol. 6, no. 1, 2001, pp. 32–46; M Williams, *The draining of the Somerset Levels*, Cambridge University Press, Cambridge, 1970; F Willmoth, 'Dugdale's history of imbanking and drayning: A royalist antiquarian in the 1630s', *Historical Research*, vol. 71, no. 176, 1998, pp. 281–302; S Halliday, *Water: A turbulent history*, Sutton Publishing, Phoenix Mill, Gloucestershire, 2004; S Haslam, *The historic river: Rivers and culture down the ages*, Cobden of Cambridge Press, Cambridge, 1991; T Glick, *Irrigation and Society in Medieval Valencia*, Belknap Press of Harvard University Press, Cambridge, MA, 1970. doi.org/10.4159/harvard.9780674281806; M Dobson, '"Marsh fever" – The geography of malaria in England', *Journal of Historical Geography*, vol. 6, no. 4, 1980, pp. 357–89. doi.org/10.1016/0305-7488(80)90145-0; HC Darby, *The changing fenland*, Cambridge University Press, Cambridge, 1983; M Cioc, *The Rhine: An eco-biography*, University of Washington Press, Seattle, 2002; S Ciriacono, *Building on water: Venice, Holland and the construction of the European landscape in early modern times*, Berghahn Books, New York, 2006; R Garcier, 'The placing of matter: Industrial water pollution and the construction of social order in nineteenth-century France', *Journal of Historical Geography*, vol. 36, no. 2, 2010, pp. 132–42. doi.org/10.1016/j.jhg.2009.09.003.

caused by pollution or by over-exploitation of resources, issues that produce tangibly perceived and understood effects. Erosion caused by deforestation, for example, or polluted air, can create critical public health problems.[7] In contrast, ecological processes can often operate more subtly, with the concomitant tendency to fade from human attention. Some are invisible, for example, the role of soil bacteria in aerating soil. Yet, ecological processes create the places and landscapes that we fall in love with, even though that love does not tend to extend to the process itself.

As such, ecological processes are often hard to teach as Karterakis et al. identify in their article about teaching the hydrological cycle. They describe their problem as the difference between knowledge embodied in experience and knowledge separate from experience.[8] Most people understand rainfall, in its infinitely varying permutations of duration and strength, because they experience it with their body. It has ramifications: respite from ferrying the children to sport, but the washing remains damp. Depending on your mood and the meteorological conditions prior to the rain, you may feel alternately refreshed or oppressed. In contrast, the movement of a water table is not something that can be learned through sensory experience. This can only be understood with specific training and instrumentation.

Until recent years, much academic scholarship privileged the latter, calling it 'objective' and worthy of study, and downplayed the everyday, embodied sensual experience as being 'subjective'. This objective/subjective split is not helpful to understanding how we comprehend and interact with the world around us.[9] Ecological processes, particularly the hydrological cycle, are particularly good at flouting such boundaries. Strang wrote:

For Australia, see J Tibby, 'Explaining lake and catchment change using sediment derived and written histories: An Australian perspective', *Science of the Total Environment*, vol. 310, nos 1–3, 1 2003, pp. 61–71. doi.org/10.1016/S0048-9697(02)00623-X; M Cathcart, *The water dreamers: The remarkable history of our dry continent*, Text Publishing, Melbourne, 2009; L McLoughlin, *The Middle Lane Cove River: A history and a future*, Centre for Environmental and Urban Studies, Macquarie University, North Ryde, 1985.

7   For example, M Williams, *Deforesting the earth, from prehistory to global crisis*, University of Chicago Press, Chicago, 2003; P Brimblecombe, *The big smoke: A history of air pollution in London since medieval times*, Methuen, London, 1987.

8   SM Karterakis, BW Karney, B Singh & A Guergachi, 'The hydrologic cycle: A complex concept with continuing pedagogical implications', *Water Science and Technology: Water Supply*, vol. 7, no. 1, 2007, p. 29. doi.org/10.2166/ws.2007.003.

9   For a recent Australian example of how subjective and objective are being challenged, see F Allon & Z Sofoulis, 'Everyday water: Cultures in transition', *Australian Geographer*, vol. 37, no. 1, 2006, pp. 45–55. doi.org/10.1080/00049180500511962.

> As the substance that is literally essential to all living organisms, water is experienced and embodied both physically and culturally. The meanings encoded in it are not imposed at a distance, but emerge from an intimate interaction involving ingestion and expulsion, contact and immersion. Engagement with water is the perfect example of a recursive relationship in which nature and culture literally flow into each other.[10]

It is first and only through the everyday embodied relationship with water that people begin to experience the work of ecological processes. The combination of both kinds of knowledge has the potential to lead people to new relationships with nature that are founded on a connective ethos.

It is also true that while we experience rain, for example, very few pay sustained attention to it. We know its impact but we don't, generally, have 'a relationship' with it. This is true for most ecological processes that support our lives. They remain in the backdrop until they are found wanting in some way, or depart from usual expectations. The processes that make up the places where we live remain in the background, a little like Cinderella.

Eric Hobsbawm remarked:

> What is officially defined as the 'past' clearly is and must be a particular selection from the infinity of what is remembered or what is capable of being remembered. How great the scope of this formalized social past is in any given society naturally depends on circumstances. But it will always have interstices, that is, matter which form no part of the system of conscious history into which men [sic] incorporate, in one way or another, what they consider about their society.[11]

Ecological processes are the perfect illustration of such an interstice in the histories of modernised, technologically dependent people.[12] Hobsbawn (by no means an environmental historian) is describing a 'fact' of historical enquiry, which I do not dispute. However, I use his comment to point out that forgetting our dependence on ecological processes is a glaring issue in historical practice.

---

10 V Strang, *The meaning of water*, Berg Publishing, Oxford, 2004, pp. 4–5.
11 E Hobsbawm, *On History*, Abacus Books, London, 2002, p. 14.
12 Sofoulis and Allon provide an example of a modern approach to demand management that focuses on this level. Their 'approach underscores the importance of investigating the ordinary, unspectacular dimensions of daily life and scrutinising those rituals of water use that have become, to a great extent, routine, habitual and, therefore, *inconspicuous* practices of consumption'. Emphasis in original. Allon & Sofoulis, 'Everyday water, cultures in transition', p. 47.

Figure 1.4: Cleared hillsides surrounding Walhalla, Thomas Henry Armstrong Bishop, c. 1901.
Source: Pictures Collection, State Library Victoria, Accession no. H36688.

This work pays as much attention to the processes as it does to the place. This means a detailed evaluation of four key phases of the hydrological cycle – precipitation, flow, storage and evaporation – combined with an unpacking of settlers' cultural baggage around nature. This approach to writing environmental history allows a more conscious reflection on human knowledge and understanding of the ecological processes upon which our lives depend. As 80 per cent water ourselves, we cannot escape our connection to the hydrological cycle.

A process-focused approach complements existing environmental history research, because every person, organisation, group or party has, at some level through their sensory experience, an understanding of how the natural world works and how they relate to it. Process-based environmental histories can be written regardless of the amount of formal or 'objective' learning anyone has. Lack of 'objective' knowledge is no barrier. This approach enables a more sympathetic approach to understanding the actions of settlers, because we are not judging them by the standards of a scientific discipline that didn't exist in their day.

There are many ecological processes that would have made good candidates for this study. The hydrological cycle, or the movement of water at all scales, was the preferred choice because of its mutability in structure and in time, its ubiquity and because of our complete dependence on it. Besides that, I think it is always best to write about something you love. Of those three characteristics, the ubiquity of water has come to inform this work's approach, while aspects of the hydrological cycle itself have informed its structure.

This book is a blend of insights and approaches from ecopsychology (particularly gestalt psychology), cultural history and the sibling disciplines of hydrology and ecology. My emphasis on the everyday, ubiquitous nature of water is related to the gestalt 'parent'. The gestalt school of psychology defines itself as working with 'the physical, psychological, intellectual, emotional, interpersonal and spiritual aspects of an individual, and which are considered inseparable from the individual's environment, history and culture' in a therapeutic setting.[13] History is not therapy, so I have adapted the holistic emphasis to better suit the constraints of writing a history of a community in profound ecological and social transition.[14] Echoing the gestalt emphasis on the intertwined whole, this work explores water in the physical, emotional, spiritual, economic and social everyday lives of colonial Gippslanders.

The book focuses on the ability of colonists to perceive ecological processes during the first 70 years of colonisation. This period set in train changes in the catchment that developments in the twentieth century would multiply. Laura Sewall states that there are five parts to ecological perception:

- learning to attend
- learning to perceive relationships, contexts and interfaces
- developing perceptual flexibility across spatial and temporal scales
- learning to re-perceive depth, and
- the intentional uses of the imagination.[15]

---

13 Gestalt.com.au. For an example of gestalt applied to an environmental topic, see C Cooper Marcus's now seminal book from 1995 in the field of design. C Cooper Marcus, *House as a mirror of self: Exploring the deeper meaning of home*, Nicolas-Hays Inc., Lake Worth, Florida, 2006.
14 T Griffiths & L Robin (eds), *Ecology and empire: Environmental history and settler societies*, University of Washington Press, Seattle, 1997. The introduction covers debate on how much of the environmental transformation that accompanied colonialism was active or passive, planned or accidental.
15 L Sewall, 'The skill of ecological perception', in T Roszak, ME Gomes & AD Kanner, *Ecopsychology: Restoring the earth, healing the mind*, Sierra Club Books, San Francisco, 1995, p. 204.

FOLLOWING THE WATER

Figure 1.5: Flora Gregson's painting of wildflowers from around Lakes Entrance, c. 1870.
Source: Pictures Collection, State Library Victoria, Accession no. H16561/24.

Learning to pay attention is the first step. It is the intention to behold and explore where one finds oneself with care and interest, regardless of location. The second component takes the perceiver into the details and textures, noting, for example, how vegetation changes with soil type, or the dietary preferences of animals. Developing flexibility across spatial and temporal scales might mean an interest in the patterns of migratory birds; which birds, and when and where did they come from? Perceiving depth means seeing how you are one organism among billions, counting the soil bacteria and earthworms as well as the higher-order species. Finally, the intentional use of the imagination suggests trying to see yourself as the snipe, or the ash tree. Or at least, trying to imagine what joins you.

The multidisciplinary body of literature that explores ecological perception is not easily retrofitted to nineteenth-century Australia. One reason is that Sewall's definition presupposes knowledge of geological time that was by no means a done deal in the nineteenth century. Second, it presupposes familiarity with the notion of biological diversity. This was not the era of BBC wildlife documentaries, transmitting the astonishing form and variety of species straight into people's living rooms. Instead, many people were only just coming to grips with the idea of microscopic organisms.[16] Third, the very strangeness of the Australian continent would have flummoxed even the most ardent proto-ecologist. While European exploration into the continent was introducing an enormous number of new species to science, their apparent oddity often found more detractors than admirers.[17] Fourth, it takes time to learn about ecological processes, and it must be said that patience was not a virtue displayed *en masse* by settlers. Rolls, for example, describes a problem with perceiving the differences in pasture recovery rates for areas receiving between 100 and 200 mm rainfall per year. Both look the same, but the pasture in the lower rainfall areas takes much longer to recover from grazing. By the time a pastoralist realised his overstocking mistake, it was too late.[18] Frawley describes a similar issue with the length of time it took white settlers to grasp the nature of soil fertility.[19] But, rather than rectify their impacts, most settlers simply moved on to virgin lands, which was quicker and more profitable. This combination of unfamiliarity and haste, compared to the time needed to perceive ecological processes, prevented most white settlers developing the skill of ecological perception as described by modern writers.

However, I do not claim that settlers did not possess any of these skills. It is particularly tempting to claim a lack of imagination. Patently, this was not so. They imagined a future and they created it on that landscape. What was missing was any reason or motivation to see with ecologically tinted glasses. This is the more basic reason why the notion of ecological perception cannot straightforwardly be applied to colonial Gippsland. Because of their pre-existing assumption of their separation from and superiority to nature, they had no *a priori* motivation to develop or apply the skills of ecological perception to their new homes.[20] Their world view

---

16   U Seibold-Bultmann, 'Monster soup: The microscope and Victorian fantasy', *Interdisciplinary Science Reviews*, vol. 25, no. 3, June 2000, pp. 211–19. doi.org/10.1179/030801800679242.
17   No animal created as much consternation as the platypus. A Moyal, *Platypus*, Allen and Unwin, Crows Nest, NSW, 2002.
18   E Rolls, 'More a new planet than a new continent', in S Dovers (ed.), *Australian environmental history: Essays and cases*, Oxford University Press, Melbourne, 1994, p. 26.
19   K Frawley, 'Evolving visions: Environmental management and nature conservation in Australia', in S Dovers (ed.), *Australian environmental history: Essays and cases*, Oxford University Press, Melbourne, 1994, p. 43.
20   The nature/culture split is well-traversed territory in many disciplines. D Tacey, *Re-enchantment: The new Australian spirituality*, Harper Collins, Sydney, 2000, pp. 162–4 and 174; LM Gibbs, '"A beautiful soaking rain": Environmental value and water beyond Eurocentrism', *Environment and Planning D: Society and Space*, vol. 28, 2010, p. 364. doi.org/10.1068/d9207; see also D DuNann Winter, *Ecological psychology: Healing the split between planet and self*, Harper Collins College Publishers, New York, 1996, p. 28; Strang, *The meaning of water*, p. 90; C Merchant, *The death of nature: Women, ecology and the scientific revolution*, Harper Collins, San Francisco, 1980.

limited what they could perceive.[21] Deborah DuNann Winter describes this using the example of a house. 'Our worldview', she writes, 'acts like the frame of a house: It determines the shape and coherence of the particular beliefs it supports. We see and experience the particular beliefs (walls) instead of the frame, but the frame exerts pivotal influence on which beliefs we hold and how they are related to each other.'[22] The frame of colonial Gippslanders consisted of dualism, separation and mechanism, factors that ecopsychology suggests lie at the root of most of the world's contemporary environmental and social ills.[23] In the nineteenth century, nature was often thought to be inanimate, operating rather like a clock. Humans were thought to be separate from nature, and separate from each other.

Ecopsychology differs so radically from traditional psychology because it places human wellbeing within the environment and inseparable from it. Psychology has traditionally proceeded on the assumption that the totality of a person is encapsulated within the boundaries of the epidermis. In this view we are skinned egos, interacting with other skinned egos in a formless, shapeless vacuum.[24]

Ecopsychologists challenged their discipline in the same way that environmental historians have challenged history, issuing a call to insert the environment into its discourse. Otherwise, there are no rivers to dive into, no grassy plains, no trees to climb. There are no wetlands to derive a living from, no water sprites to make offerings to. And without this living, breathing, pulsing, changing environment, any account of the human psyche or human history is flawed. Gibson and Neisser describe ecological psychology as 'a psychology that is about the complex embedded relationships of objects in constant transformation'.[25] This was a psychology that echoed the natural world of ecological processes.

---

21   Dean Radin describes an elegant experiment to test this. Playing cards were altered so that, for example, the six of spades was made to be red, instead of black. It took more than 40 viewings by subjects before they noticed the deck had been changed. The expectation of what a deck of playing cards should be determined what could be seen. D Radin, *The noetic universe: The scientific evidence for psychic phenomena*, Corgi Books, London, 2009.
22   DuNann Winter, *Ecological psychology*, p. 28.
23   DuNann Winter, *Ecological psychology*, p. 236.
24   Hillman in Roszak et al., *Ecopsychology*, p. xvii.
25   Gibson and Neisser in DuNann Winter, *Ecological psychology*, pp. 240–2. They coined the term 'ecological self'.

Much early Australian environmental history was interested in attitudes to the broader landscape. Frawley, building on Heathcote's work from the early 1970s, put forward five basic categories of response to, or five perceptions of, the Australian landscape up to the end of the twentieth century:

- colonial or resource exploitation
- national development/optimism
- scientific enquiry into nature
- ecological opposition to development ethos, and
- romantic attraction to wild and uncivilised landscapes.[26]

Of these, ecological opposition was the least prevalent during the nineteenth century, though it certainly existed, as Roe demonstrates in his analysis of park creation in the late nineteenth century.[27] Scientific enquiry also lacked influence because it was frequently made subservient to the goals of resource development and expansion.[28] The romantic tradition was also expressed, for example, in Bonyhady's analysis of the history of Fern Tree Gully.[29] But by far and away the strongest perceptions of the Australian landscape in the nineteenth century were resource exploitation and national development. Powell's extensive body of work on water largely addresses the nexus between settlers, the state and development.[30] Understanding of ecological processes and of the hydrological cycle in particular, however intrinsically fascinating to individuals, was collectively passed over unless it became useful to meet the transformative demands of colonial settlement.

---

26 Frawley, 'Evolving visions', p. 59.
27 M Roe, *Nine Australian progressives: Vitalism in bourgeois social thought*, University of Queensland Press, St Lucia, 1984.
28 JM Powell in Roy MacLeod, *The Commonwealth of science: ANZAAS and the scientific enterprise in Australasia, 1888–1988*, Oxford University Press, Melbourne, 1988, p. 249.
29 T Bonyhady, *The colonial earth*, Melbourne University Press, Melbourne, 2002.
30 JM Powell, *The public lands of Australia Felix: Settlement and land appraisal in Victoria 1834–91 with special reference to the Western Plains*, Oxford University Press, Melbourne, 1970; JM Powell (ed.), *Yeomen and bureaucrats: The Victorian Crown Lands Commission 1878–9*, Oxford University Press, Melbourne, 1973; JM Powell, *Environmental management in Australia, 1788–1914, Guardians, improvers and profit: An introductory survey*, Oxford University Press, Melbourne, 1976; JM Powell, *Watering the garden state: Water land and community in Victoria 1834–1988*, Allen and Unwin, Sydney, 1989; JM Powell, 'Snakes and cannons: Water management and the geographical imagination in Australia', in S Dovers (ed.), *Environmental history and policy: Still settling Australia*, Oxford University Press, Melbourne, 2000, pp. 47–71; JM Powell, 'Environment and institutions: Three episodes in Australian water management, 1880–2000', *Journal of Historical Geography*, vol. 28, no. 1, 2002, pp. 100–14. doi.org/10.1006/jhge.2001.0376.

Our colonial forebears were totally dedicated, possibly even addicted, to growth, as are we. Addiction theory has been reworked by ecopsychologists to ask why we continue to destroy what we depend on.[31] Addictions require the capacity to suppress or ignore information that might threaten the hold of the addiction. The major characteristics are denial, dishonesty, control, thinking disorders, grandiosity and emotional disconnection, most of which appear in some form in this story.

There is certainly no shortage of grandiose sentiments expressed by colonial Gippslanders as they lobbied to attain the infrastructure that they fondly believed would bring 'progress' to their district. Indeed, the grand infrastructure projects of canals, entrances and railways would exert the control over nature that was so vital to them. Living alongside them were the survivors of a culture who had lived in relative harmony with the surrounding environment for thousands of years, yet colonial Gippslanders denied the skills of ecological perception that the Kurnai possessed and dismissed them as ignorant savages. Finally, the harshness of colonial life suggests that there were few privations that they would not endure to achieve material and economic progress. The passion for growth meant that their ability to perceive their impact on ecological processes was either blunted or considered less importance than the progress achieved.[32] Griffiths noted in *Forests of Ash*: 'Improvement was nostalgic; it was dismissive of indigenous environmental systems; it was aggressive as well as progressive'.[33]

In contrast to the historic assumption by Western culture that nature exists for human benefit, we survive solely because of ecological processes that are in a continual state of flux. This has not been the standard view of modern Western European cultures and their colonial offshoots. Chodron outlined the conflict neatly in one crystal clear paragraph:

---

31   See, for example, C Glendinning, *My Name is Chellis and I'm in recovery from Western civilization*, Shambala Press, Boston, 1994.
32   For example, *Gippsland Times* (hereafter *GT*), 4 May 1881. 'The over hanging gums over course, were bound to be sacrificed to the exigencies of commercial progress, but we image that the Council will somewhat repent of having authorized such havoc among the ti tree.'
33   T Griffiths, *Forests of ash: An environmental history*, Cambridge University Press, Cambridge, 2001, p. 32.

> That nothing is static or fixed, that all is fleeting and impermanent, is the first mark of existence. It is the ordinary state of affairs. Everything is in process. Everything – every tree, blade of grass, all the animals, insects, human beings, buildings, the animate and the inanimate – is always changing moment to moment. We don't have to be mystics or physicists to know this. Yet at the level of personal experience, we resist this basic fact.[34]

Chodron is right about personal discomfort with processes of change. We resist ageing, death and many other forms of change strenuously. This resistance also operates at a collective level. There is a fundamental tension between the nature of our ecological reality and how we want it to be. As a species, we have enthusiastically set out to control as much of the environment as we can in order to make conditions favourable for our own survival and wellbeing, and of our economic systems. As Cicero said: 'Finally, by means of our own hands we endeavour to create, as it were, a second world within the world of nature'.[35]

We have liked to think (all evidence to the contrary as Barbara Hurd suggests in the epigraph) that this will protect us from what is beyond our control, and what has often been beyond comprehension. To facilitate this, Judaeo-Christian-based societies have insisted on the separation of humans from nature, and that this is a firm boundary. The opening epigraph alludes to this entrenched aspect of Western culture. Creating a persistent dualism, we have done our utmost to downplay our porosity, and our utter dependence on ecological processes for our survival.

To smooth out the fluctuations in ecological processes, we have turned to technology and infrastructure. This made survival less of a hit-and-miss affair, and, during the Industrial Revolution, fuelled the rise of the factories of eternal production. There are many histories that tackle this vast subject and, unsurprisingly, many of them are river based. Technology and changing rivers have gone hand in hand. Blackbourn's history of the Rhine is an excellent example, detailing the damming, dredging and draining of the river and its floodplain in the service of industry and capital.[36] His title, *The Conquest of Nature*, is an explicit enunciation of

---

34 P Chodron, *The places that scare you: A guide to fearlessness*, Element, Hammersmith, London, 2001, p. 26.
35 Quoted in JD Hughes, *What is environmental history?* Polity Press, Cambridge, 2006, p. 24.
36 D Blackbourn, *The conquest of nature: Water, landscape and the making of modern Germany*, WW Norton and Co., New York, 2005.

the theme of control and power mirrored through water.³⁷ Gippslanders followed the lead set by Europe, to the best of their financial and technical capability. They rerouted rivers, created entrances where there were none and removed vast swamps, all to facilitate their participation in the economic life of the empire.

In short, they worked to remodel the catchment's hydrology to make exportable products. They were exporting water, embodied or disguised as timber, grain, butter and other foods. Embodied water is a way of asking how much water went into the creation of a cow, fence paling or wheel of cheese, all products that colonial Gippslanders sent to market. Rather than viewing the product as being complete in itself, the concept of embodied water includes the trails of water needed to make it and thus situates the product more obviously in the web of ecological connections supporting its existence.

The study of embodied water emerges from contemporary concern about the damage that has been done to the hydrological cycle by manufacturing processes.³⁸ Globally, approximately 70 per cent of the world's water is currently abstracted for agricultural purposes, and estimates range from 500 litres to produce a kilogram of potatoes to 15,000 litres to produce a kilogram of beef.³⁹ These are modern estimates, but colonial agriculture follows the same pattern of intensification of water use through irrigation. Dams, pipes, channels and drains were all deployed in the Gippsland Lakes catchment to shore up the levels of embodied water in their crops, ready to be exported to Melbourne and beyond. Colonial Gippslanders chose an economic path of exporting a select set of water intensive products using increasingly water intensive infrastructure.

---

37 Many scholars look at power relations mediated through technologies applied to water. Their focus is upon governance and human institutions more so than the ecological processes inherent in the hydrological cycle. For a historical water example, see WE Bijker, 'Dikes and dams, thick with politics', *Isis*, vol. 98, 2007, pp. 109–23. doi.org/10.1086/512835. Common pool resource theory is a key body of work that exemplifies the interest in governance of natural resources, Elinor Ostrom's landmark publication of 1990 galvanised this field. E Ostrom, *Governing the commons: The evolution of institutions for collective action*, Cambridge University Press, Cambridge, 1990. doi.org/10.1017/CBO9780511807763.

38 There are vast variations in the estimates made to analyse embodied water. Some of these variations include type of production process, climate and geographical location and whether or not the embodied water of ingredients or components is also taken into consideration. As Gleick says, there is no rule about where to draw the line on the supply chain of goods. PH Gleick, H Cooley, MJ Cohen, M Morikawa, J Morrison & M Palaniappan, *The world's water 2008–9: The biennial report of freshwater resources*, Island Press, Washington, 2009, p. 335, Table 19.

39 R Clarke & J King, *The water atlas: A unique visual analysis of the world's most critical resource*, The New Press, New York, 2004, p. 19 for the 70 per cent and p. 33.

1. INTRODUCTION

Colonial Gippslanders were passionate believers in agriculture as the foundation of human civilisation. In one of its earliest issues, the editor of the *Gippsland Times* proclaimed:

> We must make the lands more profitable than by their existence in their pristine covering, this is the age of science and advancement and we must progress with both, or forever remain in the same state we are now in. We well know we cannot all be agriculturalist, but at the same time we know that agriculture is the foundation of progressiveness.[40]

Gippslanders were enthusiastic appropriators of water for their agricultural endeavours, so long as it was orderly and appropriately timed for their crops. In reality, the new colonial lands from which they sought their fortunes were anything but orderly. The great southern land came with a suite of environmental characteristics that stretched colonial Gippslander's capacity to adapt. Either too much or too little rain, coupled with fire, drought, foreign vegetation and animal plagues tested the colonisers' flexibility. (At least the sun still rose in the east and set in the west.) The Australian environment was one giant question mark of uncertainty, and making a living off the land was a game of chance.

This book takes a microcosmic look at this game of environmental roulette. Part One deals in more depth with the intellectual baggage that colonial settlers brought with them to Gippsland's plains and forests. Chapter 2 focuses exclusively on the idea of the hydrological cycle itself, examining its evolution from ancient Greece to the present and locating colonial Gippslanders along that spectrum. It also sets the book within the context of twenty-first-century concerns about global environmental change, and how environmental and hydrological knowledge is structured. Chapter 3 explores the world view of the settlers, probing their religious beliefs and cultural values about the nonhuman world.

The four chapters that make up Part Two focus specifically upon the study area and are structured by aspects of the hydrological cycle. This gives an in-depth exploration of how settlers perceived the specificity of the hydrological cycle in Gippsland by tracking the entry, passage and exit of water through the catchment. Chapter 4 considers the effect of precipitation in all its forms, and teases out how wetness influenced the daily life of settlers. I include flood in this chapter, rather than in the

---

40 *GT*, 21 August 1861.

following, for the reason that flood generally occurs with an 'excess' of precipitation and snow melt. Chapter 5 addresses what is now referred to as environmental flow; the water in creeks, rivers, floodplains and aquifers that sustain ecosystems. The emphasis is upon movement, and how settlers combined to use and remake the catchment's watercourses for their economic gains. I also consider the symbolic aspects of flowing water, especially from the point of view of biblical teachings. The importance of religious belief is extended into Chapter 6, which analyses the opposite to flowing waters: this chapter focuses on the 'problem' of still and stagnant waters. Chapter 7 addresses the effects of water in its gaseous state, and links dryness, drought, fire and irrigation.

The evidence used is largely unofficial in character, relatively diverse and sometimes, at first glance, fragmentary or even trivial. However, a gestalt approach suggests that a fragment can lead to a whole world of meaning. Like embodied water where each object trails a cloud of vapour behind it, these fragments of evidence, like single line entries in diaries, carry threads of meaning which connect with other threads.

My principal primary source has been local newspapers (in retrospect, an eye-destroying decision). Regional and local papers proved to be the standout source of information. They provided a wide range of information of interest to many settlers, addressing the physical, emotional, spiritual, economic and social aspects of their everyday lives.[41] They reproduced council proceedings at length, reported on the weather, assessed the progress of seasonal crops and debated politics. They covered a myriad of social events, infrastructure issues, church activities and business news. However, the subject matter meant that indexes, if they existed, could not be relied upon to guide searching. Particularly severe natural events and some infrastructure matters were indexed, but the wealth of material collected was not.[42] When the research started the newspapers had not been digitised, and accordingly, a five-year sampling strategy was chosen for the *Gippsland Times* from when it commenced in 1861, with at least two issues (and generally many more!) being read in detail for

---

41   L Morrison, 'The newspapers of Gippsland, 1855–1890', *Gippsland Heritage Journal*, vol. 6, 1989, p. 3.
42   Ian Lunt reported the same problem in his study of clearing grasslands. I Lunt, 'The distribution and environmental relationships of native grasslands on the lowland Gippsland plain, Victoria: An historical study', *Australian Geographical Studies*, vol. 35, no. 2, July 1997, p. 142. doi.org/ 10.1111/1467-8470.00015.

each month.[43] The five-year strategy assured that most weather conditions would be sampled. One extremely wet year (1871) and one extremely dry year (1898) were also read in detail. Local newspapers, such as the *Morwell Advertiser*, were used as available.

Diaries, manuscripts and memoirs from individuals and families were the second primary source. These provided the minute detail of everyday life required to balance out the more generic information gathered from the newspapers. Approximately 20 original diaries and manuscripts were used, supplemented by memoirs that have been published in the many newsletters of historical societies. What was most memorable about the diaries was their general terseness, compared to the florid language in the papers. The practical importance of water suggested by the papers is confirmed by the diaries, but it is difficult to come to any conclusion about the metaphoric or symbolic understanding of water for the individual diarists. Mostly, they treated the diary like a business engagement book rather than a space to confide feelings. There was little common ground between the writers' demographic characteristics. I have therefore refrained from conclusions of general applicability based on the diaries, except in agricultural or pastoral matters.

The third category of sources comprised almanacs, books, music and sermons. These sources were selected to flesh out aspects of daily life, and to help interpret the symbolic perception of water. Almanacs provided a source of readily accessible environmental advice about the Australian environment. Books and music, sourced from descriptions of concerts and meetings in the papers, provide insight into the symbolic interpretation of water in entertainment. Sermons were often reprinted and help to give context to spiritual beliefs.

Finally, a selection of government documents made up the last class of records consulted. These included land capability assessments, meteorological data, reports of various boards, parliamentary debates, papers presented to parliament and maps. This diversity lends strength and reliability to the conclusions drawn.

---

43 If I had my time over and started this work with digitised papers, I would still have chosen a five-year sampling strategy. Digitisation is a wonderful advance, but it also makes much more material available. A sampling strategy remains a sensible and time-honoured approach. In revising, I have undertaken keyword searches but nothing showed up that caused me to alter my original arguments.

Given that water policy in Australia seems to be perennially made in response to drought or flood right up to the present day, we have not really moved on.[44] During the course of my research, Gippsland burnt and flooded more than once. In January 2011, virtually the whole of south-east Queensland and the Murray-Darling Basin experienced extensive flooding. The city of Brisbane was shut down for days. But for the less florid language, the flood reportage could have been lifted straight from the *Gippsland Times* from 1871.

The ecological processes inherent in these events in south-eastern Australia between 2006 and 2011 were portrayed as an affront to human values, talents and aspirations. Such events do blight lives, and cause immense heartache and loss for the communities that are affected by them. No one with a functioning heart could avoid spilling tears for the homeless, bereaved and injured. But such events are not a deliberate affront, or an act of war from nature directed at humans.

To think in this way presupposes separation from the natural world and the ecological processes which make life possible. It is the antithesis of a gestalt way of thinking that presupposes connection. Learning to think connectedly with an ecological process orientation is a new task for us. In a society increasingly confronted by its poor environmental record, it is also is much needed. This book, I hope, is one contribution to that process.

---

44 D Connell, *Water politics in the Murray-Darling Basin*, Federation Press, Annandale, 2007.

CHAPTER 2

Making the circle round: Perceptions of hydrology through time

I do not know much about gods; but I think that the river
Is a strong brown god – sullen, untamed and intractable
Patient to a degree, at first recognized as a frontier;
Useful, untrustworthy as a conveyor of commerce;
Then only a problem confronting the builder of bridges.
The problem once solved, the brown god is almost forgotten
By the dwellers in cities – ever, however, implacable,
Keeping his seasons and his rages, destroyer, reminder
Of what men choose to forget. Unhonoured, unpropitiated
By worshippers of the machine.

TS Eliot[1]

Introduction

The hydrological cycle is like a Buddhist devotee, reborn in different guises over millennia. This chapter examines the five different guises that the hydrological cycle has worn for European intellectuals, beginning with the ancient Greeks and ending in the twenty-first century. It tells of the different ways humans have tried to understand what water is, where it comes from and where it goes to when it passes beyond our senses.

---

1   TS Eliot, 'Four Quartets'. This quote alludes to the pre-Christian pagan worship of rivers. For further information on this topic see T Andrews, *Legends of the earth, sea and sky: An encyclopedia of nature myths*, ABC Clio, Santa Barbara, 1998.

That the idea of the hydrological cycle has a history is profoundly important. It means we cannot understand the actions of past persons, communities or societies unless we have an understanding of which incarnation of the cycle they thought was true. This outing into the intellectual pedigree of the hydrological cycle situates colonial Gippslanders along a spectrum of intellectual development and perception, as much as it situates me, the writer. It can be hard to appreciate or understand the way people behaved in the past, especially if the results of their actions are now a problem. It is therefore vital to take the time to understand the different versions of the hydrological cycle, because this will allow a more nuanced interpretation of why colonial Gippslanders acted as they did.

Outline of the hydrological cycles

Since the time of ancient Greece and up to the present, there have been five permutations of the hydrological cycle. I have called these:

- the underground purification model
- the vertical model
- the divine design model
- the rainfall-driven, quantitative model
- the connected model.

Each reflects the concerns, considerations and biases of the era in which it was created. The first three reflect the fundamental question that bothered many scientists and natural philosophers for centuries; namely, where do rivers and springs come from? The majority of scientific thought on hydrology from the ancient Greeks until the late 1600s went to solving this question and centred around the underground purification, the vertical and the rainfall-driven models.

Briefly, the underground model theorised a mechanism based on underground channels. Water was sucked out from the ocean through holes in the ocean floor and channelled under the earth up into mountains, where it was released as vapour to condense and form rivers. The major opposing theory, also dating from ancient Greece and which has proved to be correct, is that rivers were fed by rainfall. The vertical model relied on theories of alchemy, of how one substance could be converted into another.

Ancient Greece and Rome

Hydrological manipulation is driven by the development of agriculture and urbanisation. Generally societies pursuing a hunter-gatherer lifestyle, like Australia's Indigenous peoples, had little need for large-scale hydrological alterations. It is for this reason that most scholarship on water manipulation begins with societies of the ancient Mediterranean and Middle East, where agricultural societies that could support cities developed.[2]

Because most histories of science begin with the classical Greek philosophers, the few published histories of hydrology tend also to start there. This approach overlooks the many precursor civilisations of the Middle East, and civilisations from other continents.[3] Nace, however, admits that the origin of the idea of the hydrological cycle is unknown, and that it probably predated the biblical Book of Amos, written in approximately the eighth century BC.[4]

Three versions of the hydrological cycle were debated in the period approximately 500 years before the common era: the rainfall model, the underground model and the vertical model. Thales of Miletos (640–546 BC), Xenophanes of Colophon (570–470 BC) and Anaxagoras of Clazomene (500–428 BC) all had conceptions of a cycle in which water moved ceaselessly in a loop. Of these, and 'making due allowance for the lack of substantive data 2500 years ago', Nace considers that 'Anaxagoras formed a concept of the hydrological cycle which was qualitatively correct'.[5] However, Anaxagoras's insights were eclipsed by the enduring inaccuracy of Aristotle, Plato and Pliny. They (and many others at the time and subsequently) supported the underground purification model.

---

2   For example, Phillip makes passing remarks about Sumer and Egypt. JR Phillip, 'Water on the earth', in AK McIntyre (ed.), *Water: Planets, plants and people*, Australian Academy of Science, Canberra, 1978, p. 37.
3   For a discussion of ancient Chinese debate on the cycle, see V Te Chow, 'Hydrology in Asian civilization', *Water International*, vol. 1, no. 2, 1976. Cited in SM Karterakis, BW Karney, B Singh & A Guergachi, 'The hydrologic cycle: A complex concept with continuing pedagogical implications', *Water Science and Technology: Water Supply*, vol. 7 no. 1, 2007, pp. 23–31. doi.org/10.2166/ws.2007.003. See also L Rezende, *Chronology of science*, Checkmark Books, New York, 2006, entries for c. 300 BCE and c. 100 BCE. For a brief history of Persian knowledge of groundwater extraction, see www.waterhistory.org/histories/karaji/karaji.pdf, accessed 24 January 2011.
4   R Nace, 'General evolution of the concept of the hydrological cycle', *Three centuries of scientific hydrology: Key papers submitted on the occasion of the celebration of the Tercentenary of Scientific Hydrology, 9–12 September 1974*, UNESCO, Paris, 1974, p. 41.
5   Nace, 'General evolution of the concept of the hydrological cycle', p. 43.

The vertical version competed with the underground and rainfall-driven models. Based on alchemical theories of elemental transmutation, cool air could convert into water in the ground and into rain above mountains.[6] This theory was based on the empirical observation of how liquid water evaporates and rises.

Greek empirical observation and theorising was prompted by scepticism about explanations for all events being based on divine direction. These three visions of a hydrological cycle should be seen as part of a broader imperative to ask questions about how the world worked; a response to the need to make the world understandable and potentially amenable to human control. Rupp wrote:

> The Greeks conviction that there were logical explanations for natural phenomena … had dazzling implications for the future. If events had causes other than the fickle fingerlings of the supernatural, infinite possibilities opened up for predicting, circumventing, redirecting, controlling or exploiting them. The natural philosophers were intellectual revolutionaries, star players in a splendid and startling era of cognitive epiphany.[7]

Predicting, circumventing, redirecting, controlling and exploiting is exactly where the Greeks' innovative thinking led to, potentially making the world less terrifying. From the relative comfort and safety of the twenty-first century it is easy to forget, after all, that this was a world of actual wolves, living in actual forests, near to actual villages. Vito Fumagalli notes that fear was a primary factor in determining responses to the environment.[8] This fear fostered a mentality that encouraged attempts to remake ecological and hydrological conditions to be more in line with human survival requirements. Water's capacity for destruction was well understood and was immortalised in religious teachings. Feliks has argued that the role of flood as divine punishment in the monotheistic religions that arose in the Middle East derived from actual experience. Relatively narrow river valleys along which roads and villages were strung meant that floods tended to be devastating because of the greater velocity of the water being channelled through the narrow space. This also meant

---

6   Y-F Tuan, *The hydrologic cycle and the wisdom of god: A theme in geoteleology*, University of Toronto Press, Toronto, 1968, pp. 25–6.
7   R Rupp, *Four elements: Water, air, fire, earth*, Profile Books, London, 2005, p. 2.
8   V Fumagalli, *Landscapes of fear*, Polity Press, Cambridge, 1994, pp. 1–2.

people had few places to escape to.⁹ Hillel discusses devastating floods in Sumerian culture as well as in the other great riverine-dominated empire of the ancient world, Egypt.¹⁰

Many historians do not credit the Romans with innovations about the theory of the hydrological cycle, even though they were the absolute masters of its practical alteration. (The exception is Marcus Vitruvius Pollio who wrote on the basis of his practical expertise, and came to the conclusion that rainfall was the origin of rivers and springs.)¹¹ The Roman obsession with bathing required the transport of vast quantities of water over long distances, and their engineers therefore developed a fine grasp of hydraulics.¹² In AD 100, Romans received almost three times as much water per capita as modern New Yorkers.¹³ Roman water redirection bears out the truth of Nace's assertion that lack of theoretical knowledge of hydrology was no impediment to hydrological alteration. Similarly, Wikander's *Handbook of Ancient Water Technology* gives examples from a variety of ancient Middle East civilisations that practised irrigation and drainage, amply demonstrating that practical bodies of knowledge about water manipulation existed from about 6,000 BC.¹⁴

The capacity to change hydrological processes without any theoretical understanding of the hydrological cycle is a key point, because the repercussions could be so great. The incorrect understanding of the cycle lead to environmental repercussions, notably irrigation induced salinity leading to abandonment of previously fertile land. However, this often did not appear as an immediate cause and effect that could be observed. Time lags between action and reaction can take decades to reveal,

---

9   Y Feliks, *Nature and man in the Bible: Chapters in biblical ecology*, Soncino Press, London/Jerusalem/New York, 1981; D Hillel, *The natural history of the Bible: An environmental exploration of the Hebrew scriptures*, Columbia University Press, New York/Chichester and West Sussex, 2006. Page 5 of the prologue describes his own personal experience of sudden torrents in the Negev desert highlands.
10   Hillel, *Natural history of the Bible*, ch. 3 on Mespotamia and ch. 5 on Egypt.
11   FH Chappelle, *Wellsprings: A natural history of bottled spring waters*, Rutgers University Press, New Brunswick, New Jersey, 2005, p. 27.
12   Roman technological skill in many areas was the subject of one (in my opinion) of the best scenes in Monty Python's film *Life of Brian*. John Cleese, as the all-talk, no-action leader of the People's Front of Judea asks the members, 'What have the Roman's ever done for us?' A member responds with 'The aqueduct', followed by sanitation.
13   O Wikander, *Handbook of ancient water technology, Technology and change in history*, vol. 2, Brill, Leiden, 2000, p. 48.
14   Nace, 'General evolution of the concept of the hydrological cycle', p. 40; Wikander, *Handbook of ancient water technology*.

suggesting that the capacity to learn about hydrological processes generally wouldn't have matched most of the ancient people's life expectancy. The role of long-term temporal fluxes in ecological processes is crucial in understanding ecological misperceptions.

Middle Ages

When the Roman Empire disintegrated, much of the theoretical musings of the Greeks and the practical skills of the Romans disappeared for centuries, plunging Europe into what became popularly known as the Dark Ages.[15] Around the turn of the first millennium, Europe was gradually rebuilding after the cumulative impact of invasions, wars and disease.[16] During this period, there was no resolution about how the hydrological cycle worked. The period saw, however, the origin of the divine design model, which would reach its height in the early eighteenth century, and the high point of the vertical model. The vertical model also had its heyday.

By the time the surviving Greek theoretical writings resurfaced in Muslim libraries and were brought to Europe through Spain, the ideological hold of Christianity was palpable. In this world view, God in His munificence created the earth for man's benefit. The Bible has numerous references to hydrological phenomena, but only Ecclesiastes 1:7 gives a description which hints at a cycle. These four lines became critical in the slow (some might say meandering) development of hydrology:

> All the rivers run into the sea;
> Yet the sea is not full;
> Unto the place from whence the rivers come,
> Thither they return again.[17]

---

15   The major exception was the Dutch. TeBrake's article 'Taming the waterwolf' is a very useful overview of the history of hydraulic engineering in the Netherlands. In fact, the Dutch had totally transformed the lowlands of their nation by the 1300s, a process that locked them into a never-ending cycle of maintenance and further drainage. Importantly, the Dutch would then go on to export their hydraulic expertise across Europe and its colonies in later centuries. W TeBrake, 'Taming the waterwolf: Hydraulic engineerings and water management in the Netherlands during the Middle Ages', *Technology and Culture*, vol. 43, no. 3, July 2002, p. 489. doi.org/10.1353/tech.2002.0141.
16   Fumagalli, *Landscapes of fear*, p. 17.
17   King James Bible, sourced via the University of Virginia's e-text repository.

The essential role of this quote cannot be overstated in defining how Christian men of learning would come to think about the cycle. As Yi-Fu Tuan said, Ecclesiastes 1:7 provided biblical authority for the idea that the hydrological cycle was the manifest wisdom of God.[18]

The emphasis upon God's divine design acted as a stranglehold on thought, encapsulated by the doctrine of St Augustine, which stated that the Bible is always true.[19] Most scholars scorned experiment and practical observation and rehashed Greek writings, particularly those of Aristotle and Plato who both supported the underground purification model. The adherence to the wrong Greek theories combined with the central role of the Ecclesiastes verse, which suggested there was a cycle but didn't actually provide the mechanism, led to a few more centuries of confusion.

With the ongoing interest in alchemy, the vertical model also had its supporters. The ancient Greek notion of alchemical transmutation in a linear pattern was overlaid with the medieval idea of the Great Chain of Being. Although a hierarchical concept, the Great Chain of Being taught that God had made everything for a reason and that relationships between the things of His creation were meaningful. Hence, respectful behaviour and thought towards the nonhuman was not only expected, but required. St Francis's position of respect towards animals and biblical writings which express reverence for water are examples. Of the earlier models, this one had the greatest capacity to recognise the preciousness of ecological connections mediated by water. However, it competed with the emphasis on hierarchy and power that was derived from it. In a world dominated by absolute monarchs, it was easy for those with vested interests and status to emphasise the hierarchical and authoritative aspects rather than the obligations of interlinking respect. In practice, there was a greater weight placed on the superiority of man vis-à-vis the people, animals, plants and things that were lower down on the scale.[20]

In terms of the practical understanding of the hydrological cycle, Tuan notes that the vertical emphasis downplayed winds, which create a horizontal component to the operation of the hydrological cycle. The preceding line to Ecclesiastes 1:7, upon which so much was written, refers to different wind directions. If the two had been read together, the correct

---

18 Tuan, *The hydrologic cycle*, p. 21.
19 AK Biswas, *History of hydrology*, North Holland Publishing Company, Amsterdam, 1970, p. 136.
20 Tuan, *The hydrologic cycle*, p. 73.

physical interpretation might have appeared much earlier. But because of the hierarchical nature of the biblical world view, with God in His heaven and a step ladder of other beings below Him, these horizontal elements and connections were obscured.[21]

Renaissance and the Enlightenment

Nace suggests that there were three key factors for the renaissance in hydrology that coincided with the Renaissance. The first was the improved skills in tools and measurement techniques. Second, the intellectual rigidity of the Church was breaking down, and the third was the translation of the rediscovered Greek texts.[22]

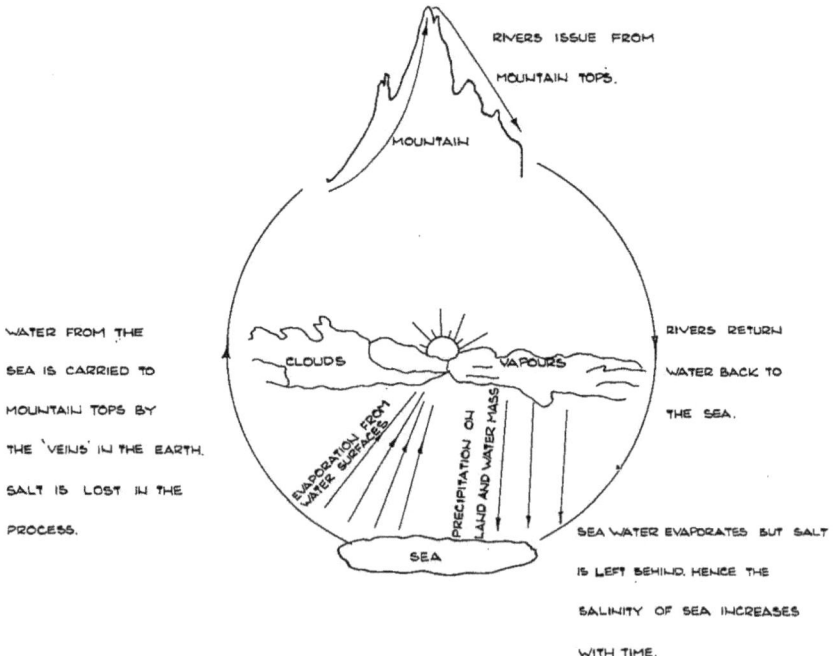

Figure 2.1: Leonardo da Vinci's view of the hydrological cycle.
Source: AK Biswas, *History of hydrology*, North Holland Publishing Company, Amsterdam, 1970.

---

21 Tuan, *The hydrologic cycle*, p. 43.
22 Nace, 'General evolution of the concept of the hydrological cycle', p. 44.

While the Greeks had not solved the directional and method questions of the hydrological cycle, their works revived the interest in observation, measurement, testing and theory amongst the Renaissance scholars. For example, Leonardo da Vinci was deeply interested in experiment, and conducted a number of flow experiments amongst his other prodigious interests.[23] During the 1600s the barometer and thermometer were invented, there were experiments into rainfall and evaporation measurement, the first correct explanation of artesian wells was produced, the first European rain gauge was constructed, and Castelli published the correct formula for the relationship between velocity and stream flow discharge.[24] It wasn't until Perrault published *On the Origin of Springs* in 1674 that the underground purification model was laid to rest. His evidence demonstrated that rivers originated with rainfall, while Halley figured out the evaporative end of the cycle. The reemergence of interest in hydrology, coupled with the Renaissance rise of scientific method, ultimately produced the hydrological science upon which most twentieth-century resource decisions were made.[25]

With the basic questions now answered, scholarship was now devoted to how to better discern God's divine design. Much scholarship was devoted to understanding the earth and retrofitting a divine intent to it. In *The Hydrologic Cycle and the Wisdom of God*, Tuan noted:

> [In this era,] not only was there was no sharp distinction between natural theology and science but the scholars who wrote on the theme of the water cycle within the context of a physico-theological treatise actually contributed to it. The contribution lay largely in extending the number of physical processes and facts

---

23 Da Vinci swung between the rainfall and underground theories of the cycle. Nace, 'General evolution of the concept of the hydrological cycle', p. 45.
24 Biswas, *History of hydrology*, p. 173; D Camuffo, C Bertolin, PD Jones, R Cornes & E Garnier, 'The earliest daily barometric pressure readings in Italy: Pisa AD 1657–1658 and Modena AD 1694, and the weather over Europe', *The Holocene*, vol. 20, no. 3, 2010, pp. 337–49. doi.org/10.1177/0959683609351900.
25 The following chronology with early examples of scientific investigations in many aspects of water's properties is sourced from Rezende, *Chronology of science*: 1590 – Galileo Galilei invents the thermoscope, forerunner of the thermometer; 1637 – Descartes describes how rainbows appear as arcs; 1643 – Evangelista Torricelli invents the barometer and describes atmospheric pressure; 19 September 1648 – Blaise Pascal and Perier Pascal show that atmospheric pressure drops at high altitudes; May 1652 – Pascal states his law of hydraulics in *Treatise on the equilibrium of liquids*; 17 September 1683 – Antonym van Leeuwenhoek describes bacteria he has seen under the microscope; 1686 – Edmond Halley describes trade winds and suggests the air currents are caused by solar radiation; 1698 – Guillaume Amontons shows that water boils at the same temperature and doesn't increase in temperature after reaching boiling point and proposes the boiling point of water as a way to fix a temperature scale; 1714 – Daniel Fahrenheit invents the mercury thermometer.

that can be subsumed in a unified scheme. For it was this unity – the beautiful economy of means and ends – that illustrated the wisdom of God.[26]

Tuan uses John Ray's immensely popular book *The Wisdom of God*, first published in 1691, as his preeminent example. It was still in print in the 1820s, giving John Ray better sales than most authors could ever dream of. The book explained how various physical phenomena had been designed by God for human benefit.[27] For example, the vastness of the sea was a particular conundrum, when it would appear that *terra firma* is of much more obvious benefit to *Homo sapiens*. The answer put forward was that the sea provided water vapour, which became rain, providing drinking water and creating rivers.[28] Ray employed Perrault for this part of his argument, and then moved onto Halley, to explain how mountains fitted into the grand scheme. The growing understanding of the physical path of water through landscapes was put to moral and theological use. Unlike in the fourth version of the cycle, there was no essential split between the science of the writers and their faith in Christianity.

Colonial Gippslanders sit at the tail end of the divine design model and cross over into the quantitative model. They were devout Christians on the whole and the notion of an earth designed for them to improve had considerable appeal, as Chapter 3 will consider in more detail.

Eighteenth and nineteenth centuries

With the basic parameters of the hydrological cycle established, research in the eighteenth century turned towards developing a more refined understanding of its components, primarily a range of advances in the understanding of surface water. Jean-Claude De La Methiere (1743–1817) noted how rainfall is disposed of in different ways: as direct runoff to streams, as evaporation and transpiration, and storage as soil moisture or deeper as groundwater.[29] It was conclusively shown that water flows faster towards the surface of a stream than at lower depths, and Dalton

---

26 Tuan, *The hydrologic cycle*, p. 4.
27 For an example of this in relation to Australia, see Cathcart's discussion of the inland sea in *The water dreamers: The remarkable history of our dry continent*, Text Publishing, Melbourne, 2009, p. 101.
28 Tuan, *The hydrologic cycle*, pp. 10–12.
29 Biswas, *History of hydrology*, p. 279.

finalised his theories about vapour pressure, which allowed estimates of evaporation to be made for the first time. Dalton's modified theory is still used.[30] Mathematicians, fluid mechanicists and hydraulicists from Europe dominated this research.[31]

The nineteenth century saw a continuation of the trend of attempting to master individual technical problems that the hydrological cycle presented. In particular, the contribution of the French engineer Henry Darcy solved a major barrier. In 1856, Darcy published his findings on how water travels through porous media, thus laying the foundations of groundwater research.[32] In the nineteenth century there was also a strong correlation between engineering, the industrial revolution and hydrology, as suggested by this chapter's epigraph. 'Many works examined relationships between precipitation and streamflow out of necessity for engineers designing bridges and other structures.'[33] The relevance of this for Gippsland was that bridges represented significant and ongoing expenditure to communities, and any advances in knowledge in this area could mean the difference between keeping communities connected. Gippsland and the remainder of colonial Australia followed the pattern for increasingly large dams to provide clean water to its burgeoning settlements.

Twentieth century

The rainfall version of the cycle, hypothesised by Anaxagoras, proved by Perrault and increasingly refined during the nineteenth century, is what is currently accepted. The cycle is understood as the product of rational, scientific enquiry. Davie noted that hydrology's claim to be a science rests upon its fundamental theory of the water balance equation, the application of which is the basis of most field studies:

---

30   Biswas, *History of hydrology*, p. 276.
31   J Hubbart with J Kundell, 'History of hydrology', in CJ Cleveland (ed.), *Encyclopedia of Earth*, Environmental Information Coalition, National Council for Science and Environment, Washington DC, 2008 [First published in the Encyclopedia of Earth, 20 May 2007], last revised 11 February 2008, editors.eol.org/eoearth/wiki/History_of_hydrology, accessed 25 February 2009.
32   Hubbart, 'History of hydrology'.
33   Hubbart, 'History of hydrology'.

> [The water balance equation is a] mathematical description of the processes operating within a given timeframe and incorporates principles of mass and energy continuity. In this way, the hydrological cycle is defined as a closed system whereby there is not mass or energy created or lost within it. The mass of concern in this case is water.[34]

The equation can be summarised in simple terms as inputs equal outputs plus or minus a change in storage. So, precipitation as the input will equal the outputs of evapotranspiration, plus runoff, plus groundwater storage, plus or minus a change in storage of water.[35] This mathematical equation version of the hydrological cycle is what is employed in policy, planning and management decisions. It was formalised in the 1930s by the American hydrologist Robert Horton, but we can trace its antecedents back to the 1700s and it is created from the individual insights and work of all the proto-hydrologists discussed so far who sought to quantify the cycle. As Davie suggested, it is a vision based upon a numerical rendering of a single substance, in and of itself.

It is often schematically presented. Figure 2.2 is one of many available to download from any internet search.

Other aspects of the environment are subordinated to water, and interpreted only in so far as they affected the parameters of the mathematical equation. Some diagrams push this representation even further, entirely removing any representation of the environment, making the path of water an abstracted series of pipes.

This reduction of the cycle to a quantitative volume was historically unique, and served a socioeconomic purpose:

> The need for a coherent world view of water cycles only became necessary through the dire need for a treatment of resource and hazard problems – those require prediction or forecasting and for the want of such insights the hydraulic civilisations (heroic manipulators of water) met their demise in environmental stress and disorder.[36]

---

34   T Davie, *Fundamentals of hydrology*, Routledge Fundamentals of Physical Geography, Routledge, London, 2003, p. 9.
35   Davie, *Fundamentals of hydrology*, p. 9.
36   M Newson, *Hydrology and the river environment*, Clarendon Press, Oxford, 1994, p. 3.

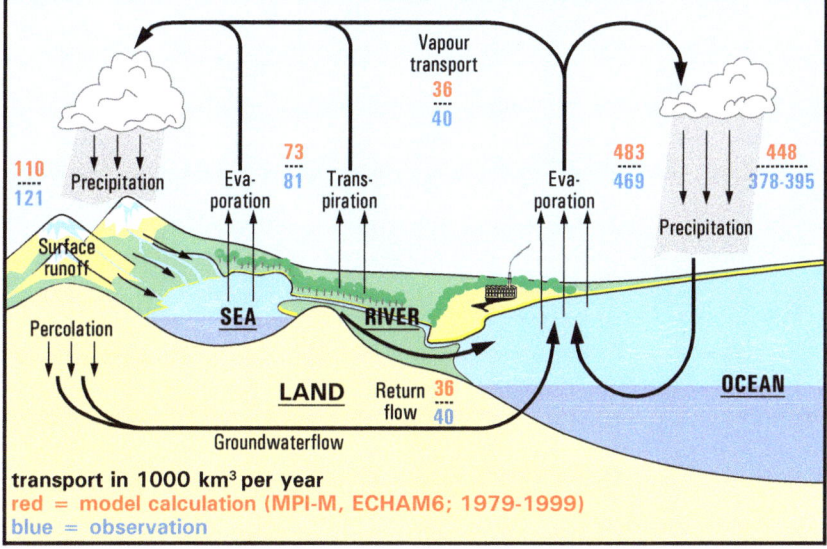

Figure 2.2: Schematic representation of the hydrological cycle.
Source: Diagram from Max Planck Institute for Meteorology (www-k12.atmos.washington.edu/k12/pilot/water_cycle/grabber2.html).

It is possible to see this process in microcosm in Gippsland, especially through the debates about the poor quality of Sale's water supply soon after the population began to grow.

This way of viewing water, as merely a quantity that can be manipulated and redirected, dovetailed with an increasing emphasis on large-scale water infrastructure. In particular, Linton pairs the timing of Horton's foundational lecture with the New Deal era of progressive projects like the Tennessee Valley authority.[37] Mega water projects built upon the wide range of state excursions into water supply, sewerage and drainage ventures in the nineteenth century. For example, the EU Water Time Project lists the start dates of major water infrastructure in 16 European cities in the nineteenth century.[38] Most of these projects were initiated to solve problems caused by industrialisation and massive population

---

37   J Linton, 'Is the hydrologic cycle sustainable? A historical–geographical critique of a modern concept', *Annals of the Association of American Geographers*, vol. 98, no. 3, 2008, p. 636. doi.org/10.1080/00045600802046619.
38   European Union Water Time Project, www.watertime.net. Click on the 'City in Time' link for further details of the historical aspects of this broader project on water governance.

growth. Many of the excellent water histories, such as Cioc's history of the Rhine, explore the nuances of these drivers at river and catchment scale.[39] Generally these histories are telling the story that TeBrake identified in his work on the lowlands Dutch, that is the inability (or refusal) of humans to exist within their local hydrological conditions. Instead, the hydrological processes are changed to suit human wants (e.g. to transport products safely, to provide irrigation water out of season, to cleanse the body and make products).

Late twentieth and early twenty-first centuries

Within the last two decades of the twentieth century, the quantitative version of the hydrological cycle was being questioned. I have labelled the fifth version the 'connected' cycle. Although there is no definite articulation of what 'it' is, there are a number of critiques that point to some common concerns.

In 1999, hydrologist Rafael Bras delivered the annual Horton lecture, in which he reflected on his career over 30 years.[40] His remarks demonstrate the emergence of the fifth version and clearly showed his commitment to its quantitative and volumetric version.

He began his studies in the late 1960s. At the time, hydrology could have been seen as a mature field but there was a 'revolution' underway, brought in by experiments in catchment modelling with computers. 'When I learned hydrology, vegetation was treated almost as a nuisance term', he says. He gives several other examples of complete turn arounds in received wisdom. For example, 'in the late 1970s and early 1980s Eagleson suggested the then radical idea that the biosphere, the climate, the hydrology and the soil were in a synergistic waltz'.[41] The assertion that water should be viewed as being interconnected with elements of the environment was a problem for a discipline that had, up till then, defined itself as primarily a quantitative and volumetric science. This was a shift from an engineering paradigm to an environmental paradigm.

---

39   M Cioc, *The Rhine: An eco-biography*, University of Washington Press, Seattle, 2002.
40   RL Bras, 'A brief history of hydrology', *Bulletin of the American Meteorological Society*, vol. 80, no. 6, June 1999, pp. 1151–65. doi.org/10.1175/1520-0477-80.6.1151.
41   Bras, 'A brief history of hydrology', pp. 1151–65.

There are other critiques of the hydrological cycle, notably arising out of its geographical birthplace. There is an inherent bias towards streamflow, which occurs globally in a limited geographic range. The proto-hydrologists like Perrault all lived in Northern Europe, where the presence of surface water all year round is the norm, not the exception. The application of a model based on this to other parts of the globe has had serious consequences:

> The hydrologic cycle upholds a long-standing Western prejudice against aridity, by which places (and often the people inhabiting the places) lacking 'sufficient' rainfall, or subject to 'violent swings' in seasonal and annual precipitation must be regarded as deficient, abnormal and in need of hydrological correction.[42]

The entire history of, say, Australia's Murray-Darling Basin could be summed up in this one sentence. It might reasonably be said that the European experience of evenness and reliability in hydrological conditions is the exception to the rule. Yet it was that expectation that colonial settlers brought with them to the continent with the greatest hydrological variability on the planet.

As an input/output model, the equational nature of the hydrological cycle carries a gain or loss mentality. The question is who or what gains and who loses? Infiltration of water into the ground is only a loss from the point of view of someone who wants to know how much water to take from a stream. Again, the emphasis upon streamflow is explicit. Streamflow accounts for only a small portion of total water, and the emphasis upon what Malin Falkenmark calls blue water fails to recognise other pathways and other connections.[43] In the Australian context, Aplin has noted that the volume of water stored in soil is greater than all surface waters. He further remarked:

> Many Australian ecosystems and individual plant and animal species are well adapted to variations in moisture availability. Many ecosystems can be accurately described as 'stop-go', going into a form of suspended animation under drought conditions

---

42   Linton, 'Is the hydrologic cycle sustainable?', p. 640.
43   M Falkenmark & C Folke, 'The ethics of socio-ecohydrological catchment management: Towards hydrosolidarity', *Hydrology and Earth System Sciences*, vol. 6, no. 1, 2002, pp. 1–9. doi.org/10.5194/hess-6-1-2002; M Falkenmark, 'Freshwater as shared between society and ecosystems: From divided approaches to integrated challenges', *Philosophical Transactions: Biological Sciences*, vol. 358, no. 1440, Freshwater and Welfare Fragility: Syndromes, Vulnerabilities and challenges, 29 December 2003, pp. 2037–49.

and exploding into life when rain falls or floodwaters arrive. One likely conflict in water management, then, is between the human desire for a steady, reliable, year-in, year-out supply of water and the dependence of ecosystems and species on variability.[44]

European-inspired agricultural practices were profoundly threatened by the stop-go nature of the Australian hydrological cycle, and this has set the stage for much of the degradation of Australian catchments.

Appreciation of the connectedness of hydrology to other ecological processes was very much a 1990s concern, as evidence mounted of cumulative impacts on natural ecosystems. In 1987 the United Nations World Commission on Environment and Development chose to open their famous Brundtland Report using the image of the blue planet as a meditation on global connectedness:

> In the middle of the twentieth century, we saw our planet from space for the first time. Historians may eventually find that this vision had a greater impact on thought than did the Copernican revolution of the 16th century, which upset human self image by revealing that the Earth is not the centre of the universe. From space, we see a small and fragile ball dominated not by human activity and edifice, but by a pattern of clouds, oceans, greenery and soils. Humanity's inability to fit its doings into that pattern is changing planetary systems, fundamentally.[45]

The Brundtland Report led to the Agenda 21 conference in Rio. The conference report has 40 chapters, all dealing with some aspect of interrelationship between humanity and its environment. With so many chapters, there was an inescapable conclusion. No aspect of social life or the environment could be considered in isolation anymore, and certainly not water, the subject of Chapter 18. Unlike the quantitative version of the cycle, the connected version emphasised much more than just the quantity of water and the direction of its movement.

Because water moves, and because it is integral to the proper functioning of so many other ecological processes, the hydrological cycle is symbolic of a web-based understanding of humanity and nature. This is partly

---

44 G Aplin, *Australians and their environment: An introduction to environmental studies*, 1988, Oxford University Press, Melbourne, reprinted 1999, both references are from p. 423.
45 United Nations World Commission on Environment and Development, *Our common future*, Oxford University Press, Oxford, 1987, p. 1.

due to water's remarkable range of physical and chemical qualities. Its capacity to move materials and hold others in suspension, to dissolve substances and change form all contribute to its connected nature. It is not possible to understand hydrology without also having an appreciation, at minimum, of disciplines like botany, chemistry, mathematics, ecology, engineering, geomorphology and zoology. Yet, as Bras so bluntly said, this viewpoint was very new in hydrology's long career. Vegetation was a nuisance!

The fifth version of the hydrological cycle is part of a changing view of the world that places interconnection between species, places and things as being of paramount importance. 'Ecologists use connectivity to describe how animals and plants live in interconnected relationships, across multiple spatial and temporal scales.'[46] The two-way relationships – for example, between the ti-tree and the swan, or the molecule of phosphorus and the plankton – are just as important as the things themselves. This includes humans. This is a way of perceiving the environment that would not have come easily to the European migrants that colonised Gippsland. Their world view, which Chapter 3 enlarges upon, was not based upon an idea of connection. In contrast, Indigenous Australians did have such a view. Rose wrote:

> Water, like country, is dreamed into existence. Dreamings created relationships that structure obligations of care, and that constitute webs of reciprocity in the created world ... Rockholes, soaks, wells, rivers, claypans, water holding trees, billabongs, springs and other localized water sources form part of the subsistence geography of country and almost invariably part of the sacred geography as well ... Along with 'owning' country came owning the water, it was a right. However as noted above, it comes with obligations of care.[47]

An ethic of caring for the connections that sustain life is a hopeful prospect, and one that the connected version of the hydrological cycle embodies.

---

46   J Weir, 'Connectivity', *Australian Humanities Review*, no. 45, 2008, p. 153.
47   D Rose, 'Fresh water rights and biophilia: Indigenous Australian perspectives', *Dialogue*, vol. 23, no. 3, 2004, p. 36.

Conclusion

This chapter has demonstrated that there is no such thing as 'the' hydrological cycle. A tour of a few millennia of intellectual history demonstrates human propensity to interpret natural phenomena according to cultural precepts and values.

Ancient Greek philosophical attempts to impose orderly theories on nature were a reaction to the mercurial whims of their pantheon. The vertical model reflected the rigid social boundaries that marked ancient and medieval societies. The eighteenth-century natural theologians, considering themselves to be the apotheosis of God's creation, produced a hydrological cycle that was all about themselves. Nineteenth-century industrial growth fostered an abstract quantitative approach, divorcing water from its ecological context just as so many European people were shorn from their own. Finally, in the late twentieth century, a new version of the hydrological cycle emerged that attempted to reinstate the cycle with the other ecological processes it is in relationship with.

This book does not argue that the fifth 'connected' version of the hydrological cycle is the correct one. It simply recognises it as a product of its own times, where the desire for ecological sustainability conflicts in a multitude of ways with the desire for progress and growth and the idea that nature is a resource. Both world views slog it out daily in local councils and state government planning and environment departments. We live on a daily basis at the conjunction between the fourth and the fifth version of the hydrological cycle.

Hydrologically inspired history therefore recognises and makes explicit the base values and world views of peoples in their catchments, and in their time frame. Or, to echo DuNann Winter, it shows the frame and the walls of the house. Chapter 3 delves more specifically into the frame and walls that constructed the world of colonial Gippslanders.

CHAPTER 3

The earth's thoughtful lords?[1] Nineteenth-century views of water and nature

To some degree all immigrants straddle two worlds, for they carry two memories in their heads and two lands in their hearts.

Richard Broome[2]

Even the dominant grasses, which so attracted the pioneer pastoralists, remain unnamed in their records.

Ian Lunt[3]

Introduction

No child is born into this world a blank slate.[4] Even if she were, her culture would immediately set to filling the void. No matter on what continent or in what time a child is born, she enters a mutually reinforcing web of custom, belief, knowledge and behaviour – her gestalt. Over time, this shapes the child's version of reality.[5]

---

1   Title is adapted from a line of Wordsworth's poem 'The Excursion', cited in WG Hoskins, *The making of the English landscape*, Pelican, Middlesex, 1978, p. 217.
2   R Broome, *The Victorians: Arriving*, Fairfax, Syme and Weldon, McMahons Point, NSW, 1984, pp. xii.
3   I Lunt, 'The distribution and environmental relationships of native grasslands on the lowland Gippsland plain, Victoria: An historical study', *Australian Geographical Studies*, vol. 35, no. 2, July 1997, p. 141. doi.org/10.1111/1467-8470.00015.
4   N Newton Verrier, *Coming home to self: The adopted child grows up*, Gateway Press, Baltimore, 2003. Chapter 2 called 'Adoption and the Brain' discusses recent research in brain neurology that demonstrates that newborn babies can recognise the sound of their mother's voice, among other skills. See also AM Paul, 'How the first nine months shapes the rest of your life: The new science of fetal origins', *Time*, 4 October 2010, pp. 28–34.
5   L Goodison, *Moving heaven and earth: Sexuality, spirituality and social change*, Pandora Press, Hammersmith London, 1992, p. 31.

She will learn how to dress and to speak, and how to relate to others different to her in age, gender and social grouping. She will learn about the properties and characteristics of the physical world she dwells in, and the levels of significance ascribed to them in her culture. Mostly, this learning is done implicitly: through language, through observation and participation in daily life, and informal instruction by family and friends. In societies where literacy was confined to an elite, this osmotic style of learning by experience was the norm.[6] It should be remembered that primary schooling is an invention of the late nineteenth century and only became compulsory in Victoria in 1872.

For a white child born in the nineteenth century in Europe or colonial Australia, this socially constructed world included what we might call 'the common water'. Experiencing thirst, a child learns that water is important to survival. If their family was afflicted by a water-borne disease like cholera, they learnt that it had to be clean water. They would see its cleansing properties at work scrubbing floors or clothes. They could observe the manufacturing uses of water, perhaps seeing horseshoes being plunged into water. Adults around them might complain of the cost of road freight, or they might live near a canal or river. They would attend the baptisms of their younger siblings, observing the use of holy water. Children in wealthy families might observe the construction of garden fountains. All this knowledge – the common water – is acquired without ever setting foot inside a school house.

The purpose of this chapter is to explore this intimate, ubiquitous relationship with water. 'Interactions with water', Strang notes, 'take place within a cultural landscape which is the product of specific social, spatial, economic and political arrangements, cosmological and religious beliefs, knowledges and material culture, as well as ecological constraints and opportunities'.[7] Strang's anthropologically inspired description echoes the discussion of gestalt psychology put forward in Chapter 1. Both emphasise the specific cultural nuances that are the product of the historical time and place, and recognise the interconnecting influences between each aspect.

---

6   Prof B Rawson, pers. comm., 26 May 2010; we were discussing ancient Rome. See also SM Karterakis, BW Karney, B Singh & A Guergachi, 'The hydrologic cycle: A complex concept with continuing pedagogical implications', *Water Science and Technology: Water Supply*, vol. 7 no. 1, 2007, pp. 23–31, doi.org/10.2166/ws.2007.003, quoted in Chapter 1 and which made the distinction between knowledge embodied in experience and knowledge acquired by formal learning.
7   V Strang, *The meaning of water*, Berg Publishing, Oxford, 2004, p. 5.

This chapter seeks to define the key mutually reinforcing beliefs and experiences about water held by nineteenth-century migrants. Using DuNann Winter's analogy of the frame of a building from Chapter 1, I suggest that there are six key parts to the framework of colonial Gippslanders' beliefs around water. I argue that they:

- had a set of religiously derived beliefs about the nature of reality, which included teachings about the relative importance of landscapes, waterscapes and water
- believed in the entwined goals of 'progress' and 'transforming the wilderness'
- were familiar with contemporary literature, poetry and music, which echoed these ideas
- had practical experience of illness related to water
- were principally engaged in agriculture and pastoralism
- were accustomed to transformed landscapes, including trained rivers, canals and swamp drainage.

Together with the understanding of the hydrological cycle laid out in Chapter 2, this led to a consensus that permanent, moderate flowing water in defined channels was the ideal. Over nearly 70 years, emigrants to the Gippsland Lakes Catchment (GLC) took individual and collective action to make the landscape conform to this ideal. This chapter will examine these mutually reinforcing beliefs about water; Chapters 4 to 7 will examine how the beliefs translated into action in the catchment.

Water symbolism in Christianity, or biblical hydrology

While notions of the universal are often suspect, there are two reasonable claims for universality in water.[8] The first is its essentialness to survival, and the second is its resulting elevation to divinity amongst many cultures. From Iris, goddess of the rainbow in Greek mythology, to Tiddalik, the

---

8   Goodison, *Moving heaven and earth*, pp. 30–2.

frog in Kurnai cosmology, human cultures have regularly incorporated water, water-related phenomena and aquatic species into their religious systems.[9] Christianity is no exception.

Christianity contains a common stock of master-narratives or myths, deeply rooted in stories that Europeans had themselves inherited from the Judaeo-Christian and Classical traditions with which most in the nineteenth century were familiar and largely took for granted.[10] Ely described these religious undergirdings:

> Proclaimed from pulpits, and taught in Sunday schools, but implied also in the solemn declarations and formal oaths which were integral to the routines of legal, political and business life, was the idea of a society living under divine as well as human law, a divine law, furthermore, spelled out in terms of fearsome sanctions for disobedience. Many saw these sanctions as laid down in what was sometimes called God's law book – the Bible.[11]

Even the least zealous would have absorbed the basic teachings that begin with Genesis. Christianity's creation story is quite explicit, the earth was a designed world for human benefit:[12]

> So God created man in His own image; in the image of God He created him; male and female He created them. Then God blessed them, and God said to them, 'Be fruitful and multiply; fill the earth and subdue it; have dominion over the fish of the sea, over the birds of the air, and over every living thing that moves on the earth.'[13]

---

9   See T Tvedt & T Oestigaard (eds), 'A history of the ideas of water: Deconstructing nature and constructing society', *History of water, Series 2, The ideas of water from antiquity to moderm times*, IB Tauris, London, 2010, pp. 1 and 16–21.
10 G Davison, *Narrating the nation in Australia*, The Menzies Lecture, 2009, Menzies Centre for Australian Studies, Kings College, London, University of London, 2009, p. 3. Davison's discussion is mostly focused on nationalist myths, but the frame of Judaeo-Christianity applies equally to environmental matters.
11 R Ely, 'Australian Federation, religion and James Bryce's nightmare', in *Intellect and emotion: Perspectives on Australian history; essays in honour of Michael Roe*, jointly published by Centre for Australian Studies, Deakin University, and Centre for Tasmanian Historical Studies, University of Tasmania, Geelong, Vic., 1998, p. 155.
12 A number of writers have pointed out that the anthropocentric view predates Christianity, e.g. R Sheldrake, *The rebirth of nature: The greening of science and god*, Century, London, 1990; and G Seddon, *Landprints: Reflections on place and landscape*, Cambridge University Press, Cambridge, 1997, chs 20 and 21. My point is that a nineteenth-century settler with minimal education would be unlikely to know this.
13 *The Holy Bible*, New King James Version, Thomas Nelson Inc., 1982.

While there is now a strong challenge to this belief, it was the common view in the nineteenth century.[14] Much of the opposition to Darwinian theory centred around its challenge to the concept of a divinely designed earth.[15]

While Cannon suggests a post-Darwinian break in Christianity, evidence from the study area does not support this.[16] If settlers did not genuinely believe the Christian narrative, it is hard to imagine why they would spend time away from home and divert money away from farms and businesses to support the churches. One might argue that the importance of religion rose for those who were moving into potentially perilous new lands. Davison has shown that the theme of Genesis and Exodus provided moral legitimacy to the process of colonisation, thus making the daily struggles of farm life seem meaningful.[17] Why else was one of the first acts of a new community to build a church?[18]

---

14 C Ponting, *A new green history of the world: The environment and the collapse of great civilisations*, Vintage Originals, London, 2007, pp. 119–21. Many Christians are also questioning this tenet of faith, e.g. the Archbishop of Canterbury released a statement in May 2005, which instructed all members of the Anglican Church to take the view that the environment is now core business for men and women of faith (Lecture given on 8 March 2005, University of Kent, www.archbishopofcanterbury.org/sermons_speeches/050308.htm, accessed 24 June 2005, site discontinued). For an account of neo-pagan critiques, see G Harvey, *Contemporary paganism: Listening people, speaking earth*, New York University Press, New York, 1997; and L de Angeles, E Restall Orr & T van Dooren (eds), *Pagan visions for a sustainable future*, Llewelleyn Publications, Woodbury, Minnesota, 2005.
15 This also echoes the divine design notion of hydrology. See A Moyal, *A bright and savage land*, Penguin, Ringwood, Vic., 1993, p. 142.
16 M Cannon, *Australia in the Victorian age: Life in the cities*, 3rd edn, Penguin Books, Ringwood Victoria, 1988, p. 77; R White, *Inventing Australia*, Allen and Unwin, North Sydney, 1981, pp. 68–9. This is not to say that the impact of Darwin's theory was not debated in Gippsland. Scientific challenges to Christianity were debated in the catchment's various churches, e.g. 25 July 1885 the *Bairnsdale Advertiser* reported a lecture on geology and the Bible being given by Rev. Morton in the Presbyterian Church. As Oelschlager notes, the Western world had no notion of the deep past, or a Palaeolithic past until the middle of the nineteenth century when Darwin and Lyell set the cat amongst the pigeons, and destroyed the idea of the world of being 4,000 years old. M Oelschlager, *The idea of wilderness: From prehistory to the age of ecology*, Yale University Press, New Haven, 1991, p. 6. See also I Keen, 'The anthropologist as geologist: Howitt in colonial Gippsland', *The Australian Journal of Anthropology*, vol. 11, no. 1, 2000, p. 79. doi.org/10.1111/j.1835-9310.2000.tb00264.x.
17 Davison, *Narrating the nation in Australia*.
18 Maddern counted 58 Anglican churches established in the whole Gippsland area, i.e. not just the catchment, between 1856 and 1900. IT Maddern, *Light and life: A history of the Anglican Church in Gippsland*, Enterprise Press, Sale, n.d., Appendix 1.

Figure 3.1: Church at Sth Yinnar. A typical example of the small community-built churches that dotted the landscape.
Source: Author.

Figure 3.2: Tambo Valley, illustrating the gentle slopes that were so sought after.
Source: Author.

Diary evidence from the GLC shows little evidence of agnosticism or atheism. George Auchterlonie made a point of recording his vow to God in his diary. The full entry for 22 February 1869 reads:

> Finished the roller. Weather stormy. Reading Dodridge's Rise and progress. Have resolved today to Seek to become a true follower of Christ. May My Creator enable me to keep this resolve and may it never be forgotten by me, amen.

Caleb Burchett was instrumental in setting up church services in Poowong, commencing in a tent on his selection. The Rev. John Watts preached the first service to a large crowd in uncomfortable heat on 17 February 1878. Margaret McCann was a strict Methodist, living on a Stradbroke sheep farm before, during and after the Federation drought. Her diary records regular attendance and involvement in church activities. Miss Caughey of Traralgon was an assistant Sunday school teacher, while George Auchterlonie liked to record his impressions of various preachers in his diary.[19] The daughters of surveyor William Dawson, who married into the Macleod clan of Bairnsdale, were described as 'wonderful women, scrupulously honest and very religious – Chapel every morning and prayers and readings with the servants – who spent their lives caring for others'.[20]

---

19   Reminiscences of Caleb Burchett, SLV, MS 8814, MSB 436; Diary of Margaret McCann, SLV, MS 9632, MSB 480; Diary of AM Caughey SLV, MS 8735, MSB 434; Diary of George Glen Auchterlonie, CGS, 4060; *East Gippsland Historical Society Newsletter*, vol. 1 no. 3, letter from Frederick Gray to Mr Lewis, dated 4 August 1855, describing his life at Lindenow. The letter concludes by asking Mr Lewis who the new preacher is in his home village in England and for his opinion of him.

20   P Macleod, *From Bernisdale to Bairnsdale: The story of Archibald and Colina Macleod and their descendants in Australia, 1821–1994*, the author, Nar Nar Goon North, 1994, p. 114. Another example is Samuel Lacey who founded an engineering works in Sale, recording in his diary how grateful he was to get off the ship in Melbourne. He wrote on 10 June 1868: 'We had the unspeakable pleasure of going to Chapel today'. Cited in SL Lacey, *The Laceys of Gippsland: The history of a pioneer firm 1870–1970*, the author, Sale, n.d., p. 9. According to his son, Samuel Lacey also derived some professional meaning from the Book of Kings, Lacey, *The Laceys of Gippsland*, pp. 50–2. See J Hibben, 'The Disher family in the nineteenth century', BA Hons thesis, University of Melbourne, 1978, p. 18 for an extract of a letter to Mrs Disher praising her religiosity.

Further evidence of Gippslanders' religiosity is found in the newspapers. The local press regularly reported on religious activities, made analogies with Bible stories and echoed biblical language.[21] In one example, the paper said:

> A correspondent remarks: The ways of pedestrians are hard in the good town of Sale, and their boots succumb quicker here to the rough pebbles strewn along its paths than anywhere else … Wisdom crieth aloud for some asphalt or cheap tar pavement to be laid down, but her voice may be likened unto the voice of the wild ass of the wilderness, which no man regardeth.[22]

In 1874, the paper described councillors Ross and Leslie of Sale Borough Council as penitent Adams.[23] There are regular reports on the visits of various temperance lecturers, and on drinking offences, that were strongly religious in tone.[24]

This trend continued throughout the whole of the nineteenth century. In later years, the *Gippsland Times* printed whole sermons in the religious column, and these occasionally used water metaphors to make their point. The following example combines a number of common expressions that wend their way through Chapters 4 to 7:

> We shall never do any good, either for ourselves or others, if we set about our work in a dull despairing way, as though, after all, it were of no use, and as though we ourselves and others were mere insects not worth working for … To make the river flow freshly and sweetly across the plain, the springs must be high among the hills. To make man clear the stagnating currents of his life, his faith and hope must be among the heights of Heaven … So I would say to you, never miss the opportunity of contemplating the character, of studying the biography, of imitating the example of a man who was greatly good. Such a man sheds a light around him

---

21 For example, *Bairnsdale Advertiser*, 2 May 1885, Cunninghame correspondent: 'Mr JD Stocks held divine service at the state school … He pointed out most forcibly how prone we are to sin, and that God is angry with the wicked every day'; *Gippsland Farmer's Journal* (*GFJ*), 1 July 1887, included a full sermon on purgatory and judgement based on Lux 16:22 and 23.
22   *GT*, 11 March 1876.
23   *GT*, 22 December 1874.
24   *GT*, 12 October 1887 for temperance; *GFJ*, 10 February 1887, 'Another illustration of the ruinous effects of over-indulgence in drink was afforded at the Warragul police court' when WF Page was tried for stealing £2 15 s; *GFJ*, 2 September 1887, opening of the Temperance hall in Traralgon. '[Mr Groom MLA] though the inculcation of temperance principles amongst the young was most desirable, and he would especially remind them that so long as they kept temperate and acted through life in a straightforward manner, they could not fail to get on in the world.'

which transfigures the world, as a ray of sunshine transfigures the wet foliage of a tree. Often in life we are like a traveler by night on dark, bad roads. We are in danger of being lost in bogs and quagmires, of falling among thieves.[25]

All of these examples suggest a common metaphorical language derived from Christianity and its antecedents in the eastern Mediterranean landscapes. Even if Cannon is correct, a lip service attitude still entails attendance at services. Absorbing attitudes about water from descriptions in the Bible did not need an unquestioning belief in creationist beliefs, nor particularly devout faith. So what does the Bible say about water?

While the different brands of Christianity placed different emphases on aspects of faith, in essence it was a rule-based, hierarchical and dualistic system. In the Bishop of Melbourne's sermon delivered at St Paul's in Sale in February 1884, this insistence on rules and obedience was quite clear. The kingdom of Heaven was available to all believers 'unless there was a stubborn, obdurate resistance … to His divine will'.[26]

Dispenza describes the dualism inherent in Christianity:

> The principal prayer of Western civilization, instead of helping to establish the bond we are all seeking between our Source and us, actually has been keeping us utterly and eternally apart. The Lord's Prayer is not about spiritual union with God; it is a heartbreaking hymn of separation … The first six words tell the entire story of unbridgeable separation. There is a God, and he is in heaven. He is not here with us or in us. On the contrary he is far off in a place called heaven, which in the cosmology of the Middle Ages is high above us, in the sky beyond the clouds, beyond the stars.[27]

---

25   *GT*, 18 June 1896.
26   *GT*, 18 February 1884. The bishop went on to say that he believed that death did not end the 'probationary period' and, in particular, the children of 'thieves and prostitutes' who were 'tainted with vice and revolting wickedness' would continue their 'probation' at God's leisure. The key elements were a belief in an omniscient male God who dwelt in Heaven, and a set of rules about what constituted Christian behaviour. The rules were founded upon the notion of the Fall, and humans were inherently sinful. The rules existed to save humans from themselves. Adherence to the rules in their entirety would gain an eternal life with God and his son, Jesus, in Heaven. Failure to obey God's precepts lead to the opposite, damnation in Hell.
27   J Dispenza, *God on your own: Finding a spiritual path outside religion*, Josey Bass, San Francisco, 2006, pp. 70–1.

This separation is played out well beyond the location of God vis-à-vis mankind. It extended to a range of other relationships, including man and woman, high and low, light and dark, good and evil, mind and body, white and black, above and below.[28] It also applied to water in the landscape, with the two major dualisms being wet versus dry and movement versus stillness.

It is not that they are dualisms *per se*, it is their polarisation that is especially important. Weir calls this hyper-separation.[29] Hyper-separation describes a mindset that emphasises difference rather than continuity and rigidly holds those differences apart. Additionally, one part of the dualism is judged as more socially desirable than the other. Heaven/man/high/light/good/mind/white/wet/flowing is preferable to earth/woman/low/dark/evil/body/black/dry/still.[30] Hyper-separation allows for no greyness, no elasticity, no shifting between categories. No multiple relationships and very few connections.

The Bible is littered with references to water that display this dualistic perception. For example, there are 67 uses of the word 'water' in Genesis alone. It is in Psalms 59 times and in Deuteronomy 22 times. The word 'river' receives 26 references in Exodus and 18 in Joshua. Words relating to evaporation are not as numerous but still present. 'Dry' is mentioned most in Proverbs, with seven appearances, and six in Joshua. The word 'wither' is in Psalms six times.[31]

How water was depicted in these occurrences creates a form of moral geography, presenting a consistently polarised portrayal of water. Rivers, springs, rain and dew are depicted positively, while floods, hail, frost, drought and swamps are not. Clouds and mist can go either way; for example, clouds can obscure the sun, which was often portrayed as a symbol of God, or they bring God's love in the form of rain.

Christian religious literature encourages believers to value permanent flowing rivers and to shun marshes and deserts. The Garden of Eden had perennial streams and presumably, as Adam and Eve had no need for clothes, a delightfully mild climate. When they were expelled from

---

28   Strang, *Meaning of water*, p. 90.
29   J Weir, 'Connectivity', *Australian Humanities Review*, no. 45, p. 154. See also J Weir, *Murray River Country: An ecological dialogue with Traditional Owners*, Aboriginal Studies Press, Canberra, 2009, pp. 48–50.
30   Goodison, *Moving heaven and earth*, pp. 13 and 26.
31   This is from a search on the King James version of the Bible, using the University of Virginia's e-text repository.

paradise, they entered a desert-like wilderness with nary a stream in sight. The Book of Revelation is quite explicit in the association of God, His virtue and love with rivers and springs. Chapter 7:17 reads 'For the Lamb which is in the midst of the throne shall feed them, and shall lead them unto living fountains of waters; and God shall wipe away all tears from their eyes'. In Revelation 22:1–2 the meaning is even more explicit: 'And he shewed me a pure river of water of life, clear as crystal, proceeding out of the throne of God and of the Lamb'. Feliks notes that flowing water in rivers and springs is used as a motif of resurgence of faith, of the triumph of obedience to God's will:

> The righteous man is compared to a 'tree planted by the water, that spreadeth out its roots by the river, and shall not see when heat cometh, but its foliage shall be luxuriant, and shall not be anxious in the year of drought, neither shall cease from yielding fruit'.[32]

Obedience was a smart strategy given the behaviour of the Old Testament God. Dispenza has described him as 'angry, terrifying, demanding, occasionally helpful, vengeful, annoyed, hard to please, sometimes merciful, usually merciless – and as volatile as a megalomaniac on a rampage. If you do not do exactly as he says, he will smite you and you will die'.[33] This psychopathic version of God was fond of using elements of the hydrological cycle as a tool of punishment, further reinforcing a dualistic perspective.

God's greatest act of vengeance was the Flood, sent as punishment for deviation from his moral laws.[34] Then there was drought. Isaiah 19:5–10 is an explicit and depressing vision of the consequences of God's displeasure:

> And the waters shall fail from the sea, and the river shall be wasted and dried up. And they shall turn the rivers far away, and the brooks of defence shall be emptied and dried up: the reeds and flags shall wither. The paper reeds by the brooks, but the mouth of the brooks, shall wither, be driven away, and be no more. The fishers shall also mourn, and all they that spread nets upon the water shall languish. Moreover, they that work in fine flax, and they that weave, shall be confounded, And they shall be broken in the purposes thereof, all that make sluices and ponds for fish.

---

32 Jeremiah 17:8 quoted in Y Feliks, *Nature and man in the Bible: Chapters in biblical ecology*, Soncino Press, London, 1981, p. 199.
33 Dispenza, *God on your own*, p. 71.
34 Genesis 6:5–12 does not specify precisely the behaviour that caused such displeasure, only that it was evil in the Lord's eye and was violent.

The Book of Jeremiah also gives an account of a terrible drought, sent as punishment for the wickedness of the people in forsaking God. While the obedient worshippers will experience the blooming of the desert as prophesied by Joel, Zecharaiah and Ezekiel, the sinful will be punished with thirst, desiccation and withering. Could there be any clearer dualism than that?

Bogs and marshes are used to symbolise being spiritually lost. Psalm 69 says:

> For the waters are come in even until the soul. I am sunk in deep mire, where there is not standing; I am come into deep waters, and the flood overwhelmeth me. I am weary with my crying; my throat is dried …

Here is a spiritual wilderness described as being drowned in a bog. John Bunyan, author of the famous book *Pilgrim's Progress*, continued the metaphor. The 'slough of despond' into which Christian falls became a metaphor that has been worked over and over again, as Giblett shows.[35]

Migrant settlers thus had a strong set of stories about the moral values of water in different places and phases of the hydrological cycle. Some waters were good and some were bad, natural disasters were divine punishment and this was presented as a universal truth. Along with their physical belongings, they uncritically transported these beliefs to their new country.

Water in literature, poetry and music

Emotional and spiritual engagement with nature is often facilitated through the arts. In this section, I demonstrate from newspapers reports that settlers in the GLC participated in artistic pursuits that represented water. Art, literature and music reinforced the perceptual disconnect between migrant settlers and the hydrological cycle, because they were authored by Europeans who reproduced the moral geography and dualisms of Christianity. They related to water from the same environmental context as the scientists who developed the concept of the hydrological cycle. Hence, the images and metaphors they employed in their art came from their empirical experience of permanent flowing rivers fed by high

---

35   R Giblett, *Postmodern wetlands: Culture, history, ecology*, Edinburgh University Press, Edinburgh, 1996, ch. 7.

rainfall. The use of these images and metaphors provided a similar view of water as found in the Bible, and for the atheist/agnostic (there must have been some) would have performed the same function as the Bible.[36]

Nineteenth-century European arts were dominated by romanticism, a reaction to the ugly impacts of rapid industrial expansion.[37] It celebrated and isolated nature, and blanked out certain aspects of human changes to the landscape: 'Urbanisation, industrialization, parliamentary acts of enclosure, the impact on the rural poor of legislation against vagrancy or poaching – all these could be excluded from consideration'.[38] The combination of hydrological determinism and nostalgia for a lost rural idyll flavoured much of the music, poetry and literature that colonial Gippslanders enjoyed, and that provided encouragement for the colonial desire to make Australia an English echo.[39]

Artistic pursuits provided a much-needed break from the difficulties of settlement, where travel was arduous, company infrequent and freight exorbitant. Synan provides an example of how music helped to soften the mishaps of travelling. A passenger on board a steamer stranded at the bar of the entrance to the Lakes recollected that: 'In the engineer we had a first class violinist who played all kinds of songs and dances, and the shouts of merriment resounded through the forest on that memorable night'.[40] Settlers could join Mechanic's Institute's libraries or the small circulating libraries run by local newsagents and booksellers, like Louis Roth, a bookseller, stationer and fancy goods seller in Sale.[41] Alternately, there was a healthy flow of travelling performers through the region and many local concerts were performed, usually to raise funds for building

---

36   There is only one diary whose writer appears to be largely uninfluenced by religious belief. Diary of Duncan Johnston, Ensay, 1882, CGS, 00317. Duncan was the son of Thomas Johnston who had the first hotel at Little River, which later became Ensay.
37   J Beattie, 'Exploring trans-Tasman environmental connections 1850s to 1900 through the imperial careering of Alfred Sharpe', *Environment and Nature in New Zealand*, vol. 4, no. 1, April 2009, p. 44.
38   M Andrews, *Landscape and Western art*, Oxford History of Art, Oxford University Press, Oxford, 1999, p. 167.
39   This appears to be so regardless of age group. Some of the most popular fairy tales were also derived from the same cultural milieu, e.g. Snow White and the magic mirror, Beowulf and the monster of the deep, Narcissus and Echo, The Frog Prince, Swan Lake/The Twelve Wild Swans.
40   P Synan, *Highways of water: How shipping on the Lakes shaped Gippsland*, Landmark Press, Drouin, 1989, p. 39.
41   *GT*, 15 February 1897. Roth had a reading library for 1 s a month: 'Latest works of Rider Haggard, Ethel Turner Fielding, Marie Corelli, etc, may be had'.

community facilities.⁴² Concert programs were often reproduced in local papers, thus providing a reasonable insight into the popular music tastes of GLC settlers. Musical talent was highly regarded in such small and isolated communities. Copeland singled out the otherwise forgotten saddler Davy Small and Sergeant Allison who had excellent voices and were active in many choirs and performances in Warragul.⁴³ Music was a significant part of Jessie Login's memories of early Sale, when she described the bark hut home of the Duncan family:

> The garden in front of them was a dream of beauty, gay with flowers, the drawing room a bit of Edinburgh transported. On the grand piano the sisters divinely played their duets, and composed music in this primeval forest at Kelvin Grove.⁴⁴

Two worlds in their head, indeed.

Music was one area where women could participate on a slightly more equal level. Gippsland's papers obscure women, except for reports on fundraising, organising tea meetings or performing as part of organisations like literary societies. These became more numerous in the later years, coinciding with both population growth and progressive social movements.⁴⁵

---

42   See, for example, *GT*, 23 October 1869, hospital fundraiser concert, attendance down due to wet weather. Watery songs were 'See our Oars with Feather'd Spray', 'The White Squall'; *GT*, 19 January 1881, advertisement for a lecture in the Victoria Hall, Sale on the songs and music of Ireland by Rev. DF Barry in aid of St Benedicts Monastery in Inverness Scotland; *GFJ*, 2 June 1887, report of the Blind Asylum singers concert: songs were 'The Queen of the Earth', 'Crowning of the Sea'; *GT*, 11 August 1886, ad for the Moonlight concert fundraiser for Gippsland hospital. A trio called Glorious Apollo to be performed by Messrs Tindall, Ingelton and Futcher.
43   H Copeland, *The path of progress: From the forests of yesterday to the homes of today*, Shire of Warragul, Warragul, 1934, p. 237.
44   Quoted in C Daley, *The story of Gippsland*, Whitcombe & Tombs for the Gippsland Municipalities Association, Melbourne, 1960, p. 42.
45   For examples of women performing in public, see *Bairnsdale Advertiser*, 27 June 1885, Ladies Benevolent Society Monster Tea Meeting: Mrs Mudie sang 'Steer My Bark to Erin's Isle', Mrs JF Stuart sang 'Queen of the Earth', Miss Odgers sang 'Bend in the River', 'and was considered to be the gem of the evening'; *GT*, 19 July 1886, Concert at Heyfield for the COE church, Miss Cross from Maffra sang 'Alton Water'; *GT*, 23 July 1886, Miss Horstman sang 'Beside the Sweet Shannon'; *GT*, 11 March 1897, St Patricks Day concert, Victoria Hall Sale, Duet called the 'Silver Rhine' to be sung by Mr P Cox and Miss A Cox.

Figure 3.3: A bark hut in the Neerim area, Archibald J. Campbell, 1877.
Source: National Library of Australia, Accession no. 644224.

The reference to water in song titles occurred often enough to become notable as the newspaper research continued. Due to the ephemerality of sheet music, many of the words of the songs have proved very difficult to track down. However, some have come to light, and some guesses can be made from the titles alone as to what kind of water imagery they conveyed. One of the most popular songs was 'Come Over the Stream Charlie', significant to the many Scots in the audience. At the Queen's birthday holiday celebrations in 1876, the Caledonian Society members performed a song called 'Children of the Mist'. Both these songs suggest the importance of Scottishness and its link to waterscapes. 'Sunshine and Rain' was also a highly popular song. Ada Crossley, Gippsland's rival to Nellie Melba, was reported in the *Gippsland Times*, on 11 January 1897, saying it was one of her favourite songs. The words to 'Sunshine and Rain' are a classic cheer-up tune:

> Lift your eyes to yon Daygiver
> Look up higher, hoping still
> Tho the rain is on the river
> The sun is on the hill.
> Tho the rain is on the river
> The sun is on the hill.

The song also reflects a bias towards sunshine and also perpetuates dualism of high and low through the high hill versus low valley symbolism.

Music was a significant part of tea meetings, and the selection of music often conveyed this dualistic nature of thinking. The Presbyterian Tea Meeting reported in the *Gippsland Times* of 27 October 1876, organised by Mesdames Login, Sprod, Hutchison, Miller, Blanch, Coupar, Bearup and Campbell, and the Misses Login, Gresley, Law, Leslie, Monger, Coupar and Campbell, had a heavy musical component. Attendees sang a hymn called 'There's a Light in the Valley', which combined both a high/low dualism with the light/dark one. In addition, they also sang Psalm 100, Psalm 24 and another hymn called 'Ring the Bells of Heaven'.

Images of water in popular music were echoed in literature.[46] Perhaps the earliest and most direct literary allusion comes from William Odell Raymond. He squatted on the Avon River, and the settlement that grew up around his run was called Stratford. He also built the Shakespeare Hotel for his housekeeper, Mrs Woods.[47] Shakespeare's popularity ensured that at least some common water references would make it into the popular repertoire. Lucy Bell recalled playing Nerissa in the Merchant of Venice, while John O'Connor credited his education to his father who would read 'Byron, Moore, Shakespeare, Milton and all the great writers'.[48]

Beyond Shakespeare, there is some evidence of discussions of literature that involved water imagery. For example, on 12 April 1870 the *Gippsland Times* published an article on the life and work of Tennyson, in anticipation of the new poem 'The Quest for the Holy Grail'. In the late 1880s the Traralgon Literary Society formed. The *Gippsland Farmer's Journal* provided a list of pieces read at the October 1887 meeting, which included two poems with watery themes: 'Traces of Ocean' and 'Bingen on the Rhine'.[49] Giblett has discussed extensively the role of nineteenth-century literature in reinforcing the negative attitudes towards swamps, particularly citing a number of Dickens's publications.[50] As one of the most widely read serialised novelists of his day, Dickens knew his market. Less well-known writers also employed similar kinds of water imagery to convey their stories. For example, the *Gippsland Times* on 17 April 1891 started a serialised new story called the 'Secret of the River' by Miss Dora Russell. Carried over several weeks, this was a combined murder/love triangle story employing a large number of water metaphors, not least of which was the riverside location.

---

46 For example, just in poetry alone, Wordsworth's *The River Duddon* and *The Prelude*, which describe ice skating on a lake; Elizabeth Barrett Browning's 'A Musical Instrument' using imagery of Pan, reeds and rivers; Walter Scott's poem 'The Lady of the Lake', Longfellow's 'Maidenhood', Milton's *Comus*, includes the rebirth of Sabrina as a spiritual goddess of the river and well as *Paradise Lost* with its four rivers; Samuel Taylor Coleridge's 'Kubla Khan', 'Where Alph, the sacred river, ran/ Through caverns measureless to man', and his poem about pollution of the Rhine called 'Cologne'; Alfred, Lord Tennyson's 'The Lady of Shallot'; Robert Browning, 'Child Rolande to the Dark Tower Came'; and TS Eliot, whose poem using rivers symbolism is the epigraph to Chapter 2.
47 M Watson, 'William Odell Raymond', *Gippsland Heritage Journal*, vol. 2, no. 2, 1987, p. 35.
48 Memoirs of John Joseph O'Connor, SLV, MS 10409, MSB 208, p. 5; Memories of the early settlement of Narracan by Lucy Bell, in RM Savige, *History of the Savige family*, the author, Frankston, 1966, p. 123.
49 *GFJ*, 13 October 1887.
50 Giblett, *Postmodern wetlands*, pp. 14–15.

Individually, none of these references amount to much, but, taken collectively, they demonstrate that colonial Gippslanders regularly heard and read a wide variety of European-modelled water metaphors in their public and private entertainments. The celebration of a lush, well-watered rural landscape encouraged colonial readers to look around their local catchments and judge them poorly. More than anything, this survey demonstrates the metaphorical nature of water, with the use of water imagery to describe states of being. In Chapters 4 to 7, I discuss examples pertaining to each aspect of the hydrological cycle.

Trained hydrologies

Figure 3.4: An example of trained hydrology, channels constructed in Walhalla.
Source: Melissa Smith.

The GLC's new residents, almost all deriving from a European nation, were familiar with a variety of altered landscapes and trained hydrologies. This was particularly the case for British migrants. According to Simmons, the British landscape was the most heavily cleared in all of Europe.[51] What

---

51 IG Simmons, *An environmental history of Great Britain: From 10,000 years ago to the present*, Edinburgh University Press, Edinburgh, 2001, p. 152. 'In 1800, England and Wales were the least wooded area in Europe with an estimate total of less than 2 million acres (810 000ha of woodland).'

most took as the natural rural landscape was in fact a product of hundreds of years of conscious change, and the nineteenth century ushered in an even more accelerated process of transformation.[52] The alteration of ecological processes, from the domestic scale upwards, would have been widely accepted as a beneficial part of daily life.

Hydrological manipulation was a key activity in the transformation of ecosystems and landscapes. Many of the hydrological interventions carried out by Europeans all over the globe involved manipulating the wet/dry and the moving/still dualism. Water histories vary according to the place and the philosophical bent of the author, but all of them tell a story of increasing attempts to control water's natural variability and fluidity. River histories are abundant, reflecting their importance as well as their permanence and visibility in the landscape.[53] River regulation and swamp drainage practices are solely designed to create uniformity – in depth, length, course and available moisture. This was particularly important in England and other countries in Europe that relied on extensive canal systems to transport goods. Hoskins says of the canal system:

> Not only did they bring stretches of water into country lacking in them … with consequent changes in bird and plant life, but they also brought – mostly for the first time – aqueducts, cuttings, embankments, tunnels, locks, lifts and inclined planes, and many attractive bridges, and they greatly influenced the growth and appearance of many towns.[54]

---

52   John McNeil's environmental history of the twentieth century obviously makes considerable comparisons to the rates of change in the nineteenth century. JR McNeill, *Something new under the sun: An environmental history of the twentieth-century world*, WW Norton and Co., New York, 2001.
53   For European examples of river histories, see P Ackroyd, *Thames: Sacred river*, Chatto and Windus, London, 2007; M Cioc, *The Rhine: An eco-biography*, University of Washington Press, Seattle, 2002; D Blackbourn, *The conquest of nature: Water, landscape and the making of modern Germany*, WW Norton and Co., New York, 2005; S Haslam, *The historic river: Rivers and culture down the ages*, Cobden of Cambridge Press, Cambridge, 1991. For American examples, see R White, *The organic machine: The remaking of the Columbia River*, Hill and Wang, New York, 1995; D Worster, *Rivers of empire: Water, aridity, and the growth of the American West*, Pantheon Books, New York, 1985. For Australian examples, see R Longhurst & W Douglas, *The Brisbane River: A pictorial history*, WD Incorporated, Brisbane, 1997; L McLoughlin, *The Middle Lane Cove River: A history and a future*, Centre for Environmental and Urban Studies, Macquarie University, North Ryde, 1985; J Roberts & G Sainty, *Listening to the Lachlan*, Sainty and Associates, Murray-Darling Basin Commission, Potts Point, NSW, 1996; G Seddon, *Searching for the Snowy*, Allen and Unwin, St Leonards, NSW, 1994; D Connell, *Water politics in the Murray-Darling Basin*, Federation Press, Annandale, 2007; N Burningham, *Messing about in earnest*, Fremantle Arts Centre Press, Fremantle, 2003; P Sinclair, *The Murray: A River and Its People*, Melbourne University Press, Melbourne, 2001; Royal Historical Society of Victoria, *River of History: Images of the Yarra since 1835*, Royal Historical Society of Victoria, Melbourne, 1998.
54   WG Hoskins, *The making of the English landscape*, Pelican, Middlesex, 1978, p. 217.

The engineered waterscape became commonplace.[55] The practice of creating water meadows was widespread in rural areas, as was aquaculture. Hoskins notes that canals were the purveyors of a new world of products that broke down the particularity and regionalism of areas – for example, by introducing standardised building products. The railways produced even more change in social and physical landscapes, and were wider in their spread.[56] No emigrant could have escaped a stark contrast in his or her head between Great Britain and the GLC. As Broome notes in the epigraph that heads this chapter, they inhabited two psychic worlds. Gippsland, with its bogs, terrible roads, periods of extended drought, bushfires and the ubiquitous eucalypt, did not compare well to cool, green, familiar and civilised Europe.

Rivers and floodplains made a logical choice for early settlement patterns. In *The Historic River*, Haslam describes the many ways in which Europe's rivers from the fourteenth century onwards were used. Grain milling was an important activity around which urban settlements grew. Safe crossing places, fish and birds for protein, building materials, and water for irrigating crops all played a part.[57] Many riparian plants had medicinal or domestic uses – for example, the bark and leaves of *Alnus glutinosa* made a tonic and astringent and *Valeriana officinalis* is still used to treat anxiety.[58]

---

55 See also S Halliday, *Water: A turbulent history*, Sutton Publishing, Phoenix Mill, Gloucestershire, 2004.
56 Elizabeth Gaskell's novels, dramatised by the BBC, centre around the social upheaval caused by the arrival of the railway into Cranford in the 1840s.
57 There were also religious reasons behind some of the locations. Guillerme demonstrates that in northern France the rivers on which cities were founded were usually associated with gods and goddesses, e.g. the Thara River was diverted by a 2-kilometre-long canal when they could have chosen the closer Liovette River. The name Thara derived from a Gallic god Taranis, the god of light. There was a similar situation with the Iton River, which was diverted to surround the city of Evreux. Most of these cities in later years developed a mythology where the first Christian bishop or martyr does some kind of battle with the local pagan deity, often a river spirit, and drives them forth from the land. AE Guillerme, *The age of water: The urban environment in the north of France, AD 300–1800*, Texas A&M University Press, Texas, 1988, pp. 9–14.
58 Haslam, *The historic river*, pp. 83–6. The dried root and flowers of meadowsweet (*Filipendula ulmaria*) treated diarrhea, influenza and ulcers. *Nasturtium acquaticum* (leafy stems) acted as a stimulant, diuretic and antipyretic. The fresh leaves of mint (*Mentha aquatica*) are, like valerian, still well known for their ability to soothe digestive complaints. The willow (*Salix alba*) was the forerunner of aspirin, giving pain control and fever reduction. Riverside rushes were used to make lights, which were usually all the poor could afford. Making them was a long process. The largest and longest reeds had to be cut, soaked, peeled, bleached in the sun and then dipped in hot fat, all of which would provide light for approximately an hour. One-and-a-half pounds of processed rushes could produce 1,200 hours of light, which was apparently in 1778 sufficient for a family for a year.

Because of rivers' social, economic and spiritual significance, social conflicts often coalesced around them. Fishing, boating, swimming, hunting, water supply, sanitation, rubbish dumping, tourism, city building and crop growing don't usually exist in the same sentence, but they regularly exist in the same limited geographical space. This naturally creates conflict between different groups. Waterscapes are, therefore, also powerscapes, reflecting how some classes can impose their water vision and dominate access to resources.[59] Cioc's work, *The Rhine: An Eco-biography*, is a prime example of the kinds of powerbroking, engineering and conflicts that coalesced around rivers. Given the Rhine's status as the principal trade river in Europe, these themes are writ large:

> The Rhine's emergence as Europe's preeminent commercial river was hardly the work of nature alone. Generations of planners, industrialists and civil engineers all contributed to this achievement. Together, they straightened the river's channel, constrained its floodplain, regulated its flow, and profoundly manipulated its ecosystems, all with the goal of making it obey human visions of safety, efficiency and productivity.[60]

The fights over drainage schemes for flood plains, bogs, marshes and other low-lying land in Europe tended to revolve around competing visions of productivity as well. Peasants with common rights often fought back to retain their access to lands, but most were dispersed. Both Darby and Williams detail the class struggles that accompanied large-scale wetland drainage schemes in Lincolnshire and Somerset in the early modern period.[61] Reclamation was widely promoted as reclaiming a savage nature, and frequently invoked biblical precedent to justify the changes. Outside of Britain, similar schemes and attitudes could be found in the Netherlands, highland Scotland, Amiens, Venice and other parts of northern Italy, southern Spain and even portions of Switzerland.[62]

---

59   Karl Wittfogel ignited this field with his work on power and irrigation practices in ancient Mesopotamian civilisations. K Wittfogel, *Oriental despotism*, Yale University Press, New Haven, 1957.
60   Cioc, *Rhine*, p. xi.
61   HC Darby, *The changing fenland*, Cambridge University Press, Cambridge, 1983; M Williams, *The draining of the Somerset Levels*, Cambridge University Press, Cambridge, 1970.
62   P van Dam, 'Sinking peat bogs: Environmental change in Holland 1350–1550', *Environmental History*, vol. 6, no. 1, 2001, pp. 32–46; PJEM van Dam, 'Ecological challenges, technological innovations: The modernisation of sluice building in Holland 1300–1600', *Technology and Culture*, vol. 43, no. 3, July 2002, pp. 500–20. doi.org/10.1353/tech.2002.0144; W TeBrake, 'Taming the waterwolf: Hydraulic engineerings and water management in the Netherlands during the Middle Ages', *Technology and Culture*, vol. 43, no. 3, July 2002, pp. 475–99. doi.org/10.1353/tech.2002.0141; A Tindley, 'The Iron Duke: Land reclamation and public relations in Sutherland 1868–95', *Historical*

It would therefore be a rare migrant who had not heard of or seen a canal, a drainage scheme or a dam. From the massive to the domestic scale, control and order were the principles that nineteenth-century Europeans applied to landscapes and waterscapes. Migrants to Gippsland were familiar with trained hydrologies at every level and scale. What we would now admiringly describe as a 'wild' river was, to them, a wasted river.

Water-borne illness

It is impossible to understand the history of hydrological interventions without considering beliefs about human health, disease and the role of water. The experience of water-borne disease was common, and shaped responses significantly.

Devastating diseases like malaria and cholera were thought to be inhaled from poor air arising from swamps and marshes. Miasmatic theory contributed to an understanding of health in the context of the environment, and the hydrological cycle fared the worse for it. Still waters were singled out as the source of miasma. Germ theory would ultimately replace miasmatic theory, but not before great drainage projects were undertaken.

The connection of illness to the environment originated in ancient Greek medical theories. One of the most significant was the theory of humours, developed by Sicilian-born Empedocles (c. 491–432 BCE), which dominated Western medicine until the beginning of the scientific revolution. In this theory, the human body was controlled by four elements, earth, air, water and fire, which were acted upon by two basic forces, love/unification and hate/division. It is not hard to see here another aspect of dualism. Each force acted upon the element in the body and could increase or decrease its qualities. Water was classified as cold and moist and produced phlegm, while earth was considered cold and dry and produced melancholic black bile. Good health occurred via the appropriate balance

Research, vol. 82, no. 216, 2009, pp. 303–19. doi.org/10.1111/j.1468-2281.2007.00441.x; J Larkcom, 'The floating gardens of Amiens', *Journal of the Royal Horticultural Society*, vol. 121, no. 9, 1996, pp. 540–3; D Speich, 'Draining the marshlands, disciplining the masses: The Linth Valley hydro engineering scheme (1807–1823) and the genesis of Swiss national unity', *Environment and History*, vol. 8, 2002, pp. 429–37. doi.org/10.3197/096734002129342729; T Glick, *Irrigation and society in medieval Valencia*, Belknap Press of Harvard University Press, Cambridge MA, 1970. doi.org/10.4159/harvard.9780674281806; S Ciriacono, *Building on water: Venice, Holland and the construction of the European landscape in early modern times*, Berghahn Books, New York and Oxford, 2006.

of elements and forces. As Glacken shows, this model can be traced to the physical environment in which Empedocles lived. Cold and moist are linked in a Mediterranean climate. In the tropics, hot and moist is a much more likely combination.[63] Because of the widely held belief that the environment affected health, this imbued certain landscapes with beneficial qualities and others with negative ones. The history of malaria is a helpful window on this concept of medical geography.

The mechanism for the spread of malaria was a mystery until Sir Ronald Ross discovered that the vector was the anopheles mosquito, gaining a Nobel Prize for his work in 1902.[64] Until then, the ague claimed countless victims. Many places in Europe with extensive wetlands had high rates of infection, especially Rome. Because of its location by the Tiber's lowland marshes, malaria killed at least four popes and, in 1623, an outbreak profoundly disrupted the election of Pope Urban VIII. One-tenth of the assembled cardinals succumbed and at least 40 of their retinue died. Urban himself lay in bed for two months, preventing his accession to arguably the most powerful job in Europe. Because of his experience, he encouraged overseas priests in botanical prospecting, which led ultimately to the development of quinine from the Peruvian chinchona tree.[65] Pick estimated that in the nineteenth century approximately 15,000 Romans died from malaria per annum, but he is unable to provide a figure for numbers affected. The desire to prevent malaria drove Garibaldi's attempts from 1875 to divert the Tiber completely.[66] Thus the power of miasmatic disease shaped world history and the actions of religious and national leaders.

For the majority of people living near seasonally inundated floodplains, there was no cure from shaking fever. All anybody knew was that it came with the summer heat, and that anyone rich enough to retreat to higher grounds had a better chance of escaping it. For example, death rates from malaria in south-east coastal marsh areas of England were three or four times higher

---

63   C Glacken, *Traces on the Rhodian shore: Nature and culture in Western thought from ancient times to the end of the eighteenth century*, University of California Press, Berkeley, 1967, pp. 10–12.

64   Malaria is a parasite that, when it reproduces, causes red blood cells to explode. This detritus depletes the capacity of liver and spleen to perform their own function. Alternating sweats and chills combined with delirium lead to death for countless thousands over millennia. F Rocco, *The miraculous fever tree: Malaria and the quest for a cure that changed the world*, Harper Collins, Great Britain, 2003, p. viii.

65   Rocco, *The miraculous fever tree*, pp. 36–48 tells the story of the papal election, who fell ill and why that shifted the balance of powers.

66   D Pick, 'Roma or morte: Garibaldi, nationalism and the problem of psycho-biography', *History Workshop Journal*, vol. 57, no. 1, 2004, p. 12.

than non-marsh areas. The Fens, the Thames, the coastal marshes of Essex, Kent and Sussex, the Somerset levels, the Ribble district of Lancashire and the Holderness of Yorkshire were all particularly afflicted.[67]

In Australia, a similar practice of escaping to higher ground turned parts of the Blue Mountains and the Dandenongs into summer holiday areas for the colonial elite. Some lobbied, generally unsuccessfully, for Gippsland's mountain towns to follow suit. In the GLC, mosquitoes could be prevalent in low-lying areas. Brodribb was unusually frank about their impact:

> The mosquitoes were dreadfully annoying, both day and night, to us and the horses. I have seen the horses, when hobbled out, completely covered in them and at times they were so irritating they would gallop about in the hobbles and roll on the ground; and although we had beautiful clear water to perform our ablutions in, we dare not approach the creeks and take off our clothes, even to wash our faces and hands, because of these formidable enemies.[68]

The name malaria derives from the Italian word meaning bad air, which reflects the traditional association of low-lying swamps with stagnant waters. The potential for the ague was a factor employed in the pro-drainage arguments in the East Anglian Fens.[69] Draining was the first order solution to removing the sources of malaria, thought to be the noxious smells arising from swamps, sewers and cesspits, followed by adequate sewage treatment. In Gippsland, there are references in the papers to the supposed healthfulness of certain places. For example, in 1885 the *Bairnsdale Advertiser* praised James Stirling's attempts to establish alpine meteorological stations in the catchment:

> This work will do good service if it only succeeds in educating the denizens of crowded and dusty towns to the existence and eligibility of the 'summer sanitoriums' to be met with in these highlands ... [the mountains] have 'the extreme grandeur and sublimity of the landscape; the freshness and variety and etherical purity of the air on our highest peaks and tablelands, with their most exhilarating and invigorating influence'.[70]

---

67 M Dobson, '"Marsh fever" – the geography of malaria in England', *Journal of Historical Geography*, vol. 6, no. 4, 1980, pp. 377–8. doi.org/10.1016/0305-7488(80)90145-0.
68 WA Brodribb, *Recollections of an Australian squatter 1838–1883*, John Ferguson in association with the Royal Australian Historical Society, Sydney, 1978, p. 39 [First pub in 1883 by John Woods and Co., 13 Bridge St, Sydney].
69 Darby, *Changing fenland*, p. 123.
70 *Bairnsdale Advertiser*, 11 April 1885. There were two follow-up notes from different correspondents also lauding Stirling's research in the following week.

Figure 3.5: The now thickly overgrown Mosquito Creek in the Moe Drainage Scheme.
Source: Author.

For comparison, here is a quote from the summer of 1890 describing a polluted pool of water and its assumed influence on health:

> The inaction of the council will very shortly bring about a boom of typhoid. The fever germinating pool at the corner of Foster and York Sts has, during the last few days, been emitting a stench that promises to shortly put a stop to traffic across the lake.[71]

Here is the high/low dichotomy at its clearest. Elevated land was healthy, low-lying land was not.[72]

While malaria was wrongly attributed to polluted wetlands, there were other water-borne illnesses that contributed to the strength of miasmatic theory. Typhoid fever and cholera were major killers during the nineteenth century. Like malaria, their causes were unknown until the development of bacteriology, although John Snow proved transmission by water for cholera between 1849 and 1855.[73] Until then, many diseases were attributed to miasma, and were a major concern for colonial Victorians.

Chappelle has noted that for most of history, the search for clean drinking water has been a dominant force:

> If you combine the natural tendency of water to pick up sediment, solutes and microorganisms with the mounds of filth generated by human and animal habitation, you can begin to appreciate how rare clean, drinkable water has been for much of human history.[74]

Shunning water as a drink, even while recognising it as essential for agriculture, was based in the common knowledge that unless you had access to a clean artesian spring, there was a very good chance that drinking the local water supply would make you sick. In 1885, the *Bairnsdale Advertiser* was blunt about the quality of local water. 'Give us water?',

---

71 *GT*, 17 January 1890.
72 For example, see the *GFJ*, 2 June 1887. 'One advice I would give to all resident on the land in South Gippsland is, to build on the highest point they can, providing it is not too much inconvenience; because in the flats the fog is much heavier than on the heights, and the atmosphere is more bracing the higher you ascend.'
73 V Smith, *Clean: A history of personal hygiene and purity*, Oxford University Press, Oxford, 2007, p. 273.
74 FH Chappelle, *Wellsprings: A natural history of bottled spring waters*, Rutgers University Press, New Brunswick, New Jersey, 2005, p. 10.

it asked rhetorically. 'Dr Youl condemned beer drinking, but he'd hold a very different opinion if he had to drink the water here, and unless rain comes soon, Terry's and Castlemaine will supercede water.'[75]

As noted in Chapter 2, these hygiene problems were a significant driver for the emergence of the quantitative version of the hydrological cycle and the creation of large-scale water supply infrastructure. The nineteenth century saw a range of moves at both personal and political levels to address the issue of hygiene. According to Haslam, 'in England it took a dandy, Beau Brummel, in the early nineteenth century, to make personal cleanliness fashionable again'.[76] Florence Nightingale pushed personal cleanliness wider, and included the cleanliness of towns, armies and hospitals among her achievements. Bathing increased in significance, with toilets and bathrooms being incorporated into housing designs and public buildings.[77] In London, Joseph Bramah was installing toilets in private houses from 1778 and his company continued in operation until 1890. The interestingly named Thomas Crapper was his main competitor.[78] Cities began to invest millions into sewerage infrastructure. London's sewer system is possibly the most well known. The money for it was voted by parliament during June 1858 when the Thames was so rank smelling that parliamentarians were forced from their chambers, handkerchiefs pressed to noses. The Bill became law in a speedy 18 days![79]

In colonial Victoria, public health, water supply and drainage in Melbourne were major policy issues as well. Dunstan notes how, as early as 1848, Melbournians were being taxed by 'the want of drainage, the filthy condition of narrow streets, courts and alleys, the prevalence of stagnant pools of water, the habit of slaughtering animals in the city proper and a large unhealthy swamp on the east side of the city known as Lake Lonsdale'.[80] One almanac memorialised the 1854 cholera outbreak

---

75 *Bairnsdale Advertiser*, 25 March 1885, Lakes Entrance correspondent. For other examples, see *GT*, 17 January 1890, on water supply at the Sale school; *GT*, 10 February 1890, on poor quality drinking water at Prospect.
76 Haslam, *Historic river*, p. 3. Also see R Hobday, *The light revolution: Health, architecture and the sun*, Findhorn Press, UK, 2006, pp. 72–3.
77 L Wright, *Clean and decent: The fascinating history of the bathroom and the water closet*, Classic Penguin, England, 2000.
78 Halliday, *Water: A turbulent history*, p. 132.
79 Halliday, *Water: A turbulent history*, p. 145.
80 D Dunstan, *Governing the metropolis: Melbourne 1851 to 1891*, Melbourne University Press, Melbourne, 1984, p. 121.

in London and the rallies in Melbourne it inspired.[81] The awful suffering involved in death from cholera or typhoid created an atmosphere conducive to social and governmental reform. Dunstan describes an emerging apparatus of public health governance in colonial Victoria, which principally centred on populous Melbourne. However, the Town and Country Police Act, extended to the Borough of Sale on 17 May 1863, included a significant range of public health concerns, which the police were required to enforce. As local government spread throughout the GLC, its legislation allowed for the employment of nuisance inspectors, health officers and night men.[82] This extension of bureaucratic control was not without its power struggles between rival projects (witness the acrimony generated in Sale about water supply in the 1870s, discussed in Chapter 5), but the sheer scale of ill health prevailed over opposition. Its importance is reflected in the fact that the 1858 opening of the Yan Yean water supply was regularly memorialised in colonial almanacs.

The experience of ill health related directly or indirectly to water was a powerful force. Generally it shaped major interventions in the hydrological cycle, prompting large- and small-scale drainage operations, centralised water supplies and, later, sewerage disposal systems. These alterations proceeded from a widespread perception that flowing water was clean, which reflected an understanding of the powers of dilution.

Agriculture and pastoralism

Australian colonial history is replete with the stories of pastoralism and agriculture. They competed with each other politically and morally for access to land and for the allegiance of settlers and politicians. Even though they were 'at each others' throats', both engage in what is generally a triumphal narrative. Hugh Copeland, for example, called his 1934 history of the Shire of Warragul *The Path of Progress*. Both agriculturalists and pastoralists were making the country productive against great personal, economic and environmental odds. An example of this narrative is Linn's *Battling the Land: Two Hundred Years of Rural Australia*.[83] The title says it all.

---

81 *Calvert's Illustrated Almanac for Victoria, 1859: being the third year after the Bissextile, or Leap Year*, 6th edn, Printed and Published by W Calvert, Neaves Buildings, Collins St East, Melbourne, 1859, pp. 15 and 18.
82 Dunstan, *Governing the metropolis*, p. 132.
83 R Linn, *Battling the land: Two hundred years of rural Australia*, Allen and Unwin, St Leonards, NSW, 1999.

Neither can be successful without making demands on the hydrological cycle; however, they do so in quite different ways and have different impacts. The cumulative impact of the 'war' on rural landscapes, waterscapes and the species that inhabit them has challenged this view of Australia's past, as has a growing understanding of the sophistication and skill of Indigenous land management practices. George Main has argued that colonial settlers were engaged in turning the landscapes and waterscapes into a giant rural factory, where indigenous species have been removed to make way for modern industrial agriculture.[84]

The success of this economic transformation depended in large degrees upon the use of water as an input. That is, water conceptualised as a resource, a raw material that is to be value added, stripped of its poetry, beauty and ecology. In the GLC, a direct verbalisation of the view of nature as a resource input came from the visiting Duke of Manchester in 1880. At a banquet given in his honour by the Mayor of Sale, JJ English, the Duke said that his travels were 'to gratify a long formed desire to make himself acquainted with the position of the colonies, and to learn something of the vast natural resources represented by their extensive commerce'.[85]

The Australian colonies were meant to be an integral part of an empire that facilitated the flow of material goods and cultural ideas for the benefit of white Europeans. To achieve this, both agriculture and pastoralism required water. Water was essential for the creation of those tradeable products, and, given the appalling condition of most colonial roads, water was a more efficient and cheaper method of transport where rail was not available. As Worster notes, this is a viewpoint that strips water of any religious, cultural or social worth of its own, so that it becomes merely something quantifiable, to be turned into units of grain or pounds of meat.[86] This kind of resource management thinking remains the current paradigm, and has come under fire from a range of writers.[87] Viewing water as a resource is one of the key components of the quantitative

---

84   G Main, 'Industrial earth: An ecology of rural place', PhD thesis, Centre for Resources, Environment and Society, The Australian National University, 2004.
85   *GT*, 29 November 1880.
86   Worster, *Rivers of empire*, p. 31.
87   For example, J Macy & M Young Brown, *Coming back to life: Practices to reconnect our lives our world*, New Society Publishers, Gabriola Island, Canada, 1998; J Seed, J Macy, P Fleming & A Naess, *Thinking like a mountain: Towards a council of all beings*, New Society Publishers, Gabriola Island, Canada, 1998; T Rozak, *The voice of the earth*, Simon and Schuster, New York, 1992; C Glendinning, *My name is Chellis and I'm in recovery from Western civilization*, Shambala Press, Boston, 1994.

approach to hydrology. Such a utilitarian view of the environment was the norm in the nineteenth century, and was facilitated in part by the Christian notion of man's dominion over the planet.

It was always the intent of the British Government that the Australian colonies should be agricultural in nature. Instead, the mixed bag of convicts, military and free settlers faced an environment that was better suited to wool growing. Pastoralism prevailed as the principal industry until the gold rushes of the 1850s. Map 6 shows the extent of squatting runs in 1857. While pastoralists had less need to clear or drain, there were significant impacts on soil, through erosion and compaction, and on biodiversity from overgrazing.[88] Lunt says that of the approximately 600 square miles of prime grazing native grassland between Sale, Stratford, Heyfield, Boisdale and Maffra, there are only tiny fragments left along rail reserves and in cemeteries.[89] Maps 8 and 9 illustrate vegetation loss between 1750 and 2005.

Figure 3.6: A fragment of grassy woodland, photographed along the road between Glengarry and Cowarr.
Source: Author.

---

88 G Aplin, *Australians and their environment: An introduction to environmental studies*, Oxford University Press, Melbourne, 1988, pp. 362–6.
89 I Lunt, 'Snakes Ridge views', *Gippsland Heritage Journal*, no. 13, 1993, p. 37.

The agricultural sector began to grow from the 1850s onwards. Men who realised that they were never going to make their fortune on the diggings turned instead to farming, seeking to supply domestic and international markets.

Figure 3.7: Depictions of Maffra from 1882. These montage scenes showing progress were common.
Source: Pictures Collection, State Library Victoria, Accession no. IAN13/05/82/68.

The settlement of Gippsland was directly linked to the experience of drought and over-exploitation of pastoral lands. Within a mere 32 years, Sydney's Cumberland Plain was filled and exhausted. The discovery of the passage through the Blue Mountains by Blaxland, Wentworth and Lawson was announced in the *Sydney Gazette* on 12 June 1813.[90] While

---

90   Despite this important discovery, it took some considerable time for any great movement of Europeans beyond the Sydney area. According to Perry, in 1821 less than 1 per cent of the population was living west of the range. The delay was partly to do with the Governor's reluctance to encourage expansion outside the 'natural prison' that the topography of the Cumberland Plain represented. Second, official policy was to encourage intensive farming rather than broad-scale grazing. However, a series of poor years caused by a combination of drought, overgrazing and attacks by caterpillars forced a change. Governor Macquarie finally allowed some stock to be moved to Bathurst in 1818, and in 1820 allowed further temporary grazing to take place south-west of Sydney. For a number of reasons, expansion was delayed until the Order in Council of 25 November 1820 marked 'the real beginning of the great outward movement of pastoralists and graziers which continued into

there were some delays, the emerging, wealthier pastoralists went looking for better pastures, and thus commenced a wave of movement. Every bad season prompted fresh exploration, with apparently little reflection on the part of the pastoralists about their role in degrading the country.[91] The combination of water shortage and overgrazing was repeated, turning attention southward to the future state of Victoria in the late 1830s.

There was some contemporary recognition of the role of overstocking in fuelling land exploration. *The Colonist* of 9 July 1840 was blunt on this point when discussing the exploration of Gippsland and the relative merits of Angus McMillan and Count Strzelecki's explorations: 'To the graziers whose runs or stations may have been overstocked, and indeed to the public at large, the discovery of this country was an event of considerable importance'.[92]

Other writers have detailed the competitive nature of exploration in Gippsland.[93] Who was first is not relevant to this story. What is important are the explorers' perceptions about water. The two earliest sources are Angus McMillan's letter to his employer, Lachlan McAlister, after an exploratory trip in early 1840, and Strzelecki's account.

McMillan did not find the journeying easy, especially attempting to cross the rivers and wetlands. He viewed the landscape with economically tinted glasses, exemplified in his description of the country around the Mitchell

---

the thirties and forties'. TM Perry, 'Climate, caterpillars and terrain: A study of grazing expansion in New South Wales 1813–1826', *Australian Geographer*, vol. 7, no. 1, 1957, pp. 3–7. doi.org/10.1080/00049185708702319.
91   Governor Brisbane, who held office from 1821 to 1825, gave a substantial impetus to this push for fresh land by allowing expansion in all directions and by instituting a ticket of occupation, which permitted grazing only and which protected the ticket holder from encroachment by other squatters. This ticket system was a delaying tactic developed by Brisbane until decisions were made on the preferable method of land disposal. Perry, 'Climate, caterpillars and terrain', p. 8.
92   Reproduced in full in 'Primary sources: McMillan's letter to the Colonist', *Gippsland Heritage Journal*, vol. 3, no. 1, 1988, pp. 39–41. Hume and Hovell were the first to explore the Port Phillip District, as Victoria was then called, in 1824. A decade would elapse before the combined factors of drought, overstocking and capital availability made Gippsland a serious prospect.
93   See P Morgan, *The settling of Gippsland: A regional history*, Gippsland Municipalities Association, Leongatha, 1997, p. 30; C Dow, 'Tantungalung Country: An environmental history of the Gippsland Lakes', PhD thesis, Monash University, Melbourne, 2004, p. 41. Angus MacMillan met the Omeo squatters James MacFarlane and George McKillip in June 1839. At this time, McMillan either met or heard of the explorations of Edward Baylis, Andrew Hutton and Walter Mitchell who had travelled south of Omeo and on to the Lakes. Edward Baylis reached the Lakes in November 1838, and was forced to return to the Monaro because his party had run short of rations. Andrew Hutton, who was an employee of William Morris of Nungatta on the Genoa River to the east of the Lakes, camped at Lakes Entrance. He alleged an attack by Aborigines, which forced him to abandon the cattle. A different account of this, sans violence, was given some years later to Alfred Howitt, by a Brabralung elder named Tulaba. Where they both agree is the impediment to stock movement that the waters represented.

River floodplains as 'the most delightful country I eve [sic] beheld, well adapted for cattle, sheep or cultivation'.[94] McMillan was on a business trip, and his single task was to find the best pastoral land. This meant fresh green grass (those same species they never recorded) and plenty of water for stock to drink. The abundance of both made the obstacles of wetlands and difficult river crossings seem less problematic. He sounds, in fact, almost incredulous when he wrote: 'Country still improving, if it is possible to do so'.[95]

Water's presence on a run in safe and accessible quantities materially helped the production of stock numbers. In the annual cycle of production, water was used at every stage. Access to clean water was critical for drinking water for stock and station workers. Accordingly, competition for runs with reliable surface water sources was fierce. Any map of the location of squatting homesteads is in effect a proxy map of water resources.

Figure 3.8: Eugene von Guerard's painting of Angus McMillan's Bushy Park, 1861, depicting a small wetland in the foreground with cattle drinking.
It is also notable for its illustration of the flat, lightly wooded, open country so favoured by settlers.
Source: National Library of Australia, Accession no. 1857989.

Aside from drinking water, there were three key processes associated with the squatting industry that required water as an input. These were sheep washing, boiling down and tanning, with boiling down having perhaps the worst water-quality impacts. Boiling down generated large amounts of organic residue, which were usually dumped in the nearest creek.[96]

---

94 'Primary sources: McMillan's letter to the Colonist', p. 40.
95 PD Gardner, *Through foreign eyes: European perceptions of the Kurnai tribes of Gippsland*, 2nd edn, Ngarak Press, Ensay, 2004, p. 40.
96 This rather gruesome process created a valuable, exportable product called tallow, used in soap and candle making, which continued until the advent of refrigeration. Australian Academy of Technological

The construction of washpools also had an impact on stream morphology, with consequent erosion and incision of streams that permanently changed their form, process and ecological function.[97] There is also good reason why tanning is classed as a noxious industry and is subject, these days, to strict impact assessment procedures. The environmental effects of all these have largely missed the notice of historians.

Agriculture was equally dependent on water as an input but in different ways. Pastoralists sought out flowing surface waters, while agriculturalists depended more on rainfall. More importantly, it was the distribution of rainfall throughout a season that had the greatest impact on crop production. Too much or too little rain at the wrong time could spell disaster, as could two other forms of precipitation, hail and frost. Gippsland farmers grew a wide variety of crops, and some took considerable risks in experimenting with plants like flax, chicory and arrowroot. Fruit and vegetables were either homegrown or supplied by local (often Chinese) market gardeners operating around the larger towns.

Little is known about domestic and small irrigation practices in the catchment in this period. In a photograph of the Chinese market garden at Walhalla, it is possible to see some handmade channels diverting water to crops. There is some evidence to suggest a link between the presence of Chinese market gardeners and the development of irrigation for crops in the 1880s, and it is certainly feasible that domestic gardeners may have learnt from the Chinese.[98] Map 7 marks the location of Chinese market gardens in Sale, very close to Flooding Creek. There were also many ex-diggers who were familiar with the construction of water races, so there may have been small-scale domestic irrigation of gardens. Irrigation was seen as insurance against the fickleness of the hydrological cycle, which could not be relied upon to generate the steady flow of sellable crops that the settler economy demanded.

Sciences and Engineering, *Technology in Australia 1788–1988: A condensed history of Australian technological history and adaptation during the first two hundred years*, University of Melbourne, Melbourne, 2000, ch. 1, pt 4. Knowledge of the distribution of boiling plants in Gippsland is rudimentary. Adams refers to a plant close to Port Albert, and this is confirmed by the listing for Arthur King, JP, in the 1851 Victoria Directory for Country Districts and small towns. N Cox, 'Residents of Gipps' Land 1851', *Gippsland Heritage Journal*, vol. 8, 1995, p. 48. Buckley had one on his run, which remained in operation for years; for example, he recorded boiling down on 25 January 1849, for several days in early July 1849, 10 October 1853 and on 1 March 1870. Porter refers to a large plant at Flooding Creek, but, unhelpfully, does not give a reference. Daley mentions one at The Heart, with sheep yards that were south of the Friendly Society's grounds, Daley, *Story of Gippsland*, p. 42. The Heart was close to Flooding Creek and it is unclear whether this is the one Porter referred to.
97 Pers. comm., Sara Beavis, 30 September 2011.
98 *GT*, 18 November 1865, 29 December 1876.

This was backed by a steady stream of claims about the moral desirability of agriculture. In a quote echoing the Lord's Prayer, the *Gippsland Times* declared, in its first year of publication, that 'it is agriculture which spreads the great and bountiful table at which the mighty family of civilized man receives his daily bread'.[99]

This kind of sentiment was common during the land selection era, when squatters were pitted against farmers. To his cost, Governor Gipps had opposed the wholesale spread of squatting because it reduced Crown options, the options of future citizens of the colony and created greater expense. He understood the British Government's preference for agriculture, which it saw to be morally superior to pastoralism because it produced settled, cleared, permanent and hopefully self-supporting areas.[100] The aim of substituting an English-style rural yeomanry whose work ethic would subdue the wilderness was enshrined in legislation commencing in 1865.

Stephen Legg has examined in detail the comparative successes and failures of selectors in South Gippsland.[101] Those who were lucky enough to secure flattish, moderately elevated, lightly wooded and well-drained selections had the easiest route to 'progress' because such lands required less investment of capital and sheer sweat to make them 'productive'. Map 4 shows a detail of the topography of northern Gippsland, showing the sharp change from hills to flats. Map 5 shows how the earliest runs focussed on the rivers in this gently undulating country. John O'Connor recalled the difficulty of forest land:

> So in all we had about 900 acres of dense forest, in no place could you see more than fifty yards ahead of you … The settlement [of Narracan] began on the outskirts of this appalling forest.[102]

Legislation thus supported the wholesale destruction of Gippsland's vegetation, as selectors were compelled to clear and fence as part of their conditions to attain freehold title. Drainage was not a condition, but selectors did include it in their estimations of improvements.[103] While

---

99 *GT*, 11 December 1861.
100 J Powell, *Environmental management in Australia, 1788–1914*, Oxford University Press, Melbourne, 1976, p. 23. Powell went on to note that Gipps was cognisant of the government desire, but realised its impracticality. Instead, he attempted a compromise position and pleased no one.
101 SM Legg, 'Arcadia or abandonment; the evolution of the rural landscape in South Gippsland, 1870–1947', MA thesis, Dept of Geography, Monash University, 1984.
102 Memoirs of John Joseph O'Connor, p. 11.
103 As far as I am aware, no study has been undertaken of land selection files to ascertain the extent of local drainage practices.

collectively considering the native land and waterscape of the GLC as nothing special, settlers enthusiastically set about utilising its individual resources. Unable to perceive the forest, wetland or grassland, they saw potential products: sleepers, palings, ceiling joists, maize, sheep, skins, oats, cows. And yet, the operation of the hydrological cycle was essential to the production of them all.

Figure 3.9: Gregory's Cottage, Cooper's Creek, Walhalla Road, c. 1885.
Source: Pictures Collection, State Library Victoria, Accession no. H40181.

The assumed social, cultural and moral benefits of agriculture regularly received support in Gippsland papers.[104] For example, in 1876, land around the former Heart run was released for selection. The *Gippsland Times* published a letter that heavily criticised the size of the lots being released, suggesting that they were unaffordable for the small family farmer. 'NG' (not Nehemiah Guthridge the mayor, who wrote in the following issue saying it wasn't him but that he agreed with NG's sentiments) wrote how such a decision could undermine the yeoman, utopian ideal. The large size of the lots would privilege men of capital, or pastoralists:

---

104 Goodman provides a succinct summary of these benefits. D Goodman, *Gold seeking: Victoria and California in the 1850s*, Allen and Unwin, St Leonards, NSW, 1994, pp. 110–12.

> For it was a dream of a rural population settled upon the rich meadow lands of the Heart, a dream of waving corn fields and thriving homesteads, of golden harvests and merry autumn gatherings. A dream too, of a well to do prosperous yeomanry, with troops of stalwart children crowding round the cottage doors, or merrily wending their way to parish church or school; a dream of such a prosperous future for North Gippsland, that its great reproach in the eyes of strangers, an indifference to agricultural enterprise, would be for ever removed, and that we should be able to adopt as our own, with some title to its use, that grateful motto of a rural people 'Speed the plough!'[105]

NG clearly links small farms, family and Christianity with qualities of rootedness and prosperity. He also transplants the halcyon, seasonal imagery of Northern Europe, apparently not recognising the existence of drought and flood and fire. Speeding the plough was a common image, and explains the endless observations of rural correspondents on the number of acres under crop in each district.

The importance of this agrarian imagery as an ideology is noted by Goodman. The gold rushes profoundly disturbed the agrarian ideal, as rural men rushed off to dig up streams, and gardens and farms lapsed into disarray.[106] Conservatives deplored the effects of the gold rush because it uprooted people from traditional agrarian ways of life. Radicals, on the other hand, employed the agrarian and yeoman ideal and the supposed moral benefits that accrued from it to bolster their opposition to the squatter's pastoral empires. Either way, the image of the cultivated farm with its frill of English flowers was seen as the epitome of progress, but it could not be achieved without water. Water was the foundation for both the mythical country landscape and the actual country landscape.

'Progress' in 'taming the wilderness'

If colonial Gippslanders had a secular faith, they believed in progress. The creation of towns, schools, churches, mines and farms were all the more significant for the 'primeval' conditions from which they were wrung.[107]

---

105 *GT*, 14 March 1876.
106 Goodman, *Gold seeking*, p. 109.
107 *Gippsland Illustrated*, 1904, p. 6. First facsimile edition by Kapana Press, Bairnsdale, 1987, first published by the Periodicals Publishing Co., Melbourne, 1904.

Figure 3.10: The A1 Mine at Gaffney's Creek, Thomas Henry Armstrong Bishop, c. 1901.
Source: Pictures Collection, State Library Victoria, Accession no. H36688.

Figure 3.11: New Public Hall, Sale, FA Sleap engraver, 1879.
Source: Pictures Collection, State Library Victoria, Accession no. IAN21/02/79/21.

These included untrained rivers that flooded and dried up in the same year, extensive wetlands, steep ridges and gullies, thick forests and scrub, fires and snakes, and the Indigenous inhabitants, who did not see the land as a wilderness to be tamed.

The notion of wilderness is a dualism that sets up a split between wild and civilised, stagnation and progress, habited and uninhabited. This is what Oelschlager describes as 'a bifurcation that the human story lies in our triumph over a hostile nature'.[108]

> Our prevailing definitions of wildness and wilderness preclude the recognition of nature as a spontaneous and naturally organized system in which all parts were harmoniously interrelated: in consequence, human kind has believed itself compelled to impose order on nature.[109]

Colonial Gippslanders, indeed almost all European migrants, shared this metanarrative about their new home. This broader definition of wildness naturally included the hydrological cycle. At base, the imposition of order meant mitigating variability, deflecting the force and damage of floods, making wetlands into dry lands and blunting the severity of drought in any way possible.

The ideology of progress was well established before the majority of settlers arrived in Gippsland. In eighteenth-century Scotland, a host of organisations dedicated to the idea formed including the Society of Improvers (in 1727), the Edinburgh Society for Encouraging Art, Science, Manufactures and Agriculture (established in 1755), the Select Society (1754), the Philosophical Society (1737) and, most prestigiously, the Royal Society of Edinburgh, begun in 1783.[110] Their establishment was the consequence of increasing technologies and the discoveries of science that allowed the elite to imagine a life without ecological limits. It became possible to think of human history as the story of advance, not decline.[111] In particular, *a la* Adam Smith, this advance was predicated on the increasing desirability of individual wealth, prosperity and the acquisition of consumer goods, which in turn promoted capital investment in production.

---

108 M Oelschlaeger, *The idea of wilderness: From prehistory to the age of ecology*, Yale University Press, New Haven, 1991, p. 2.
109 Oelschlaeger, *The idea of wilderness*, p. 8.
110 CWJ Withers, 'On georgics and geology: James Hutton's "Elements of Agriculture" and agricultural science in eighteenth-century Scotland', *Agricultural History Review*, vol. 42, no. 1, 1994, pp. 38–48.
111 C Ponting, *A new green history of the world: The environment and the collapse of great civilisations*, Vintage Originals, London, 2007, pp. 124–7.

It is perhaps easiest to illuminate the progress philosophy of nineteenth-century migrants by illustrating what it was not:

> If there is to be real and sustainable progress, it must be a continuing enhancement of life for the entire planetary community. It must be shared by all the living, from the plankton in the sea to the birds above the land. It must include the grasses, the trees and the living creatures of the earth. True progress must sustain the life giving purity of the air and the waters. The integrity of these life systems must be normative for any progress worthy of the name.[112]

A definition of progress that included ecological processes and species other than humans would have been if not unthinkable, then certainly unacceptable to the majority. There were a minority of colonists who expressed regret at the environmental changes they wrought, whilst simultaneously maintaining it was all for the greater good. At a dinner in his honour in 1864, Surveyor WT Dawson regretted that:

> in the course of improvements now being effected by the Borough Council, he saw the stumps of trees that he recollected as the giants of the forest, that were venerable from age before the white man ever cast eye on the continent of Australia – now being removed to afford greater facilities for the busy traffic of the township of Sale.[113]

Ecological changes were noted but, as Synan summed up: 'It was puzzling, but scarcely anyone took time to inquire or to care'.[114] More often than not, the changes were lauded:

> I confidently believe that the time is not now distant when the few gaps in the virgin forest, made by the present settlers will extend over the whole lower district, when in place of somber gum forest and tea tree scrub, the make shift hut and half cleared paddocks, the eye of the travellors will rest for miles upon waving cornfields and well stocked homesteads of a happy and prosperous race of farmers.[115]

---

112 T Berry, *The dream of the earth*, Sierra Club Books, San Francisco, 1998, p. 82.
113 *GT*, 10 June 1864, quoted in P Synan, *Gippsland's lucky city: A history of Sale*, City of Sale, Sale, 1994, p. 30. His summary, which I have quoted, was about the death of forests on the plains around Sale up to around 1870 and is also from p. 30. Other examples of regret include C(K) Barrett, *A busy life: The reminiscences of Charlie Bartlett born at Providence Ponds, near Stratford, Victoria, in 1890*, Stratford Historical Society, Stratford, 1980, p. 11. Reminiscences of Mr McTaggart, first station master at Warragul, reported in Copeland, *Path of progress*, p. 143. Poetry by James Brown or James Brown Alton reproduced in *Coach News*, vol. 15, no. 1, September 1987. Reflections on species loss by Allan MacLean, *Argus*, 28 January 1905, Early Gippsland.
114 Synan, *Gippsland's lucky city*, p. 30.
115 Dr Hedley at a dinner to honour Angus McMillan, reported in *Gippsland Guardian*, 7 March 1856, cited in J Adams, *From these beginnings: History of the Shire of Alberton*, Alberton Shire Council, Yarram, 1990, p. 30.

Their assumption of their own superiority separate from nature, and their superiority to other ethnic groups, was made possible by their religious beliefs. It was also supported by a consumer-based industrial revolution, which made the colonies seem like just so much raw material.[116]

There is abundant evidence from the newspapers to confirm this. Progress (preferably of white well-born men, and then women) was their pre-eminent theme. Correspondents were either lamenting the lack of it or celebrating its achievement. In a letter to the editor, a member of the Church of St Paul's criticised Bishop Moorhouse for describing Gippsland in terms that made the place seem like it was full of 'savages'.[117] Progress meant different things to different people, but generally it never meant ecological preservation and protection. In its first year of operation, the *Gippsland Times* championed the twin ideas of progress and taming the wilderness. In an editorial about the possible opening of the Lakes, the editor lamented the feuding between various rival groups about whether it should be rail or a permanent entrance. He went on to warn readers that unless they resolved their differences and worked cooperatively 'the country which gave such promise of future prosperity, will gradually dwindle and decay, returning back to the non-productive state in which it was first discovered'.[118]

In 1887, the Glenmaggie correspondent associated progress with ease of travel, and therefore access to entertainment and learning.[119] Daniel Wilson of Longford also supported the importance of local opportunities for learning. In April 1890, he expressed regret at the cessation of Sale Mechanic's Institute's winter lectures, calling it a retrograde step 'in the march of intellect'.[120] In the same issue, notice was given of a new journal called the *Scientific Australian*, 'a monthly journal of industry and instruction'. Newspapers regularly reported on activities at Mechanic's Institutes, and they were always listed in directories.[121] The newspapers

---

116 D Garden, *Droughts, floods and cyclones, El Ninos that shaped our colonial past*, Australian Scholarly Publishing, North Melbourne, 2009, p. 21.
117 *GT*, 21 March 1877.
118 *GT*, 18 December 1861.
119 *GT*, 1 July 1887.
120 *GT*, 30 April 1890.
121 *Butler's Woods Point and Gippsland General Directory* of 1866, p. 68, reported that the Sale Mechanic's Institute has 'Colonial and English papers and periodicals; and the library contains about 1000 volumes of good books. The Committee get up a course of about twelve lectures yearly, which are delivered by gentlemen in the neighbourhood gratis'. It also employed a librarian.

make it clear that progress, learning and industry were valued in the community, and readers were encouraged to attend local events and institutions to better themselves and better Gippsland.

Frequently progress was associated with new businesses started, public works, building activity, harvest yields and population growth.[122] This was perhaps most brazenly summed up by the *Gippsland Farmer's Journal* in its first year of operation from Traralgon. Its first issue declared that its motto was 'ADVANCE TRARALGON'. Six months later, and employing the dark versus light dualism, it trumpeted:

> 'LET THERE BE LIGHT'. Traralgon is moving swiftly in the progressive path: now tradesmen are coming in rapidly, and building is brisk on all sides.[123]

Reporting on the borough statistics in 1876, the *Gippsland Times* said: 'These figures indicate an increase in the material prosperity of the town, and, though the progress has not been so rapid as to be startling, it is nevertheless satisfactory to know that Sale is progressing'.[124] Public works, which frequently involved management of water, were very much a part of this ideal of progress. Perhaps the sole claim to innovation in Australian hydrological development that the GLC can claim is the successful development of groundwater for town supply in Sale. Certainly John King of Nambrok thought so, referring to it in toasts at the banquet for the Duke of Manchester, calling it indicative of enterprise and public spirit.[125] Swamp drainage, too, was seen as a sign of progress, banishing the potential for illness and creating productive agricultural lands from watery wastes.

---

122 *Bairnsdale Liberal News and North Gippsland District Advertiser*, 14 May 1879, reporting on a lecture given by the government statistician in Melbourne, Mr HH Hayter, on the 'Colony of Victoria: Its progress and present position'. *GT*, 19 March 1896, a lengthy article discussing government returns on agricultural development across the state. It gave detailed on expenditure on improvement for each shire, but the opening statement lamented how far Gippsland was lagging behind two northern shires, Numurkah and Bonang. The development of various industries engaged a progressive rhetoric. For example, the opening of cheese and butter factories were huge developments locally, and were celebrated as such.
123 *GFJ*, 27 January 1887 and 2 June 1887. Capitalisation in original.
124 *GT*, 27 January 1876.
125 *GT*, 29 November 1880.

For others, it meant bringing Christianity to a hitherto uncivilised land, sometimes at the business end of a gun. Brodribb in *Recollections of an Australian Squatter* described his approach to the Kurnai at the founding of Port Albert:

> We gave the overseer full instructions how to act should they ever make an attack upon them again. Not to fire over their heads but to shoot at their legs, and if hard pressed, to kill one. My experience tells me, under such circumstances, half measures will not do. Show them at once you are determined not to trifle with them and above all, keep them at a distance. They are never to be trusted in a new and unexplored country.

Leaving their overseer behind, Brodribb and his party set out to explore the GLC. He concluded that 'it was altogether a new country, and only inhabited by savages'.[126] Savages inhabit wildernesses, while civilised white men transform wilderness and savages into productive farms and docile Christians.

There was hardly any change in attitude in three decades. Frederick Hagenauer wrote in 1874:

> Many people acquainted with the manners of the Blacks regarded it as almost an impossibility that they could be induced to settle down at one place, to take to industrial habits, become civilized and above all to become Christians. Successive annual reports, however, have clearly shown that by God's blessings, with much patience and persevering labour a great change has taken place.[127]

Hagenauer tells us much about the ideal citizen in his contrast between the Kurnai and what he has been attempting to instruct them in. As late as 1891, the primeval wilderness imagery was being used at the highest levels. The *Gippsland Times* reported the Minister for Lands would exercise forfeiture powers on unimproved selections 'in their primeval condition' in order to keep up with demand.[128]

---

126 Brodribb, *Recollections of an Australian squatter*, pp. 33 and 38.
127 Letterbooks of FA Hagenauer, vol. 2, 1874, NLA MS 3343.
128 *GT*, 11 February 1891.

Implicit in progress is an assumption of movement, a linear progression from one lesser condition to a more advanced and desirable one. John O'Connor said of his father's determination to take up a second selection: 'No one could put him off it, and no one could understand why he should drag his family back to the awful wilderness again, and I can remember my Mother being in tears at the thought of it'.[129] All of the quotations so far in this section echo movement, regardless of whether it was personal growth, as in the case of Daniel Wilson, or actual, as in Traralgon's building boom. Most importantly, the dualisms of movement and stillness, progress and stagnation, and cultivation and wilderness were assigned values of desirable and undesirable respectively.

The movement of water will be a particular focus in Chapter 5. For the time being, it is enough to suggest that colonial Gippslanders projected their cultural assumptions onto the waters of the catchment. For people who had given up every familiar notion of home, family and community, it would not do to stand still. The enormous gamble of squatting or selecting in the difficult environment of colonial Gippsland meant that progress was imperative. Migration was a statement of hope, an aspiration to a better life. In retrospect, it doesn't appear so benign, given the assumption of domination over native people and appropriating resources. But taken in their own terms, a better life meant individual landownership, and the substitution of a basically British approach to rural landscape. As McCleary and Dingle memorably say: 'In the manner of the time, [Gippsland] would be a man made landscape, with woman made trimmings'.[130]

Conclusion

The epigraphs of this chapter suggest the conflict between the imagined world of Europe, with its green pastures and flowing brooklets, and the daily experience of rural life in the Australian bush, replete with snakes and supposedly melancholy gums. In the previous chapter, I demonstrated how colonial Gippslanders had a view of the hydrological cycle that blended aspects of the divine design version and the hydraulic version. The title of this chapter, derived from a poem by Wordsworth, captures

---

129 Memoirs of John Joseph O'Connor, p. 4.
130 A McLeary & T Dingle, *Catherine: On Catherine Currie's diary*, Melbourne University Press, Melbourne, 1998, p. 2.

this blend. The word 'lord' suggests their own surety that they were made in God's image and thus were lords of the earth, as decreed from on high. The thoughtful aspect suggests their increasing hydraulic knowledge, gained through study and application. Convinced as they were of the moral integrity of the colonial project, and possessed of an increasing amount of technical skills, they did genuinely perceive themselves as the earth's thoughtful lords.

This chapter has demonstrated a 'common water' that existed in the hearts and minds of the GLC settlers. This was based upon the combination of several factors, namely:

- a religious tradition that valued surface water above all, and that ascribed high moral value to cultivation compared to other land uses
- a literary and musical tradition that arose in a physical environment where surface water was the norm
- regular exposure to water-borne diseases
- settlers who came from countries with a long tradition of altered waterscapes
- settlers who were principally engaged in economic activities that were entirely dependent upon surface water.

This combination would, in the long run, have substantial ecological ramifications for the catchment. In the following four chapters, I detail the actions of the GLC settlers that would help to create these intractable problems. I look in detail at their understanding of hydrology, and each chapter will address one of the four key components of the hydrological cycle.

CHAPTER 4

'Notwithstanding the inclemency of the weather':[1] The role of precipitation in the catchment

There appears no termination of this unprecedentedly wet season, and the residents think that Jupiter Pluvius has taken up his quarters for good and all.[2]

Introduction

Colonial Gippslanders preferred moderate, permanent flow. For this to occur, steady and reliable rainfall is needed. If there is anything in Australia that is reliably unreliable, it is rainfall. The epigraph from the *Gippsland Times* in 1870 tells a story of seasonal bewilderment, of people whose expectations about rainfall are thrown firmly out of kilter.[3]

---

1   *GT*, 31 December 1870, Crooked River correspondent. The full quote read 'Not withstanding the inclemency of the weather, great and manifold have been the festivities held in commemoration at Grant'.
2   *GT*, 1 November 1870.
3   For other examples of bewilderment, see *GT*, 11 September 1876, Maffra correspondent: 'In adverting to the weather, I must say that the last three days have been so extremely warm as to suggest that old Father Time had dispensed with one of the seasons and thrown us right into the heat of summer'. Settlers were generally like the Dargo correspondent in August 1886 when s/he said that 'We are getting a little rain, though the showers are like angels visits, few and far between; still just enough to show that the clerk of the weather has not quite forgotten us'. Some of the examples cited below at fn. 4 could also have been included here.

Chapters 2 and 3 demonstrated how migrants from Europe were encouraged by their culture in multiple ways to expect regular rainfall, lush greenness and permanent surface water as the norm. The mountainous parts of Gippsland come as close as possible to this as the Australian mainland can manage – for example, Erica averages 1,100 mm of rain a year. Yet the catchment experienced significant variability in rainfall, inducing otherwise devout Christians to talk in mythical and pagan terms, as in the epigraph.[4] The mythological references that surface in the newspapers during unexpected weather suggests how much settlers felt themselves to be at the mercy of environmental processes that they did not understand.

This chapter concentrates on the wet part of the hydrological cycle. Section one explores what colonial settlers knew about precipitation, provides an analysis of rainfall variability and discusses the roles of meteorological science and traditional weather lore in the nineteenth century in shaping settler's ecological perceptions. Section two focuses upon the practical impacts of rainfall and flood on social and economic life. Finally, the last section probes the emotional aspects of rain through an examination of the figurative language of precipitation used by settlers.

---

4   For other pagan allusions, see *GT*, 26 January 1875, Rosedale correspondent: 'If there were any worshippers of Moloch living in our midst, they could have had a place for sacrifice last Thursday or Friday by patronizing the Rosedale School'; *Gippsland Mercury* (hereafter *GM*), 15 March 1877, Upper Maffra correspondent: 'we hope that Jupiter Fluvius will soon favour us with his much desired gifts. We have not experienced such a dry season in the last 8 years'; *GM*, 17 February 1877, also referring to Jupiter Fluvius; *GT*, 29 December 1876, report of a fundraising bazaar: 'Here was everything to tempt the disbursement of ready cash, which began to flow apace into the ready hands and pockets of the Hebe-like forms which thronged the building, inviting all and sundry to win fabulous articles with infinitesimal outlay'; *GM*, 11 August 1877, Hygeia – in reference to the death of a man who'd worked on Good Luck Creek for 16 years and who died of natural causes; *GT*, 22 December 1874, 'Now that the Rosedale Council have got the plans passed for the bridge over Shiel's gully on the Sale and Rosedale Rd, traveling ratepayers will not be displeased to see the works progressing. This ugly little pinch is a kind of Charybidis to the Scylla of the cutting'; *GT*, 3 February 1876, Charles Berry is warned to study the tale of Tithonus. Tithonus was granted immortality because a goddess fell in love with him, but he forgot to ask for his beauty to be preserved; he aged terribly, wished to die but couldn't and was turned into a grasshopper. Berry would become 'a mere political grasshopper'. The report in fn. 1 also referred to Terpsichore, the muse of dance; *GT*, 22 March 1870, report of the Sale cricket team's trip to Port Albert: 'old Sol smiled the next morning, on a party of ten knights of the willow … trudging to the bathing house, towel in hand … Old Father Neptune might have allowed his grim visage to relax into a smile, when he saw the merry we disported ourselves in the briny element'; *GT*, 11 November 1881, Rosedale correspondent, complaining about the weather on the Prince of Wales's birthday, which forced the suspension of the cricket match between Sale and Rosedale. 'Whether the transit of Mercury the day before over old Sol's disc has so disgusted that noble liminary that he declined since to show his face, or from other reasons not known outside the realms of science … about dinner time (vulgo) tears began to flow from the clouds in earnest, and all hopes were abandoned by those who had entered the cricket ground, as of some more ancient visitors to Dante's Inferno of the prospect of any amusement.'

Knowledge, measurement and prediction

The birth of meteorology

The modern nightly weather bulletin complete with five-day forecast is the exception to the pattern of human experience. The settlement of the Gippsland Lakes Catchment (GLC) took place at the midpoint of a great transition, when traditional weatherlore and the emerging science of meteorology uneasily cohabited. Jankovic, in his history of English weather, notes that over the eighteenth and nineteenth centuries there was a transition from qualitative, unique and local weather to quantitative, systematised and national weather.[5] This quantitative emphasis corresponds with similar developments in hydrology, detailed in Chapter 2.

Before the twentieth century, European understanding about precipitation was in a constant state of flux. Europe was late to the practice of rainfall recording.[6] Data from the recording rain gauge (probably invented by Christopher Wren in 1663) furthered questioning about weather patterns. Scientists began to appreciate that there was a big difference between measuring rainfall on an ad hoc basis, and a systematic approach over a wide geographical area. In 1677, Richard Townley of Townley Hall, Lancashire, commenced the first continuous observations of rainfall. Followed by others across England, the practice of keeping cumulative records extended both knowledge and questions. For example, in *The Natural History of Selborne*, Gilbert White said that his six-and-a-half years of observations were insufficient for a conclusion about a mean. White's brother-in-law took this observation to heart and kept records for 59 years.[7] Hence, the establishment in the United States in the nineteenth century of a geographically dispersed network of continuously recording stations was a major milestone.

---

5   V Jankovic, *Reading the skies: A cultural history of English weather, 1650–1820*, Manchester University Press, Manchester, 2000, p. 9. Also WEK Middleton, *A history of the theories of rain and other forms of precipitation*, Oldbourne, London, 1965, especially comments on the United States.

6   Other cultures, notably 'India around the fourth century BC, Palestine in the first century AD, China in the thirteenth century, and Korea in the fourteenth century', had all developed methods to accurately measure rainfall. Benedetto Castelli (the 'father' of hydrology) made the first rain gauge in Europe in approximately 1639. However, this gauge could not record, and it was Christopher Wren who is attributed with this invention. The earliest recorded description is from 1663. From then on, rain gauges are the subject of considerable design attention. Biswas gives details of the various types invented across Europe. AK Biswas, *History of hydrology*, North Holland Publishing Company, Amsterdam, 1970, pp. 231–4.

7   Cited in Biswas, *History of hydrology*, p. 248.

Across Europe, the establishment of Royal Societies 'by first class men' fostered the fledgling sciences of meteorology and hydrology, but Biswas's comprehensive history of hydrology indicates that the science of precipitation was nowhere near as sexy as flow, velocity and discharge.[8] William Heberden first noticed that rainfall gauges at different elevations produced different results, but incorrectly concluded that it was due to 'some hitherto unknown property of electricity'.[9] These differences are due to orographic effects. Water vapour cools as it rises and condenses into clouds and, where conditions are suitable, releases precipitation.[10]

Rainfall was obviously a factor in the debate on the origin of rivers and springs. There were three theories on the origins of rivers, which echoed the models of the hydrological cycle. Theory one held that rainfall was the source of rivers. Theory two was alchemical – the conversion of air to water created rivers – while theory three opted for recirculated sea water as the mechanism.[11] This was not conclusively settled until Perrault published his now classic work *On the Origin of Springs* about the Seine in 1674. Perrault's findings were confirmed in 1715 when Antonio Vallisnieri published his analysis of measurements in the Italian Alps.[12] However, the real purpose of these works was to settle the flow issue, and were not about precipitation itself.

Middleton concurs with this view of rainfall research, noting that the field was thin until the turn of the nineteenth century. Luke Howard commenced his classification of clouds in the first quarter of the century. An understanding of cloud types was a great help to predicting precipitation.[13] The great nineteenth-century advances in understanding precipitation were based on advances in thermodynamics, 'the branch of physics that deals with the conversion from one to another of the various forms of energy and how these affect temperature pressure, volume, mechanical action and work'.[14]

8   JG Symons quoted by Biswas, *History of hydrology*, p. 249.
9   He published in 1769 and did his work in London. Biswas, *History of hydrology*, p. 277.
10 Pers. comm., Sara Beavis, 30 September 2011.
11 Biswas, *History of hydrology*, p. 183.
12 Biswas, *History of hydrology*, p. 254.
13 Middleton, *A history of the theories of rain*, p. 147.
14 *Encarta Concise English Dictionary*, Pan MacMillan, Sydney, 2001.

The nineteenth century was marked by robust scientific debate about the phenomenon of precipitation, as demonstrated by some of the ideas that were later demonstrated to be wrong. This included the widespread notion that electricity played a role in cloud formation and other rainfall phenomenon: that water dissolves in air, that water vapour moves independently of the air surrounding it, and that while air could ascend, there was apparently no way for it to descend.[15] Air was also the subject of similar intellectual debate, as is shown by Walker.[16]

In Australia, meteorology was establishing itself, and the climatic variability stimulated debate. 'The science of Meteorology is not, as yet, in a sufficiently advanced state to enable any reliable theory to be formed to account for abnormal seasons, either of excessive drought or excessive rainfall in any country', wrote FS Peppercorne in *On Australian Meteorology and Hydrography* in 1879.[17] Starting in Melbourne and Sydney, observatories were set up, and a network of weather observation stations followed the frontier of settlement.[18] Under the charge of the government astronomer, these networks were the manifestation of scientific meteorology. Leading scientists attempted to make emerging scientifically derived weather patterns available. Ellery from the Victorian Observatory allowed his 'Plain rules for foretelling the weather' to be published in Mullen's *Almanac* in 1884.[19] The catch was that each farmer had to own both a barometer and a thermometer for the plain rules to be useful. While many settlers might have lauded the development of a meteorological network, the fact remained that the information they received from it was generally too little and too late.

---

15 Middleton, *A history of the theories of rain*, pp. 148–51.
16 G Walker, *An ocean of air: A natural history of the atmosphere*, Bloomsbury, London, 2007.
17 FS Peppercorne, *On Australian meteorology and hydrography*, Coupland Harding, Hastings St, Napier, New Zealand, 1879, p. 1.
18 K Douglas, *Under such sunny skies: Understanding weather in colonial Australia, 1860–1901*, Bureau of Meteorology (Metarch Series), Canberra, 2007; A Moyal, *A bright and savage land*, Penguin, Ringwood, Vic., 1993, p. 31, also ch. 8.
19 *S Mullen's Victorian Almanac for 1884*, (Being Second Year after Bissextile, or Leap Year) and Astronomical Ephemeris, containing all necessary information, reducted to the meridian and latitude of Melbourne expressly for this work; also, Departmental, Postal, Telegraph and Mail information, notable events during past year, original medical notes, farming and gardening, and a mass of miscellaneous information.

Weather rhetoric and reality in the catchment

In Gippsland, a network of weather stations slowly grew from Sale outwards from 1870 onwards, but, as shall be seen, the weather in Sale was no reliable guide to the rest of the catchment. Farmers needed other ways to understand the weather.

From the earliest reports by explorers, Gippsland earned a reputation for having a 'good' climate. There had been enough rain to make the catchment look like a pastoralist's version of Heaven. In 1835, George MacKillop wrote of Strathdownie (now Benambra) that it:

> contains 60000 acres of as fine land as I have seen anywhere in the colonies. The Strath is well watered by a large stream running through the middle of it and according to the native who was with me the climate was bland all the year round.

Writing soon after, Strzelecki said: 'Everywhere Nature seems to have most liberally enriched this part for the benefit of man', an excellent example of the divine design philosophy discussed in Chapters 2 and 3.[20] Angus McMillan could barely believe how lush the grasses were. The news of this magnificent new land spread quickly, and the idea of Gippsland as a mild and pleasant climate spread with it.

Settlers used adjectives like 'fine', 'moist' or 'salubrious'. Dr King of Ballarat said at his farewell party that: 'He had come to Sale in the first instance to recover his health, and … he had found here a fine climate … about the finest climate in the colony. The rivers were equal to the climate'.[21] Yet a comparison of rainfall data with the newspaper regional correspondents reports show the startling variability that was experienced within the catchment.

---

20 Cited in P Morgan, *The settling of Gippsland: A regional history*, Gippsland Municipalities Association, 1997, p. 95. An excellent example of the human-centred approach to human–nature relations.
21 *GT*, 7 October 1881. During his convalescence, King helped to establish a fish acclimatisation society.

Table 4.1: Mean monthly and annual rainfall in mm across measuring stations in the GLC.

	Sale	Stratford	Bairnsdale	Morwell	Maffra	Tanjil Bren	Lakes Ent
Jan	46.5	50.8	60.2	54.2	51.8	89.4	55.9
Feb	48.4	54	50.2	39.9	40.4	89.2	41.5
Mar	54.5	59.3	67	45.9	54.1	102.7	54
April	47.9	49.8	50.2	58.4	45.8	136.9	62.8
May	46.3	48.4	54.4	53.6	47.6	190.7	65.6
June	49.7	53.6	58.4	60.2	46.4	148.2	61.9
July	41.5	40.1	50.2	66.1	38.6	157.9	54.9
Aug	44.4	46.6	48.9	63.1	40.3	174.9	50.3
Sept	55.2	53	56.8	80	51.1	171.8	59.2
Oct	63	67.7	70.2	79.3	60.8	175.2	63
Nov	57.4	56.7	64.4	75.1	58.1	171.6	71
Dec	52.4	57.8	67.7	67.6	59	140.9	70.8
Year	607.5	637.4	698.8	740.7	594.1	1,742.3	711.2
Yrs of data	74	74	68	23	95	29	41

Source: Bureau of Meteorology, 17 July 2007.

Table 4.1 provides information on mean monthly and annual rainfall. The geographic variability is obvious comparing Tanjil Bren at an altitude of 838 metres with Stratford, lying on the flattish red gum plains along the Avon River. Stratford's annual average is 637.4 mm, while Tanjil Bren receives nearly three times that amount, 1,742.3 mm. There are also times when rain was almost completely absent. For example, the statistics for Sale show that the driest June ever recorded was in 1886 when only 6.6 mm of rain was recorded. In 1895, Sale residents experienced their driest November, with 0.8 mm in the gauge.[22] In contrast, the nineteenth century held the record for wettest month in seven out of the 12 months. The year 1873 was particularly notable, notching up a saturating February and November with 220.7 mm and 206.7 mm respectively.[23]

---

22   Monthly Climate Statistics for Sale, Station No. 085133, generated from Bureau of Meteorology (BOM) online on 17 July 2007.
23   Monthly Climate Statistics for Sale, Station No. 085133, 1879 was the wettest May (228.6 mm), 1896 was the wettest June (136.4 mm), 1897 was the wettest August (104.9 mm), 1880 was the wettest September (148 mm), 1886 was the wettest October (153.5 mm).

These distinct differences explain the regular comparisons in primary sources about the wetness in certain areas, like Warragul, and the complaints of dryness in places like Bairnsdale and Stratford. Rainshadow areas also exist at Licola, the area around Maffra (the twentieth-century Macalister Irrigation District) and Swifts Creek.[24] We can see the rainshadow effect in operation in this quote from 1876:

> There has been abundant rainfall at Moe and Morwell, and some at Traralgon, but very little has reached further east – not even to Toongabbie which locality generally rejoices in a moist climate from its proximity to the ranges.[25]

Despite the evidence of actual variability in place and time, many still clung to a vision of how they would like certain seasons to be. The *Gippsland Times* Upper Maffra correspondent wrote in September 1876:

> I am delighted to be able to convey to you, even if in borrowed words, the fact that our luxuriant and genial fields assume once more their wonted verdancy. It is to be hoped that a genial summer will maintain the pleasing prospect.[26]

Perhaps the Upper Maffra correspondent was a newcomer, for summers are rarely genial and luxuriant. These were, after all, what Michael Cathcart called wet country people, with 'water coursing through their industries, their farms, their buildings, their class relations, their faiths and superstitions, their songs and their games'.[27]

Traditional weather lore

Before modern meteorology, Europeans had developed elaborate systems for understanding and predicting the weather. These were based on the surviving theories of ancient Greece, particularly Aristotle's *Meteorologica*, or, more typically amongst the less educated, on intergenerational oral traditions. The emphasis here is on the latter. The majority of migrants

---

24 Agriculture Victoria, Victorian Resources Online: East Gippsland, vro.agriculture.vic.gov.au/dpi/vro/egregn.nsf/pages/eg_climate, accessed 26 October 2017. 'Rainshadows caused by the ranges occur in the Mitchell and Tambo River valleys, and on the plains. Rainshadow effects are evident when comparing average annual rainfall of 659mm at Tabberabbera in the Mitchell River Valley in the bordering high country which receives an average of 1080 per annum.'
25 *GT*, 4 July 1876. See also *Bairnsdale Advertiser*, 18 April 1885, noting less rain in Bairnsdale than west of Sale.
26 *GT*, 25 September 1876, Upper Maffra correspondent.
27 Cathcart uses this term specifically in relation to the First Fleet, but I think it can be appropriately used for any migrant from northern Europe. M Cathcart, *The water dreamers: The remarkable history of our dry continent*, Text Publishing, Melbourne, 2009, p. 9.

to Gippsland were not highly educated members of the upper classes. To them, reading Aristotle may have seemed more like punishment than pleasure. The majority of people relied upon a detailed oral tradition of weather lore, along with other aspects of environmental knowledge.[28]

European traditional ecological knowledge was highly honed. It was, in fact, a great skill possessed by rural people, rather than the vague and possibly magical art that sensationalising folklorists and sceptical scientists have since portrayed it.[29] As Hatfield points out in her work on domestic plant medicine in England, 'many [of the rural working classes] could tell the time accurately without a watch, and predict the weather without listening to reports or recording pressure changes'.[30] Time could be told by observing the opening and closing of certain flowers.[31] Observation of animal behaviour was also a significant part of weather lore, as were rainbows:

> Rainbow in the morning, sailors take warning;
> Rainbow at night, sailor's delight.

Lee and Fraser interpret this saying in the light of the position of the observer:

> Throughout much of North America and Europe, weather systems tend to move from west to east. Because the rainbow is seen opposite the sun, a morning rainbow means that rain is falling to the observer's west and thus is likely to arrive soon ('sailors take warning'). Conversely, a later afternoon rainbow, ('night' here is obviously poetic, not literal) means that the rain is to the east and is receding from the observer. Because many rainstorms occur in the afternoon, the rainbow is often (and quite fittingly) taken as a sign of fair weather to come.[32]

---

28 Jankovic, *Reading the Skies*, p. 2.
29 L Watson, *Supernature: A natural history of the supernatural*, 3rd edn, Sceptre Books, Great Britain, 1986, p. 257. Watson described an experiment that analysed the chemical composition of mistletoe (a well-known 'magical' plant) and found that the levels of compounds varied significantly across a year as well as during the day. This confirmed the herbal lore that suggested mistletoe, along with many other plants, should be harvested only at certain times of the day or year to achieve the best potency.
30 G Hatfield, *Memory, wisdom and healing: The history of domestic plant medicine*, 2nd edn, Sutton Publishing, Gloucestershire, 2005, p. 5.
31 A tiny vestige of this kind of knowledge was passed on to me by my grandfather, Phillip Jacka. In the front lawn of their home in Victoria Park in Perth was a tiny plant with pink flowers he called Four O'Clock's because they would start to close their flowers at around this time every summer afternoon. This plant was a type of oxalis, but it shares its common name with *Mirabilis jalapa*, a native of Peru. *Mirabilis* opens in the evening and flowers through the night.
32 RL Lee & AB Fraser, *The rainbow bridge: Rainbows in art, myth and science*, Pennsylvania State University Press, Pennsylvania, 2001, p. 29. I learnt a version of this from my grandfather that substituted 'red sky' for 'rainbow'.

Such traditional knowledge is highly valuable, but also locationally specific. Could settlers translate their old models of weather and climate knowledge onto a foreign land that bore little, if any, resemblance to the world they left behind? How did someone from Galashiels, or St Erth, come to grips with Australian weather?[33] Under Australian conditions, traditional English weather lore could not be relied upon, contributing to the sense of disorder and chaos that many settlers felt.[34] With no familiar plants or animals, there was little opportunity for translation of their traditional knowledges into the Australian environment. Additionally, the presence of Indigenous custodians of weather lore and ecological knowledge rubbed up uncomfortably against their own sense of being the epitome of God's plan. Viewing Aboriginal knowledge as anything other than superstition threatened that view.

Given the climatic variability, the promise of systematised predictions offered by the newly emerging science of meteorology would have seemed ideal. At the same time, the delivery of useful predictions by meteorologists to country farmers was dogged by time delays. Additionally, every farmer would have had a different definition of 'useful' depending on their crop and the stage of its growth. A potato farmer from Thorpdale would differ from a hops grower at Bairnsdale. There was a tension between the desire for generalised predictions based on systematised observations and the reality of the isolated farmer, say, somewhere north of Bairnsdale. In practice, farmers had to make up their own minds about when it might rain, using whatever came to hand and seemed to work.

---

33   I selected these two towns as they were the birthplaces of two female ancestors who migrated.
34   A number of writers have commented on the feelings of alienation and strangeness in settlers caused by the reversed seasons and the radically different flora and fauna, e.g. Moyal, *A bright and savage land*, pp. 131–2; R Broome, *The Victorians: Arriving*, Fairfax, Syme and Weldon Associates, McMahons Point, NSW, 1984, p. 26; E Rolls, 'More a new planet than a new country', in S Dovers (ed.), *Australian environmental history: Essays and cases*, Oxford University Press, Melbourne, 1994, pp. 22–36.

Living between rhetoric and reality

A GLC farmer might read about meteorological debates in supplements in the local papers or in libraries, but it seems unlikely that they would find them helpful on a day-to-day basis.[35] Given that meteorological science offered little of practical value to settlers, what other tools did they have at their disposal to figure out this perplexing world? There were four ways:

- their own recorded observations
- almanacs
- the possible survival and adaptation of traditional weather lore
- astronomical theories.

The most abundant evidence comes from observations that settlers made themselves. Every diary consulted recorded their experience of the weather and made comparisons among years. In her comparative study of British and Australian almanacs, Perkins discusses an Australian sheet almanac held in the British Library. Its advice for October and November acknowledged that only personal observation by the farmer would be of use and that no book could hope to fulfil their need for advice.[36] Every diary consulted made weather observations, to varying degrees of sophistication. McLeary and Dingle note that Catherine Currie, a selector's wife at Lardner in West Gippsland, recorded everything including rainfall.[37] On her selection near Lake Wellington, Isabella MacLeod also made regular notes on the weather in her diary. On 5 June 1873, it was 'dull-damp' and the following day was 'rainy and miserable all day'.[38] The timing, quantity and duration of rainfall was a matter of strategic importance to these newly establishing farmers. While most settlers who kept diaries always made a passing note of the weather, very few of them took measurements. More significantly, the weather observations are interpreted in the light

---

35 For an example, see *Castner's Rural Australian*, December 1875, p. 8; and *GT*, 1886 Supplement. Hutton and Connors discuss the role of the newspapers in reporting the activities of scientific societies and in starting 'nature columns'. D Hutton & L Connors, *A history of the Australian environment movement*, Cambridge University Press, Cambridge, 1999, pp. 30–1.
36 M Perkins, *Visions of the future: Almanacs, time and cultural change 1775–1870*, Clarendon Press, Oxford, 1996, p. 175. doi.org/10.1093/acprof:oso/9780198121787.001.0001.
37 A McLeary & T Dingle, *Catherine: On Catherine Currie's diary*, Melbourne University Press, Melbourne, 1998, p. 13.
38 R MacLeod, *The Commonwealth of science: ANZAAS and the scientific enterprise in Australasia, 1888–1988*, Oxford University Press, Melbourne, 1988, p. 223.

of their current circumstances. These are *personal* measures of rainfall effectiveness, whether that might be to moisten the soil enough to start ploughing or to fill the waterholes for desperately thirsty stock.

Buckley and Auchterlonie, whose diaries have the longest runs, show evidence of attempting to figure out weather patterns. In the mid-years of his diary, Buckley tended to sum up the month. In a month that had seemed contrary, he would remark how he had never seen it so dry/wet/windy/frosty since he arrived. Gradually these individual observations accumulated. One can imagine conversations held in the stores and streets of the settlements that, with time and repetition, grew to be the new weather lore, approximations that the newspapers reflect. A good example comes from 1890:

> The remarkable heat and dryness of the Autumn is causing heavy loss in those parts of the colony where a fair amount of moisture is customary at this time of year, and is a necessity for crops produced in them. In North Gippsland, the maise [sic] and hops crops are pretty nigh ruined, especially the former, while the potato crops in other districts are suffering severely for want of moisture … Altogether the autumn is reckoned as the hottest and driest experienced for many year past.[39]

Because rainfall variability could cause such havoc, it is not unreasonable to assume that traditional methods of weather forecasting might be adapted, and in some ways attain a greater significance, in areas being newly settled. From the sources consulted, nothing conclusive can be said, although there are some clues worth following up. As noted previously, European weather knowledge involved a detailed understanding of local plants. One reference to this was found in newspapers. In its March 1876 issue, *Castner's Rural Australian* reported the research of Herr Hanneman, Director of the Botanic Garden in Proskau, Germany, on plant behaviour that could predict the weather. For example, *Convolvulus arvensis* (common English bindweed) and *Anagallis arvensis* spread their leaves at the end of wet weather, in contrast to some varieties of clover, which contract their leaves. *Stellaria media* (chickweed) and *Pimpinella saxifraga* in clear morning weather will straighten its flower and spread its leaves, and droop in rain. If *Calendula pluvialis* hasn't opened its flowers

---

39   *Morwell Advertiser*, 7 February 1890.

by 7 am, one could expect rain. Even the name of this plant is a clue to its connection with rainfall. However, the article made no references to any Australian flowers. Nor did it suggest that farmers could adapt this knowledge of European plants to the Australian context.[40] Rather, it was presented in the manner of curious facts that one might read aloud to the children at tea time.

On the other hand, David Brown, a road labourer, recalled in his memoirs an anonymous German man in Brandy Creek whose nickname was 'the weatherman'.[41] One cannot rule out the adaptation of old-style European ecological knowledge, for three reasons. First, the newspapers almost never provide sources for their weather, describing them only as 'those most capable of giving an opinion on the subject'.[42] This could be the Weatherman, or it could mean the Mayor of Sale, who in 1870 owned a thermometer, barometer and hygrometer. Second, there is the strong tradition of oral knowledge suggested by Hatfield. The almanac referred to by Perkins above advised that December was the month for gathering and drying medicinal herbs, which suggests that some parts of traditional herbal knowledge continued.[43] Third, by the emphasis in Australian history on the practical skills of the bushman. It would be logical to assume that part of the skill set of a practical bushman would include knowledge about the weather. Perhaps the constant noting of weather conditions in diaries is a part of the revamping of the old version of European weatherlore. While some settlers must have learnt some botanical information from Indigenous people, this research found no evidence of the transfer of Indigenous ecological knowledge in Gippsland.

The final method of weather forecasting was planetary observances. In this approach, the weather was caused by the movements of astronomical bodies and their phenomena, especially the moon. Most almanacs recorded phases of the moon, and there are comments throughout newspapers that indicate the broader public found moon tracking to be normal. It was in relation to weather extremes that comments were found linking the moon to the weather. For example, the *Gippsland Times* on 21 June 1880 queried: 'Whether it is owing to the approaching eclipse of

---

40   *Castner's Rural Australian*, March 1876, p. 16.
41   David Brown, Reminiscences Of Brandy Creek, CGS, 696.
42   For example, *GT*, 24 December 1886.
43   Perkins, *Visions of the future*, p. 175.

the moon, which will take place tomorrow night, or not, the temperature during the last four or five days has been marked by extreme severity'. In 1881, cricket lovers were thwarted by downfalls on the Prince of Wales's birthday public holiday match between Sale and Rosedale. The Rosedale correspondent said:

> Whether the transit of Mercury the day before over old Sol's disc has so disgusted that noble liminary that he declined since to show his face, or from other reasons not known outside the realms of science, the weather was just as provokingly aggravating in the morning as the most arrant flirt could desire in her conduct to her lover.[44]

Whether or not he believed it, the Rosedale correspondent displayed an understanding of weather prediction based on astronomical movements. Describing locally significant events in these terms suggests that many Gippslanders accepted this idea.

Almanacs are an important source for understanding colonial weather knowledge, as significant as settler diaries.[45] They are vital to understanding what information was easily available for settlers to make agricultural and social decisions. Not having an almanac would be like not having a mobile phone in the twenty-first century. Almanacs were a cross between a calendar, a business diary and a government year book. They always carried day and date information; listed the government, its ministers and departments; and provided data on essential services like postage and telegraph services. They also published varied types of scientific, meteorological, astronomical, astrological and environmental data. There were also specialty almanacs, like the *Victorian Law Calendar and Almanac*, referred to in the *Government Gazette*.[46] For the very poor, many papers printed sheet almanacs.[47]

---

44   *GT*, 11 November 1881, Rosedale correspondent. See also, *GT*, 4 January 1876, Report of a meteor.
45   Perkins, *Visions of the future*, pp. 3, 10–11.
46   *Victorian Government Gazette*, 1 December 1855, p. 94.
47   For example, the *Gippsland Times* on 1 January 1876 published a sheet almanac that gave day, date, mail info (arrival and dept of the mail steamers) and postage rates.

The *Gippsland Times* often assessed the content of different almanacs and made comments about their value for money. In 1869, the *Gippsland Times* favourably reviewed Clarson and Messina's and Dr LL Smith's.[48] Robert Thomson, MLA, kept his 1848 *Blackie's Literary and Commercial Almanac*, which fortunately survived the flood that destroyed most of the Thomson papers stored in a cellar at Clydebank.[49] The weather table on pages 18 and 19 illustrates succinctly how the moon was thought to influence the weather.

Table 4.2: Extracted from *Blackie's Literary and Commercial Almanac*, 1848.

New and Full Moon	Summer	Winter
At new or full Moon, the Moon entering the first or last Quarter at 12 noon or between 12 and 2	Very rainy	Snow and rain
... afternoon 2–4	Changeable	Fair and mild
Evening 4–6	Fair	Fair
Evening 6–8	Fair if wind at NW, rain if wind at S or SW	Fair and frosty if wind at N or NE Rain or snow if S or SW
Evening 8–10	Ditto	Ditto
Night 10–12	Fair	Fair and frosty
Morning 12–2	Ditto	Hard frost unless W, S, or SW
Morning 2–4	Cold with showers	Snow and stormy
Morning 4–6	Rain	Ditto
Morning 6–8	Wind and rain	Stormy weather
Morning 8–10	Changeable	Cold rain if wind W, snow if E
Noon 10–12	Frequent showers	Cold with high wind

---

48   *GT*, 23 and 29 October 1869. These two almanacs are very similar, as Clarson and Massina also published Smith's. The texts of the 1877 editions of both are virtually identical. The major point of difference is that Dr Smith supplemented the standard information with a range of medical information, which is itself interesting for how dampness or dryness was seen in relation to health. For other examples of reporting on almanacs, see *GT*, 1 August 1876, copy of *Bradshaw's Northern Miners Almanac* from Charters Towers area, giving crushing details; *GT*, 1 December 1876, 'We have received a copy of McKay's almanac for 1877 being the eighteenth year of its publication. It abounds in useful information'; *GT*, 18 November 1881, 'We have received a copy of the Temperance Year Book for 1882, published Messrs Dunn and Collins Melbourne, It is compiled with special reference to temperance matters, but contains a quantity of general information, including and almanack, &c. Hutchinson's Australian Almanac for 1882 is also to hand, being the 23rd year of publication. It is of the usual standard of merit, containing information upon a large variety of topics, political and social, including an educational register, medical notes, postal rates, acts of parliament in 1880–2, 7c, &c.'.
49   Other settlers with extant almanacs include the following: Alexander Hunter's journal was recorded in a book, which also had an almanac printed in the front. It's a basic almanac with weather table, moon and eclipses, English, religious and other dates, but no facts etc in the blanks. *Lett's Diary or Bills Due Book and an Almanack for 1839*, Hunter Family Papers, SLV, PA 99/64. Michael Wilson suggests that Rev. Francis Hales adapted the Churchman's Almanac for his journal. Journal of the Rev. Francis Hales (edited by Michael H Wilson), SLV, MS 12950, box 1716/14.

Immediately following the table is a series of guidance statements on weather prediction. For example:

> Dew – If dew lies plentifully on the grass after a fair day it is a sign of another. If not, and there is no wind, rain must follow. A red evening portends fine weather, but if the red spread too far upwards from the horizon in the evening, and especially morning, it foretells wind or rain, or both.[50]

The moon theory of weather prediction was accepted by the state. For example, the report in the *Government Gazette* on Alberton's meteorological records for December 1854 noted the moon's age and described the weather experienced. Smith's *Almanac* for 1877 provided a week by week weather prediction chart, but did not explain how it was derived. Given the layout of Smith's *Almanac*, it would seem likely that the predictions were derived from the moon theory. For July, the almanac claimed: 'Fair weather will probably set in with this lunation and will last till the end of the month'. Week 2 of October refers to 'electric disturbances in the atmosphere' to be followed by stormy weather. The year of 1870 was notable for the number of displays of the Aurora Australis, and the *Gippsland Times* noted that 'the farmers evidently think they prognosticate foul weather. Almost every heavy electric storm this season has been immediately followed by tempest, and rain'.[51] Knowles Middleton noted that the connection between electricity and rainfall was commonly debated in scientific circles in the nineteenth century.[52]

A comparison of almanacs published in Victoria dating from 1841 to 1895 shows that information about the moon is the most common. The phase of the moon was always given, followed by the age of the

---

50 Thomson Family Papers, NLA, MS 8600, box 2, folder 9. The remainder of weather signs were: 'Mists – A white mist in the evening, over a meadow with a river will be drawn up by the sun next morning and the day will be bright. Clouds – Against much rain, the clouds grow bigger and increase very fast, especially before thunder. When the clouds are formed like fleeces, but dense in the middle and bright towards the edges, with the sky bright, they are signs of a frost, with hail, snow or rain. If clouds breed high in the air, in thin white trains, like locks of wool, they portend wind, and probably rain. Two currents of clouds portend rain, in summer thunder. Heavenly bodies – A haziness in the air which fades the sun's light and makes the orb appear whitish, or ill defined, or at night if the moon and stars grow dim, and a ring encircle the former, rain will follow. If the moon looks pale and dim, we expect rain, if red, wind; and of her natural colour, with a clear sky, fair weather. If the moon is rainy through throughout, it will clear at the change, and perhaps the rain return. Wind – If the wind veers about much, rain is pretty sure. If in changing it follows the course of the sun, it brings fair weather; the contrary, foul'.
51 *GT*, 22 November 1870.
52 Middleton, *A history of the theories of rain*, p. 148.

moon in days. Almanacs might also give times of the moon's rise and set, hours of available moonlight and the dates of lunar and solar eclipses.[53] The 1863 Victorian *Almanac* noted equinoxes. Because almanacs recorded the passage of the moon, there was no need for settlers to do so. Thus it is rare to find lunar references in personal writings. In diaries, only George Auchterlonie mentions the moon, and this was in relation to notable weather. On 29 September 1868, he wrote: 'very warm today large circle around the moon tonight'. A month later, he wrote: 'weather most oppressively hot today, sun rose red, atmosphere of a dirty haze, moon clear'.[54]

The almanacs emphasis on the moon also reflected dependence upon its silvery light for navigation, in the age before street lighting was common. Duncan Johnston noted taking a moonlight walk with friends on one of his visits to Bruthen.[55] Most large evening events were timed for the full moon to give the greatest amount of illumination to travelling revellers. Reflecting this importance, one local history is titled *Wednesdays Closest to the Moon*.[56]

Interestingly, few almanacs provide temperature or rainfall data, even in later years when meteorology had gained more recognition as a science. Smith's almanac was an exception. The 1864 edition provided an average for each month, and readers were informed that the figure had been derived from a seven-year dataset.[57] The colonial authorities understood the value of meteorological records, as Governor Gipps forwarded instructions to La Trobe as early as 1840 on the proper way to keep meteorological journals.[58] But for individuals, the expense of equipment precluded many from keeping records. Given the monetary and time costs, it is perhaps no wonder that the moon-charting method popularised in almanacs lasted so long. Anyone can pop outside and check the phase of the moon.

---

53 I compared 23 almanacs over a date range of 1841 to 1895.
54 Diary of George Glen Auchterlonie, 29 October 1868, CGS, 4060. The following day was recorded as 'excessively warm'.
55 Diary of Duncan Johnston, Sunday 5 March 1882, CGS, 00317.
56 B Collett, *Wednesdays closest to the full moon: A history of South Gippsland*, University of Melbourne Press, Melbourne, 1994.
57 *Dr LL Smith's Medical Almanac* for 1864 and guide to mothers and nurses and persons residing in the bush, printed by Clarson, Shallard and Co., 85 Bourke St East, Melbourne and 207 Pitt St, Sydney.
58 AGL Shaw, *Gipps–La Trobe correspondence: 1839–1846*, Melbourne University Press at the Miegunyah Press, Melbourne, 1989.

It was a terrible shame that it was wrong. Neither the new sciences nor the old wisdom cut the mustard for colonial migrants. It is no wonder they invoked those fickle Greek gods when talking about the weather. The practical implications of a turn in the weather could only have been magnified by how unpredictable the weather really was.

The effect of rain

The anxiety caused by the uncertainty of rainfall was exacerbated by the detailed knowledge of its positive and negative effects. In an agricultural economy, the quantity and duration of rain had ramifications across most areas of social and economic life. This section will examine how settlers reported positive and negative effects of rainfall.

Negative effects from precipitation included destruction from flood, storms, fallen trees and frost; coughs, colds and more serious health complaints like rheumatism and pneumonia; difficulty in travelling (see Chapter 6); the need for particular construction techniques to make buildings weatherproof; prevention of mining, outdoor works on farms and in construction; and many social activities like picnics.

On the positive side, precipitation supported germination, crop growth, ploughing, dam filling, milk yield, boating, bathing, washing, stock condition and the destruction of some vermin like grasshoppers. However, the line between 'good' and 'bad' precipitation was fine as shown in this quote from spring 1886:

> The pleasant change in the weather is all the conversation one can hear at present. Plenty of grass and water for the season, yet I hear some beginning to grumble that there will be too much rain for the crops. At present the crops and pasture are both looking splendid, and promise to give a grand harvest.[59]

The line was variable, with big differences between pastoralists and farmers, followed by differences between farmers and their chosen crops. In this section I commence with flood, move onto the less overwhelming negative aspects of heavy rain and then contrast this with the discussion on the welcome aspects of rain.

---

59 *GT*, 13 October 1886, Briagolong correspondent.

4. 'NOTWITHSTANDING THE INCLEMENCY OF THE WEATHER'

Too wet

The editor of the *Gippsland Times*, in his offering on Saturday 17 December 1870, summed up the ideal notion of precipitation when he wrote:

> The weather necessarily enters largely into all our calculations whether we will or not. [In] agriculture, it plays a part of paramount importance; for the seed may be ever so choice; the soil ever so prolific, and the situation ever so favourable, but without the life giving showers of spring, and congenial heat of the summer's sun, the blade will not show, neither will appear the full corn in the ear.[60]

By December 1870, the GLC had suffered a surfeit of water, experiencing a staggering 17 floods. The words, 'life giving showers' and 'congenial heat', suggest a farming community wearied by extremes. The season had been 'unprecedented in the memory of the oldest inhabitant'.[61] District Surveyor Dawson summed up the year:

> Two high floods in a season are beyond the average, and three has also been looked upon as extraordinary. Upon occasions the rivers have never overflowed their banks at all for two, and upon occasion that I recollect, not even for three seasons, thus showing what exceptional weather we have had this year.[62]

As the district had only been colonised since 1838, this was not a long time to make judgements about the unusualness of the winter of 1870. Yet it is quite clear from the discussion that litters the paper's pages in this year that there is strong sentiment about what was normal and what wasn't.

The winter and spring of 1870 did not provide the life-giving showers that obviously would have been preferable. Instead, there were extensive snow falls, hard frosts and widespread heavy rains and floods. Nothing about 1870 was normal. Is it possible to know what would have been considered climatically and socially normal in a district that had been so sparsely settled for such a short period of time? A glimpse of normality can be seen by looking at what events and activities this prolonged experience of flooding interrupted.

---

60 *GT*, 17 December 1870.
61 Report of Surveyor WT Dawson to the Surveyor General reproduced in full in *GT*, 13 December 1870.
62 *GT*, 13 December 1870. The report was to government but the *Gippsland Times* was able to publish it. Dawson, too, appears overwhelmed by the waters, although he couches his dismay much more bureaucratically. He was also unhappy with the report, as due to 'continued bogginess of the bush lands' he had been unable to make the full range of damage inspections that he would have liked.

If any subject got the lion's share of column space that year, it was the damage to roads and bridges. The *Gippsland Times* printed paroxysms of verbiage on the condition of Punt Lane, which connected the township to the river wharf, and thence south to Port Albert (see Map 10). Punt Lane was only one of the casualties of the 17 floods experienced in 1870. There are at least two court cases where carriers sue the respective councils for the death or laming of their horses.[63]

In the Shire of Avon, the council faced a particularly difficult situation with nearly every bridge and crossing point damaged or destroyed. The council meeting of 22 August is full of memorials, correspondence and surveyors reports about flood damage. The reports also show evidence of individuals taking action to restore order, like seeking permission to cut drains to get rid of floodwaters off their lands. Unable to influence the arrival of the flood, they tried to shorten its effect. In September, all the flood-affected councils lobbied the state government to spend their promised £5,000 on the main coach road to Melbourne, 'the like of which' one of the attendees 'ventured to say could be found in no part of the civilised world'.[64]

Another indication of 'normal' for the white citizens of Gippsland was in the number of events that were poorly attended or cancelled. In September the picnic and kangaroo hunt to celebrate the successful introduction of trout to the catchment was postponed for between three and four weeks. The Ladies Benevolent Society's proposed fundraising concert survived the haranguing delivered by the rector of St Paul's, only to be postponed because of the weather. Meetings of the Hunt Club were down on numbers. Government business was delayed, with the mail coach being detained and business in the land court affected by the inability of the surveyors to keep their schedules during bad weather. The 1870 Agricultural Show in Sale was postponed due to the flood that arrived on 25 March.[65]

---

63   The first court case was against Sale Borough Council regarding a horse that took a whole day to extract from the bog of Punt Lane, and suffered so much during the process that it had to be destroyed. *GT*, 23 August 1870. The second court case was against the Avon Shire Council, discussed at the council meeting of 22 August 1870 and reported in the same 23 August edition.
64   *GT*, 2 September 1870.
65   *GT*, 2 September 1870 for the trout reference. There are regular reports throughout this whole year, and reports from 1869 of the mail being delayed across the catchment – for example, 1 June 1869 the coach from Melbourne was a whole day late. Coverage of the Land Court proceedings was in *GT*, 5 June 1869. See *GT*, 26 March 1870, for the description of the flood that postponed the agricultural show.

Figure 4.3: Flood debris piled up against the railway bridge to Walhalla across the Thomson River, 2008.
Source: Melissa Smith.

Figure 4.4: Flats at the mouth of the Latrobe River, Thomas Henry Armstrong Bishop, n.d. (c. 1894–1909).
Source: Pictures Collection, State Library Victoria, Accession no. H40967.

Figure 4.5: A Gipps Land track after rain, *The Australasian Sketcher with Pen and Pencil*, 1882.

Source: Pictures Collection, State Library Victoria, Accession no. A/S11/02/82/37.

The impact on individuals from floods can be gleaned from diaries. The King family had to beat a hasty retreat in the face of flood in November 1844. The station day books record the intent to move two huts further away from the river.[66] Mary Cunninghame wrote in 1846 that 'unfortunately our vegetable garden has been flooded by the lake rising higher than it was ever known to do before which will put us back a good deal with our gardening'.[67] George Auchterlonie noted that his neighbour had lost land to the erosive power of floodwaters. Miss Caughey wrote on 27 August 1882 that rain was heavy enough to stop church attendance because the 'fearful' roads made for a dangerous trip. The following month they planned a picnic to host local councillors debating a bridge site, which the family hoped to influence. 'I hope it will not rain after all our trouble', she wrote. The normal social and business life of the catchment was severely disrupted for weeks or months until lands dried out and repairs could be undertaken.

Part of the difficulty in understanding 'normal' rainfall relates to measurement, or rather the lack thereof. At this time, measurement of rainfall was an emerging part of hydrological and meteorological science globally, but there is scant evidence of it occurring publicly, regularly and consistently in the catchment until the later years of the nineteenth century. On 19 July 1870, the *Gippsland Times* wrote:

> The rainfall in the neighbourhood of Melbourne, according to the Argus, has only been 12.80 inches since the beginning of the year. The rains thus far during 1870 have varied much in different parts of the colony; but it is questionable if any place has been more severely visited in this respect than Gippsland. It would be well if a rain gauge were established at one of the Government offices, either here or at Bairnsdale, as records, besides being interesting, would be likely to prove useful in a country so liable to periodical inundation by disastrous floods.[68]

---

66   Diary of Miss AM Caughey, 27 August 1882, 2 September 1882, and 21 September 1882, SLV, MS 8735, MSB 434; King Station Day Books, 23 November 1844, SLV, MS 11396, MSB 404.
67   Letter from Mary Cunninghame, at Roseneath, 20 September 1846 (postmarked from Alberton) to Lilias Bonar. Thomson Family Papers, NLA MS 8600, box 1, folder 1.
68   *GT*, 19 July 1870.

Gippsland had been settled from 1838, making 30 years of flood experience amongst the white population. These were mainly caused by the sandbar at the entrance remaining closed, inundating the swampy fringes of the lakes and rivers. It was a perfectly normal part of the hydrological cycle. These relatively small events contrasted with other qualitatively different floods, notably in 1863, when a number of people drowned and the whole Bairnsdale cemetery was washed away.

Whatever normal was, floods disrupted it because they weren't predictably timed like the famous Nile floods. The *Gippsland Times* gave extensive descriptions throughout 1870 of how high the water reached and its path. Describing a flood around Narracan, the reporter noted:

> On Blind Joe's Creek the water was higher than during any portion of the winter floods, and a considerable portion of fencing has been washed away and buried, amongst other *debris*, in the scrub. At the Middle Creek the log crossing, lately constructed by order of the Rosedale Board was swept away … At Sheepwash Creek the water was very high, but no damage was reported as having been done. The river at Rosedale is running bank high, and threatens an overflow before long.[69]

This typical report sets up an implied hierarchy of river heights. There is an average winter flood, exceeded at Blind Joe's Creek. At Sheepwash Creek the water was high, in contrast to the term 'bank high', in contrast again to 'overflow'. This shorthand description assumed that people who read the paper knew the places in question, and were *au fait* with the 'ordinary' conditions there.

In some rare flood reports, numbers are offered as an indication of magnitude. When discussing the height of a river, these are usually in relation to a bridge, but there are also mentions of what is thought to be the normal river height. For example, in the 26 March 1870 issue, the Heyfield bridge was under threat from an extra 2 feet of water 'above its ordinary height'. When the paper went to press, the floodwaters were 2 feet under the Raymond St bridge. The following edition described the extent of the floods across the district, when information had been able to reach Sale. Largely there are no numbers given. One of the exceptions was the 6-foot-deep swamp opposite the English church, which attracted

---

69   *GT*, 16 October 1869, emphasis in original.

attention because some people had decided to go rowing.[70] Where there is no available landmark, a straight depth in feet is given, such as Desailly's flat being 2 feet under a week after the first rush of waters, or the depth of snowfall on the Dargo High Plains.[71] Snow depth was important because it gave a potential indication of flood severity in the spring.[72] On 9 April, the Boggy Creek correspondent wrote that the machinery at the Sons of Freedom Mine had been submerged for up to 60 hours, and that it was the heaviest rain seen in 10 years.

The rareness of measurement in the 1870 reportage confirms the importance of subjective and relative knowledge, especially when much of the population may have had minimal schooling. Everyone would have been able to visualise water up to the eaves of a barn.

Figure 4.6: Boggy Creek, J Williamson engraver, 1865.
Source: Pictures Collection, State Library Victoria, Accession no. IAN23/12/65/5.

---

70   *GT*, 29 March 1870.
71   For Desailly's flat, see *GT*, 23 April 1870; for snow, see *GT*, 3 May 1870.
72   For example, the *Illustrated Australian News for Home Readers*, reported on 20 August 1867 that John Temple Seaborne, while making his way from the Ovens, got stranded on the Dargo High Plains during snow and was so severely frostbitten he had to be taken to Grant for medical treatment. Similarly, deep snow (up to 3 feet) also reported at Woods Point and Matlock, p. 2.

Residents of Sale knew the next flood was coming from this report on 16 April:

> The drizzling steady rain of the past two days has put an effectual damper upon all the sport and enjoyment in connection with the Easter holidays. Everything now wears a most uncomfortable, cheerless aspect … There was a continuous soaking rain all yesterday, and the Thomson has again risen to within five feet of the bridge level. It is feared that the copious rainfall in the mountainous districts will bring down a volume of water sufficient to again cover all the low lying ground in the vicinity of Sale and Stratford … The swamps, supplemented by the overflowing gutters and channels, are at the present time rising rapidly, and a very little will place Sale in an inundated condition. At the hour of going to press, we have had nearly 50 hours of rain without intermission.[73]

As the floods continued through the year, the reference point for heights became the floods themselves. The flood reported on Tuesday 17 May, which had commenced with 'incessant rain' on the previous Thursday, was reported to exceed the 1863 flood mark. Until then this flood, with its high death toll, had been considered the extreme. More commonly, the description is referenced to floods occurring within 1870. There were now more landmarks by which to describe the ingress of the waters. Mr Cobain's barn was flooded up to the roof, destroying his barley store. The Greenwood family at Clydebank was rescued off the roof of their hotel, while the bark huts of the Chinese market gardeners were submerged to their ridge lines.

In this story of relative heights, only two things sank (besides spirits): the pilings of the Cunninghame St and the Nuntin bridges.[74] Notably these are all vertical measures. In the whole of 1870, there is not a single numerical reference to the breadth of space water could occupy, and only occasional qualitative ones.[75] Gippsland's settlers were very much as Cioc described people living along the Rhine:

---

73   *GT*, 16 April 1870.
74   *GT*, 17 May 1870.
75   *GT*, 29 March 1870, 'The Thompson is now in every part a swollen expanded river'; *GT*, 19 April 1870, 'The river washed over Punt Lane road at different places, flowing into the swamp on either side, and giving the whole country towards Longford the appearance of a vast lake'.

> Flood is a highly anthropocentric term, rooted in the human proclivity to think of a river as having a fixed length but no prescribed breadth, with the result that the flood plain is often used for farms and settlements as if it were not part of the river system.[76]

Sale residents appear to have forgotten that Sale's original name was Flooding Creek, a clear toponomic clue to their likely experience if they settled there. Their ecological perception of a flood was generally expressed in words that represent degrees of verticalness, such the height of Cobain's barn, rather than the breadth of the floodplain on which they settled. Height mattered because it was tied to the fate of the social, physical and economic infrastructure that underlay their culture. This emphasis on high and low will continue in Chapter 6, which considers still water lying in low in the landscape.

Other ecological knowledge learnt

It takes more than rain to make a flood. The detailed tracking of floods in the papers provides evidence for how colonial Gippslanders perceived these related ecological aspects of the hydrological cycle. In 1897, a newspaper report mentions that there were fears of a flood, because 'the southerly winds were backing up the lakes, but the rainfall in the mountains does not appear to have been heavy enough to cause any trouble'.[77] This demonstrates the understanding that more than one environmental process is at play. Additionally, newspaper evidence shows clearly that GLC settlers understood the relationship between snow, heat and flood timing, as well as the effect and implications of soil saturation. The combination of water and clay was lethal for the horse that lost its life floundering in a bog in Punt Lane.[78] Similarly, it was noted that the annual movement of stock to the high plains would be delayed for months because the ground would be too boggy.

It also interfered with the timing of agricultural operations, setting ploughing and seeding back by weeks. In August it was noted that early sown crops were suffering from waterlogging, and that everyone was holding out for a dry change in order to attempt a late sowing.[79] A few years later, the *Gippsland Mercury* aired opinions on soil moisture in a less wet

---

76 M Cioc, *The Rhine: An eco-biography*, University of Washington Press, Seattle. 2002, p. 33.
77 *GT*, 16 August 1897.
78 *GT*, 17 September 1870.
79 *GT*, 4 October 1870.

year. The Tinambra correspondent said: 'Some farmers are of the opinion that a few showers would be of great advantage now, while others think there is sufficient moisture in the soil'.[80] This probably reflects differences in land units and the ability of different soils to retain moisture. A farmer who selected on a sandy unit would clearly have a different opinion to a farmer who was lucky enough to get a good loamy soil.

That a correlation between snow melt and flood was understood is evident from comments about a kind of informal flood-warning system, of which the papers were integral. On 16 October 1869, the paper published the following warning:

> Constable Reilly writing from Warrangarra reports: 'The spring flood of the present year is now passing down the Dargo and Mitchell Rivers. The water is overflowing their banks. A heavy flood may be expected at the end of the present week about Sale and the Lakes.' This report we fear is only too true and persons whom it concerns had better take timely warning.[81]

When the first flood struck Sale, the paper recorded that the police constables had visited the common and other lower elevation areas to advise residents to make preparations to shift livestock and valuables to higher ground. The collective wish was probably for warm dry weather so that the regular cycle of farm activity could get back to normal, but this could also be a dangerous wish after such a severe winter:

> The milder weather which we experienced for the last two days has had the result of bringing down a quantity of water to the low lying lands from the upper country, evidently from melting snow. It is to be hoped the thaw will be gradual; else the probability is more floods will be experienced in this district, as the ranges have more than the usual quantity this season to be melted away.[82]

The existence of this informal flood-warning system speaks of hard learning from experience. It shows that settlers did try to institute protective systems. In the flood of 25 March 1870 in which the bridge across the Avon River at Stratford was destroyed, the *Gippsland Times* said:

---

80   *GM*, 24 July 1877.
81   *GT*, 16 October 1869. The spelling of Warrangarra could mean either the Wonanngatta River or the Wongungurra River. This practice was also discussed in J Sadleir, *Recollections of a Victorian police officer*, George Robertson and Co., Melbourne, 1913, p. 153.
82   *GT*, 9 August 1870.

This bridge has been swept away a second time. It will be recollected that it was partly re-erected at a considerable elevation above the ordinary flood level of the river, in order to prevent the reoccurrence of the disaster only a few years ago, but it seems this precautionary measure was powerless to prevent its demolition.[83]

The term 'ordinary flood level' illustrates that there was a mental benchmark. This is a kind of intellectual epigraphic record, the mental equivalent of the actual marks that floods leave in the landscape.[84] In another example, the *Bairnsdale Liberal News* reported on a flood in 1879. Comparing damage to the flood of 1876, it noted that hop growers had learnt from the earlier flood because they had bound the hop poles and not lost quite so many.[85]

Given the constant discussion of interruption to nearly every aspect of social and economic life, it is not surprising to see so few references to the other side of the balance sheet. There were only four positives noted in the *Gippsland Times*. In August 1870, the Omeo correspondent detailed the devastation to mining operations around Livingstone Creek. His tone lightened, however, at the realisation that the flood had removed 'hundreds of tons of tailings, and other debris which impeded the proper working of many river and creek claims throughout the district'.[86] The health of the downstream aquatic life in the river channel would have been severely affected, but the Omeo correspondent had no appreciation of this.

The second benefit was the boost to pasture growth. The preceding summer had been quite harsh, and the ample quantities of water seemed to make the world anew. Fresh green grass, sprouting crops and full creeks, at least in the autumn of 1870, made for positive hopes for a bumper season. Third, the full lakes and swamps were also a beacon to all kinds of waterbirds and, therefore, to the euphemistically named 'knights of the trigger'.[87] Finally, the benefit with the most far-reaching consequence was that the sandbar entrance to the Lakes was scoured open. The Lakes Entrance pilot described the conditions as superlative in the six years that

---

83   *GT*, 26 March 1870.
84   For a discussion of epigraphic marks as evidence, see N McDonald, 'On epigraphic records: A valuable tool in reassessing flood risk and long-term climate vulnerability', *Environmental History*, vol. 12, no. 1, January 2007. doi.org/10.1093/envhis/12.1.136.
85   *Bairnsdale Liberal News and North Gippsland District Advertiser*, 4 June 1879.
86   *GT*, 2 August 1870.
87   *GT*, 2 April 1870 and 30 April 1870.

he had been the pilot.[88] The force and quantity of water from the floods had scoured a deep channel whose longevity would fuel the rail versus ship debate that was such a feature of the catchment's public life.

Staying safe, warm and dry

Focusing solely on floods misses the impact of heavy precipitation that stops short of flood. There are similarities, however – notably the difficulty with travel and the interruption to the social calendar. Heavy rain accompanied by wind storms, frosts and hail were hazards that colonial settlers had to guard against. Stock could be killed by debris or weakened by frosts, while fences and other farm infrastructure could suffer damage. Duncan Johnston's barn was destroyed by a storm in September 1882.[89] Mr T White, recalling his early days in a tent on the west side of Narracan Creek, noted that the pigs were as anxious for shelter during storms as the humans. They would escape, come into the tent and shelter under the beds.[90] Occasionally, people caught outside were injured or killed by falling trees.[91]

Protecting oneself from the cold and wet was vital to maintaining good health. A variety of sources from the Narracan/Moe area all suggest that the constant rain had adverse health impacts. Colds and bronchitis were common. One teacher 'caught a severe cold walking thro [sic] the flood water to school'.[92] Councillor Michael O'Connor and his father-in-law George Holland both died of pneumonia, aggravated by the constant wet and damp conditions.[93] Thomas Butterfield, a selector at Swan Reach, died of pneumonia in 1879. It took three days before the river was low enough for someone to come and help his wife and daughter bury him.[94] Dampness also affected rheumatism.[95] Henry Meyrick developed rheumatism while spending a winter in a tent. While attempting to fetch the doctor during a flood he was drowned because of the resultant weakness and pain in his arms.

---

88   *GT*, 2 August 1870.
89   Diary of Duncan Johnston, 17 September 1882.
90   *Coach News*, vol. 10, no. 3, March 1983, p. 4.
91   See, for example, the storm reported in *GT*, 14 January 1897, where a man and three horses were killed by lightning, reserves were flooded and a boarding house at Lakes Entrance was de-roofed.
92   M Fletcher & L Kennett, *Changing landscapes: A history of settlement and land use at Driffield*, International Power Hazelwood, Morwell, 2003, p. 30. See also Diary of George Glen Auchterlonie, 12 December 1870, 'my cold very bad today coughing up a great deal of stuff from my lungs'. Despite his illness, George had carted 91 rails from the nearby bush that day.
93   John Adams, *So tall the trees: A centenary history of the southern districts of the Narracan Shire*, Shire of Narracan, Trafalgar, 1978, p. 50.
94   *East Gippsland Historical Society Newsletter*, vol. 1, no. 2, 1979, p. 9.
95   Linden Gillbank, 'Ferdinand Mueller in Gippsland', *Gippsland Heritage Journal*, no. 10, June 1991, p. 10.

Ferdinand von Mueller developed rheumatic fever, allegedly from too long a dunking in the Tambo River in search of aquatic plants. Duncan Johnston 'got a wetting' while helping his friend drive cattle to the Tambo. It was a bad day all round for Johnston, for he had also been press-ganged into standing for the shire elections.[96] Norman Gunn remembered travelling to balls in the Thorpdale South area wearing 'clean overalls over our good suits, then our good overcoat, and over it an old one to keep the saddle dry'.[97]

Figure 4.7: The floodplain between the Swing Bridge and Longford, November 2011.
Source: Author.

---

96   Diary of Duncan Johnston, 30 January 1882. He recorded other instances of getting wet while travelling on 17 March 1882, 28 March 1882 and 6 April 1882.
97   'Memories of Norman Gunn', *Coach News*, vol. 16, no. 2, December 1988, p. 11.

George Auchterlonie described a dreadful night that he spent camping out in a storm:

> Weather heavy blasts of wind & rain throughout the day At 5 p.m. the wind increased to a perfect huracane accompanied with bitter rain. I could not light a fire & my clothes were wet thru so I fastened down the tarpaulon & slung my hammock & turned in. Kept awake a long time with the dread that the tarpaulon would be blown to pieces or the wagon upset.

The following day, he said:

> At getting up found it anything but agreeable pulling ones legs out of a warm bed and shoving them into a pair of socking wet trousers in a bitter raw morning. Packed up did not take time to light a fire, swallowed my cold water break fast in a hurry & and got away for my bullocks, found them looking very miserable.[98]

Like Auchterlonie's tarpaulin, settlers had some key items that could make wet weather more endurable. Paul Cansick's lead advertising line was often 'NO MORE DAMP FEET', extolling the qualities of his locally produced shoe leather.[99] A letter from 76-year-old Henry Dendy reveals the amount of planning that had to go into waterproofing. His plans to re-roof with bark were thwarted because it was not the bark season, so he was forced to use a different solution. He wrote to a friend to send 'four lbs of litharge and a small quantity of some dark colour slate or dark stone or anything you think proper'. Combined with boiled oil, this mixture would waterproof unbleached calico and would serve as a roof until barking season.[100]

Attempts at waterproofing extended to harvested crops. The Currie family diary recorded for 17 March 1882 that 'Da started to thresh but [the rain] came on just enough to stop him'. The wet weather of 1870 forced George Auchterlonie to devote a whole day to turning the stooks because the insides were not dry.[101] A similar story unfolded in February 1886. Eight inches of rain in as many weeks meant that crops were rotting in

---

98 Diary of George Glen Auchterlonie, 8 and 9 September 1870.
99 *GT*, 29 March 1870.
100 Letter from Henry Dendy to his friend WB Andrew of Brighten, March 1876, reproduced in *Coach News*, vol. 15, no. 3, March 1988. See also Reminiscences of Reginald Murray on early coal exploration in Gippsland, in *Coach News*, vol. 13, no. 4, June 1986.
101 Diary of George Glen Auchterlonie, 13 December 1870.

the stooks.[102] The design of charcoal-fired hops kilns avoided iron roofs, as they could collect moisture and precipitate back on to the crops, which were supposed to be being dried.[103] Heavy rains were, however, more of a problem for farmers than graziers. To a grazier, heavy rain meant lush pasture and a good return.[104]

Wet and stormy weather also affected the progress of various public works. Councils and government departments preferred to schedule major works like bridges for summer and autumn.[105] The newspapers yield multiple reports of delays to works and social functions.[106] Angus McMillan gave perhaps the best description of rain stopping work, in his diary from 1864 while cutting the alpine track:

> Sat 26th March. All hands set to clearing the road at 8am Rain all last night. Mr Jones parting out clearing the track to the Upper Dargo … 3pm heavy rain, the road cutters had to stop work, 20 to 5. Mr Jones returned like a drowned rat, encountered fearful scrub, even his forehead is bleeding with the bits of leeces which are always numerous in this altitude in this damp underwood.[107]

---

102 *GT*, 10 February 1886. See also *GT*, 18 January 1897, Stratford correspondent: 'The late heavy rains did a large amount of injury to the crops that were cut and awaiting the threshers, several stacks in this neighbourhood having to be separated so as to allow the sheafs to dry. So bad have the grasshoppers been and so vicious that they had commenced to destroy window curtains before the rain destroyed them. The river here only rose slightly even though we had nearly 4 inches of rains, but the ground was so parched the heavy downfall was largely absorbed. The fruit crops though largely thinned by heavy weather will be indifferent in quality codlin moth prevailing'.
103 J Adams, *Path among the years: History of shire of Bairnsdale*, Bairnsdale Shire Council, Bairnsdale, 1987, p. 103.
104 *GT*, 20 December 1886, Rosedale correspondent; see also Diary of Duncan Johnston, Tuesday 30 May 1881.
105 *GT*, 26 January 1875.
106 For example, *GT*, 4 January 1876, the Rosedale New Year's concert on New Year's Eve was largely washed out by 'drizzling rain'; *GT*, 14 December 1881, Traralgon Shire Council engineer reported that the inclemency of the weather has slowed everything, but as it was mostly clearing, not draining, no damage was done; *GT*, 10 December 1886, a really wet week: all over the catchment, part crops ruined. Rain caused delay in several functions including the Caledonian society sports; *Walhalla Chronicle*, 27 April 1894, Eight hr day picnic: 'the weather was very unpropitious and to a certain extent marred the fun, as shelter had to be sought from the continual showers', p. 12; *GT*, 12 October 1896, Shire Engineer Avon, reports. Of the eight contracts currently in hand, at least two were delayed by rain. 'The heavy rains that have fallen since your last meeting did considerable damage to the roads in the north and east ridings. I annex hereto a list of the roads etc requiring repair, in consequence of the floods and heavy rains. I am pleased to say that no damage has been done to the bridges in the entire shire as far as I have ascertained as yet'; *GT*, 22 March 1897, wet weather dampens the St Patrick's Day sports festival; *GT*, 17 May 1897, show for the mechanics hall was 'marred by abominably cold and wet weather'.
107 Diary of Angus McMillan during 1864 Alpine expedition, 26 March 1864, SLV, MS 9776, box 268/2.

Welcome rain

So when was rain beneficial? Spring rains were vital to get crops growing:

> During the last week a welcome change has taken place in the weather, nice genial showers falling at intervals during the day, and an occasional heavy one at night. It has been a god send to the crops here, having come in the nick of time. Everything is beginning to assume a healthy and green appearance …[108]

Rain was most welcome in March and April, especially after a blistering summer, and settlers had learnt to expect it then. Its desirability was influenced by the preceding conditions, especially if there had been a series of seasons without sufficient rain. Given low rainfall over winter, Duncan Johnston was extremely happy for a week's home detention in October describing it as 'nice steady rain'.[109]

Insufficient autumn and winter rains meant patchy germination of crops. For a grazier, nicely spaced 'genial' showers through autumn would give enough pasture to sustain animals over the winter, until the combination of warmer weather and spring rains would produce the summer flush of growth. The term 'genial' is used often in newspaper reports. Its synonyms provide an image of a warm, reliable and good-natured friend, who can be trusted to show up when needed.[110] Farmers were acutely aware of the importance of rain and evaluated rainfall effectiveness continually. 'Although steady, the downpour was by no means heavy, but it was an excellent soaking rain, and as there was no wind to dry it up, the ground received the fullest benefit from it.'[111]

A closer understanding of the rainfall timing comes from the report of the Agricultural Show in 1886, held in October, after a relatively wet spring. The show was considered a success because previous dry seasons had prevented many entries for fruits and vegetables. 'Good cabbages, cauliflowers, turnips and rhubarb, no cherries too early, gooseberries too immature but still shown. Strawberries passable. Roses had suffered from recent high winds.'[112] A month later, the Upper Maffra correspondent continued this emphasis on

---

108 *GT*, 16 August 1886, The Heart correspondent.
109 Diary of Duncan Johnston, 23 and 24 October 1882.
110 See fn. 18 for an example. Also, *GT*, 1 July 1897, Stratford correspondent.
111 *GT*, 20 August 1886. For another example, see *GT*, 20 November 1876, 'Steady and continuous rain fell in the Rosedale and T districts on Thursday night and all day Friday, the smaller creeks are everywhere surcharged with the excess of water, This rain, owing to the late drying winds was greatly needed and will prove of great benefit to the young grass'.
112 *GT*, 29 October 1886.

the specificity of rainfall impacts: 'It commenced raining heavily sometime during the night, and is still continuing. It will prove most beneficial *if not too heavy*, the grass in particularly being almost burnout out of the ground with the recent heat'.[113]

Farmer and graziers knew exactly how much rain they needed for where they were at in their annual cycle. At the beginning of a new season, a soaking rain of some duration was needed to be able to start ploughing. After crops had been sown, a soft steady rain was required, enough to moisten the soil and encourage germination, but not so much that the seed might rot in the soil.[114] More importantly, they could know this without any reference to a centralised data bank of measurements.

There is evidence of settlers moving to areas that they believe had more reliable rainfall, from initially attractive lower, thinly wooded plains to areas that were higher and more densely forested.[115] George Auchterlonie and his family left the Maffra district and selected adjoining blocks at Narracan, at the head of Wilderness Creek.[116] The selectors who established Gormandale were fleeing the impacts of the drought on the flatter lands around Merriman's Creek.[117]

This must be counterbalanced against Legg's study, which illustrated that the greatest number of walk-offs were from blocks on heavy, wet forested lands.[118] It was, however, in these areas that dairying was eventually established. Dairying is a graphic example of how rainfall controlled the spread of different types of agricultural industries. Until the advent of irrigation in the twentieth century, successful dairying was restricted to areas with a rainfall average of 25 inches.[119] This explains the advent of anxious reports from the late 1880s about the quantity of milk supply to the various co-op creameries. For example:

---

113 *GT*, 26 November 1886, emphasis added.
114 *GT*, 11 November 1881, Upper Maffra correspondent: 'The last day or two we have had some nice showers, which although not sufficient to saturate the grass lands, have done an immense amount of good to the crops and both crops and gardens are now looking well'.
115 W Frost & S Harvey, 'Forest industries or dairy pastures? Ferdinand von Mueller and the 1885–93 Royal Commission on Vegetable Products', *Historical Records of Australian Science*, vol. 11, no. 3, June 1997, p. 432. doi.org/10.1071/HR9971130431.
116 Fletcher, *Driffield*, p. 17.
117 K Huffer, 'Dedicated to the pioneers of Gormandale: To those who succeeded and those who failed', Typescript, NLA Npf994.56 H889, p. 6.
118 S Legg, 'Farm abandonment in South Gippsland's Strzelecki Ranges, 1870–1925: Challenge or tragedy?', *Gippsland Heritage Journal*, vol. 1, no. 1, 1986, pp. 14–21.
119 T Dingle, *The Victorians: Settling*, Fairfax, Syme and Weldon Associates, McMahons Point, NSW, 1984, p. 115.

since the late rains, the milk supply at the Rosedale creamery has increased 300 gallons, the weekly supply now being according to the Courier, about 1000 gallons. At Glengarry the supply has increased 300 gallons a day.[120]

Brinsmead confirms the seasonality of milk production in Gippsland, stating that it was largely a spring and summer affair.[121]

The positive impacts of moderate rainfall enabled the growth of an orderly and progressive society. Daily farm and government work proceeded, development of new agricultural enterprises was encouraged and social relations based upon Church and family could be maintained through the ability to travel and socialise. These were the physical manifestations of reliable, moderate and predictable rainfall.

The affect of rain

Responses to precipitation go beyond strictly scientific and practical approaches. As demonstrated in Chapter 3, colonial settlers were familiar with the figurative use of water in the Bible and, more broadly, with the use of ecological phenomena to describe the world to each other. All aspects of the hydrological cycle are used in this way. Rain, snow, mist, frost and hail are forms of precipitation that are used imaginatively to describe states of feeling. The quote from the *Gippsland Times*, which conferred on showers of rain a vivifying capacity, is one example.[122]

The Bible and other theological writings are one source of powerful imagery about precipitation. These metaphors provide a complementary approach to understanding rain, snow, hail, frost, dew and flood. Deuteronomy 33:28 gives a largely benevolent view of precipitation, saying: 'My doctrine shall drop as the rain, my speech shall distil as the dew, as the small rain upon the tender herb, and as the showers upon the grass'. God's grace comes clothed in the form of soft and gentle precipitation, echoing the life-giving showers alluded in the *Gippsland Times*, which are ideally both moderate and regular. In another edition, the paper calls rain the providence in the clouds.[123]

---

120 *GT*, 22 February 1897.
121 GSJ Brinsmead, 'A geographical study of the dairy manufacturing industry in Gippsland, 1840–1910', MSc thesis, University of Melbourne, 1977, p. 166.
122 See fn. 58.
123 *GT*, 17 December 1870.

Middleton has pointed out the importance of Biblical scholarship throughout the centuries in reinforcing symbolic use of hydrological processes to explain faith. For example, Isidore, Bishop of Seville, used the process of cloud formation to explain how men are gathered to God:

> The clouds are to be understood as holy evangelists, who pour the rain of the divine word on those who believe. For the air itself, empty and thin, signifies the empty and wandering minds of men, and then thickened and turned into clouds, typifies the confirmation in the Faith of minds chosen from among the empty vanity of the unfaithful.[124]

Isidore's other descriptions included hail and snow, principally used to signify a lack of faith or a lack of love; mist meant confusion, whereas dew signalled purity.

As discussed in Chapter 3, God both rewards and punishes with water. One of the most important teaching stories in the Bible is the flood story.[125] Flood is clearly a method of punishment and retribution. Also, the passage of a flood is used to illustrate undesirable human characteristics. The Book of Job 6:15 says: 'My brethren have dealt deceitfully as a brook, as the channel of brooks that overflows'.[126] A flood is the antithesis of a moderate and ordered Nature. It is quite clear from the descriptions of the numerous floods of 1870 that the floods were regarded in very negative terms. The 26 March 1870 edition used descriptions such as 'a frightful rush of water', 'the resistless torrent', 'a hopeless case' and 'imminent danger'. The very number of floods in this very wet year was itself an affront to the much cherished notions of order and progress. The constant discussion of problems with roads and bridges reflects this underlying desire for permanency, order and moderation.

There is evidence that colonial Gippslanders used hydrological metaphors to describe their world in this symbolic manner. One example of local preachers following this tradition was Rev. Canon Watson, whose sermon from the words 'Go show thyself unto Aham, and I will send rain upon the earth' came from the Book of Kings (I Kings 18:1) Watson concluded that God's munificence was in direct proportion to human obedience to his

---

124 Quoted in Middleton, *A history of the theories of rain*, p. 12.
125 For a discussion of a variety of aspects of the mythological importance of flood, see A Dundes (ed.), *The flood myth*, University of California Press, Berkeley, 1998.
126 See also: 'But let judgement run down as waters, And righteousness as a mighty stream', Amos 4.

laws.¹²⁷ In the newspapers, a cloud metaphor was used to describe political relations amongst European nations in 1886, while the Cunninghame correspondent to the *Bairnsdale Advertiser* called the source of rain the 'bottles of heaven'.¹²⁸

The presence of rain at the right time was an emotional upper for many farmers. On 24 February 1886, the Upper Maffra correspondent for the *Gippsland Times* wrote that the late steady rain has lifted the 'desponding hearts' of the farmers of the district.¹²⁹ George Auchterlonie would go into some detail on his refreshed crops after a well-timed bout of rain. Margaret McCann, allowing for her general terseness, was positively effusive on 14 August 1895 when she recorded 'nice rain badly wanted tank flowing over first time for the year'. A month later, she reported that 'Wall flowers, verbena and white lilies are blooming in my garden'.¹³⁰

While rain was welcome, things could also go too far. Too much rain dampened the spirits. Writing in the last edition of the paper before New Year's Eve, the Grant correspondent of the *Gippsland Times* wrote: 'After our long and severe winter a bit of sunshine is very welcome'.¹³¹ In Chapter 3, I discussed Ada Crossley's favourite song, 'Sunshine and Rain', which valorised sunshine and was performed many times publicly.¹³²

Other types of precipitation are pressed into a similar moral symbology. In a story in the *Bairnsdale Liberal News and North Gippsland District Advertiser* in 1879, George the hutkeeper begins (with unintentional comedy): 'My life has been a sad one. It has been like a stormy dark night. Once a bright star shone up it for a short happy period; but then

---

127 Reported in *GT*, 8 September 1886. In the same issue there was also Rev. WH Gray of the Presbyterian Church, who sermonised on the text from Acts of the Apostles 1:9–11. 'And when Jesus had spoken these thing, while they beheld, he was taken up and a cloud received him out of their sight.' At Primitive Methodist Church, Rev. GH Cole chose from St Johns gospel (John 8:12), 'I am the Light of the world'.
128 *GT*, 31 December 1886; *Bairnsdale Advertiser*, 16 April 1885. See also *Bairnsdale Advertiser*, 14 May 1885, 'A cloud has come over the prospects of the Good Hope Mine as the late yield was far below expectations'.
129 See also *GT*, 29 October 1886, Stratford correspondent.
130 Diary of George Glen Auchterlonie, 7 November 1868, 'The wheat looks wonderfully improved since the rain, it has shot out to about two foot high. The Egyptian now nicely in bloom. The oats have become quite green again and growing well, the barley is quite revived'; Diary of Margaret McCann, 23 September 1895, SLV, MS 9632, MSB 480.
131 *GT*, 29 December 1880.
132 *GT*, 22 October 1896, the Glenmaggie correspondent wrote that at a coffee social and concert for the Church of England, Miss Purdue sang 'Sunshine and rain', Miss Bell sang 'Valley by the sea', Mr Haws sang 'Will of the wisp'. *GT*, 15 April 1897, Miss O'Farrell's pupils perform it at the Mechanics Hall.

heavy, angry clouds obscured it again. I was left in darkness blacker than before'. In 1870, the *Gippsland Times* uncharacteristically published two poems that provide examples. The first was called 'Beautiful Snow', and the second was called 'On Morning'.[133] 'Beautiful Snow' uses the contrast between the appearance of pure new snow and after it has been lying in the streets as the central motif to tell the story of a fallen woman. The first three stanzas set the scene of a happy town receiving the first snow of winter, with dogs chasing snowflakes, and people riding out in their sleighs. The fourth stanza begins the story of the woman's misfortune:

> Once I was pure as the snow, but I fell,
> Fell like the snow flakes from heaven to hell;
> Fell to be trampled like filth in the street,
> Fell to be scoffed at, to be spit on and beat.

The poem continues on with the enumeration of her losses, her beauty and innocence, her friendships and her relationship with her family. She faces an icy death as a homeless person. The final stanza continues both the Christian and the hydrological theme, and offers her salvation:

> Helpless and foul as the trampled snow,
> Sinner despair not! Christ stoopeth low
> To rescue the soul that is lost in its sin,
> And raise it to life and enjoyment again.
> Groaning, bleeding, dying for thee,
> The crucified hung on the accursed tree,
> His accents of mercy fall soft on thine ear –
> 'Is there mercy for me? Will He heed my weak prayer?'
> Oh, God! In the stream that for sinners did flow,
> Wash me, and I shall be whiter than snow!

The second poem is not so overtly religious, but certain key words show it to have a religious subtext. Its anonymous local poet was comparing the rise of the sun and the beauty of a dew-spangled morning to the light of God's truth. The sun, referred to throughout as a male, is kingly and banishes gloom and darkness. Dew, which is largely portrayed in the Bible as a benefit to mankind from God, reinforces the sense of the beauty of the morning.[134]

---

133 'Beautiful Snow' was published on 12 April 1870, and 'On Morning' on 10 December 1870.
134 Genesis 2:5–6: 'But there went up a mist from the earth and watered the whole face of the ground', and Genesis 27:28: 'So God give thee of the dew of Heaven, and of the fat places of the earth'. Also Samuel 17:12 and Deuteronomy 33:28 referring to falling dew.

Popular English poets employed precipitation imagery. One of the more famous is William Wordsworth, whose now classic poem with the opening line 'Into every life a little rain must fall' today adorns greeting cards. It was reproduced in the *Gippsland Times* on 23 May 1874. Shakespeare regularly used precipitation to heighten the emotional effect of his dramas. Would the opening scene of Hamlet have been quite so effective without the mist? Thomas Savige named his property at Narracan Falls after a Longfellow poem.[135] These books could be found in the 2,000-volume library of the Sale Mechanic's Institute, and the subscription reading library run by Louis Roth, a bookseller, stationer and fancy goods merchant in Sale.[136] Alternatively they could be borrowed from men like Rev. William Spence Login who, according to *Butler's Directory* of 1866, had 200 volumes in his church library.[137]

While these examples are admittedly patchy, taken collectively they indicate that colonial Gippslanders were familiar with the emotional and spiritual depictions of precipitation common in Europe, and that they continued to apply them in their new country. Each succeeding chapter will show examples relating to rivers, swamps and the absence of water, thus building the metaphorical world of water that colonial Gippslanders held.

Conclusion

Colonial Gippslanders went from a world of lush greenness maintained by regular predictable rainfall to one that fluctuated wildly; a world where it was possible to have 17 floods in one year, and to then have no rain for months on end. Because they were engaged in transferring northern hemisphere agricultural systems, reliant on seasonally specific and abundant rainfall, to such a place, the anxiety about precipitation was high. Their perception of northern European weather and hydrology as the norm emphasised the supposed abnormality of the colonial

---

135 Adams, *So tall the trees*, p. 48.
136 *GM*, 9 January 1877, Sale's library was the biggest, but other smaller institutes still had a large number of volumes available and they also subscribed to many popular journals and newspapers; *GT*, 15 February 1897, Roth advertised that 'Latest works of Rider Haggard, Ethel Turner Fielding, Marie Corelli, etc, may be had'.
137 *Butler's Directory*, 1866, p. 72.

situation. Unseasonal or heavy rain had important economic and social repercussions, highlighting the contrast between the green, wet lands of their birth and where they had come to.

This chapter illustrates the limbo that settlers were in. While some had traditional knowledge of weather, it was only relevant in Europe. The new meteorological science promised much, but for reasons of distance and disagreement, could not provide isolated settlers with the information they needed. Settlers tried to address their anxiety and uncertainty in a number of ways. They instituted systems of measurement in the hope that patterns would emerge, they adapted their building techniques to conditions as experienced, and they exchanged information with each other formally and informally. Fundamentally, though, there was nothing they could do to influence precipitation rates. Instead, they would channel their anxiety into channelling the surface waters of the catchment. Once water was on the ground, it was fair game.

CHAPTER 5
'Fair streams were palsied in their onward course':[1] The desirability of flowing waters

Land most fertile, water fresh and pure from limestone, climate unequalled, population increasing, Railway transit commands all markets. The Lakes Entrance opened to all intercolonial ports. This grand estate may be briefly described as river flats of unsurpassed richness, having depth of soil that ages formed and time cannot work out. A frontage of about six miles to the renowned Mitchell River, besides other water facilities, insuring permanent supply.

Campbell Pratt and Co., real estate agents[2]

It was said there was only two or three rivers to cross, but we found more than five or six, and all of them muddy on the banks, and never a day hardly with passing creeks that we often had to lower the drays with ropes, and carry the things on our backs … It has been a mad undertaking, but it may turn out better.

John Gellion[3]

---

1 From a poem by Allan McLean (future deputy prime minister of Australia) published in 1888, cited in P Morgan, *The literature of Gippsland: The social and historical context of early writings with bibliography*, Centre for Gippsland Studies, Churchill, 1983.
2 *Gippsland Times*, 12 February 1886, Campbell Pratt and Co.'s advertising for the sale of the subdivided Axelea Estate near Bairnsdale.
3 Letter from John Gellion, 1 August 1844, cited in *Coach News*, vol. 2, no. 3, March 1975, pp. 7–9.

Introduction

Surface waters are the most visible part of the hydrological cycle. As such, they have dominated human responses to the cycle in most aspects: physical, economic, social, emotional and spiritual. This chapter explores the interaction of flowing waters with these aspects of life in colonial Gippsland.

It is on the Earth's surface that humans have the most capacity to reshape hydrological systems. However, hydrological intervention is a choice. To first think it desirable and then act upon it requires a particular type of world view. Unlike Australia's Indigenous custodians, the colonial settlers were not content to let the continent's variable waters flow at their own will. Dow has described in detail the entwined and seasonally variable relationship between the Kurnai and the surface waters of the Gippsland Lakes Catchment (GLC).[4] Whereas the Kurnai were content to adapt to the changeability of the waters, affording the water both ecological and spiritual significance, the displacing Europeans perceived the catchment's moving waters very differently. Gippsland's rivers were both an obstacle to be surmounted and a resource to be used and exploited.

The European-inspired agricultural economy they wanted to build was profoundly mismatched to their actual environment. European farming evolved in a more regular hydrological regime, both in the amount and distribution of rainfall and the resulting flow in rivers. In Gippsland, the attempt to reproduce European agriculture was challenged by the variability in available water, as well as by predation and diseases.

The settler's major response to the 'problem' of European agriculture in a non-European land was to regularise and redirect flow, wherever technically and economically feasible. As the nineteenth century progressed, money and technology became increasingly available, and the rivers were changed to achieve permanent, moderate flow as much as possible. The capacity to effect sweeping hydrological change increased during the nineteenth century as a whole, and was supported by an understanding of hydrology that was quantitative and volumetric. This understanding was nested within a broader view of nature as a machine, or as a puzzle that could be teased out by breaking the world down to

---

4 C Dow, 'Tantungalung Country: An environmental history of the Gippsland Lakes', PhD thesis, Monash University, 2004.

its component parts.⁵ This mathematical rendering of flowing water obscured the other connections of water, especially to the other species that depended on the very variability that settlers wanted to erase.

Economic life was founded upon the metaphor of flow, and the actual flow of goods was facilitated in a large part by water transport; 'water and progress ran together'.⁶ Progress meant evening out the vagaries of flow, aiming for seasonal uniformity and even geographical distribution. To improve flow was the heart of prosperity. This idea was equally influential in drainage practices (see Chapter 6) and the same flow metaphors are found in the spiritual and recreational practices of the settlers, each reinforcing the other.

This chapter begins by examining the metaphorical aspects of flow, before moving on to a discussion of settler's physical dependency upon flowing water. This order is significant because cultural understandings of flowing waters, especially from religious teaching, buttressed their choice to act in the ways they did in the physical environment of the catchment. The examples of Sale's water supply and a discussion of the tanning industry illustrate the interdependencies between social, economic and cultural aspects of flow. The remainder of the chapter details the ways colonial Gippslanders made hydrological alterations to improve flow, in particular focusing on bridge construction, snagging and the creation of the permanent entrance.

Metaphorical flow

The historical uses of river flow mimic the diversity of definitions of the word itself. Table 5.1, sourced from Haslam's *The Historic River*, includes medical, religious, magical, recreational and ornamental ways people use rivers in addition to the more prosaic uses such as cleaning and processing goods.

---

5   C Merchant, *The death of nature: Women, ecology and the scientific revolution*, Harper Collins, San Francisco, 1980, p. 195; R Sheldrake, *The rebirth of nature: The greening of science and god*, Century, London, 1990, pp. 32–41; D Worster, *Nature's economy: A history of ecological ideas*, Cambridge University Press, Cambridge, 1994, pp. 39–44.
6   JM Powell, 'Snakes and cannons: Water management and the geographical imagination in Australia', in S Dovers (ed.), *Environmental history and policy: Still settling Australia*, Oxford University Press, Melbourne, 2000, p. 59.

Table 5.1: Historic uses of rivers in Europe.

Use	Variations
Drinking	Residents, travellers and domestic animals
Cooking	Boiling, steaming, poaching various foods
Cleaning	Clothes, people and animals (e.g. sheep-dip), homes, farmyards, stables etc., roads and other built-up surfaces
Other domestic supply	Fire engines, hydrants, heating by volcanic water (Rekyjavik), water meters etc.
Food and drink	Fishing (rivers or constructed fish ponds), waterfowl, water cress (rivers or constructed cress beds), irrigation, draining, grazing, alcoholic and other drinks
Materials (source of or processed in the river)	Osiers and withies (poles and baskets), rushes (chairs, matting and lights), reeds (thatch), strewing plants, tussock sedges (stools). Soaking of flax, hemp and fleeces, extraction of sand and gravel
Medicinal	Springs, spas, water plants, leeches
Religion	Holy wells, baptism, ritual cleansing, pilgrimages to holy rivers and springs. Choice locations for religious hermits/communities
Magic	Rain-making, rain and flood removing, fertility, other magic fountains and springs
Industry	Providing power through watermills to grind grains, olives and other plants, to fashion armour, jewellery, tanning, fulling, dyeing, making paper, gunpowder, metal products, boilers for steam engines. Processing and washing of mining ore
Transport	Boats for trade, business and pleasure
Recreation and ornamental purposes	Ornamental fountains and water gardens, lakes, river parties, fishing, walking, picnicking, camping, swimming, paddling, water clocks, punting, watering residential gardens, freshwater pearls for jewellery
Waste disposal	Sewage, factory, mine and farm effluents, non source point runoff from farms and roads, general waste, drowning unwanted animals, murder and suicide of persons
Punishment	Ducking, drowning
Defence and attack	The strategic use of river bends, marshes, moats, stakes in a river, chains, fortified bridges, forts

Source: Reproduced from S Haslam, *The historic river: Rivers and culture down the ages*, Cobden of Cambridge Press, Cambridge, 1991, p. 2.

Based on a survey of European rivers from 1400 to approximately 1800, Haslam's list is evidence of a widely shared cultural world (a gestalt, if you will) centred on the importance of flowing water in rivers, springs and wells. It demonstrates the intertwined, holistic nature of flowing water as it spread through the physical, social, intellectual, economic and spiritual life of Europe.

The word 'flow' is an adaptable one, with 10 different uses as a verb. Only one relates strictly to hydrology. This indicates how much the visual action of flowing water has penetrated the English language. Flow has become metaphorical for a range of meanings, principally revolving around movement.

- To move along in a stream: *The river flowed slowly to sea.*
- To circulate: *The blood flowing through one's veins.*
- To stream or well forth: *Warmth flows from the sun.*
- To issue or proceed from a source: *Orders flowed from the office.*
- To menstruate.
- To come or go as in a stream: *A constant stream of humanity flowed by.*
- To proceed continuously and smoothly: *Melody flowed from the violin.*
- To hang loosely at full length: *Her hair flowed over her shoulders.*
- To abound in something: *The tavern flowed with wine.*
- To rise and advance, as the tide (opposed to ebb).[7]

Flow is generally a positive description, regardless of which sphere of life it is applied to. When we describe a dancer as fluid we are describing the flowing quality of her movement, made possible by an enviable suppleness. In landscaping, blue-flowered plants with long flexible stems can appear to flow down banks. Mentally, we speak of 'the flow of ideas', or 'a stream of consciousness'. Emotionally, descriptions like 'my heart was overflowing with love' are common. Spiritually, the flow of rivers can be taken as a metaphor for life's passage, as in Alfred, Lord Tennyson's poem *Crossing the Bar*, which correlates crossing the bar of an estuary with passing into death.[8] The verb 'flow' is then modified by descriptions of speed, as in a measly trickle of water, or a raging torrent. While it is describing the speed of water, the adverb is actually pointing to meanings of abundance or lack.

Examples of flow metaphors in the newspapers surveyed suggest that Gippslanders shared this basic meaning. The slow action of water eroding stone was used to describe progress towards getting the railway in 1869.[9] The desired effect of the proposed railway was likened to a stream:

---

7   Dictionary.com (accessed 19 February 2007).
8   A Allison, H Barrows, C Blake, A Carr, A Eastman & H English (eds), *Norton anthology of poetry*, 3rd edn, WW Norton and Co., New York and London, 1970, p. 716.
9   *GT*, 19 October 1869.

A journey of six or seven hours would land a man in Melbourne from Sale while goods and cattle would be going down in almost a continuous stream, the loads being augmented at the various stages, thereby opening up and developing the great tract of country between Melbourne and Sale.[10]

The railway would be an improved river, suffering none of the variability inherent in actual water transport. Flow, money and economic growth were frequently linked, and this association remains consistent.[11] These examples demonstrate how settlers thought about the flow of water, goods, people and money, the foundation of their economic life. Importantly, these positive perceptions were reinforced by the symbolic depictions of flow in their spiritual and social lives.

The flow metaphor was used to describe social events and issues, from the growth of the Mechanic's Institute library in Sale in 1862 to the New Year's revelry at Dutson in 1891.[12] Another way was in poetry, which regularly employed the dualism of movement and stagnation. The chapter title is derived from poetry published by Gippsland resident and future deputy prime minister, Allan McLean.[13] A poem about the

---

10   *GT*, 23 October 1869, report of the Public Meeting at Rosedale. For other political uses of flow metaphors, see *GT*, 23 February 1881, comparing immigration to building a bridge across the ocean; *Moe Register*, 22 December 1888, 'waves of prosperity have passed over the metropolis, the Premier has a phenomenally bright budget to disclose, from which it is hoped vivifying rivulets will run to all parts of the country'; *GT*, 16 August 1897, Allan McLean on Federation: 'The native born in my district are sensible young men, ready to part with a substance for a problematical benefit of being permitted to drink at the springs of national life'.
11   For example, see *GT*, 4 December 1863, a correspondent from Stratford complaining how the loss of the bridge and lack of legal representation in the town is impeding the flow of money; *GT*, 5 July 1866, expressing the desirability of a flow of homesteads across Gippsland; *GT*, 12 September 1874, describing free trade policies with a positive flow metaphor; *GT*, 7 March 1876, flow of benefits expected from canal construction.
12   *GT*, 17 January 1862, describing the flow of books and periodicals arriving at the Sale Mechanic's Institute; *GT*, 5 January 1891, Dutson correspondent: 'Your numerous readers in Dutson will expect to hear from me, and though I feel more inclined to sleep than write, I suppose I must wake up to the occasion, but I must tell them first that if they make their entertainments so seductive and lasting they must not expect "their own" to go through it all, and then write an account of it without a fair amount of rest. Some of the members of our community seem to partake of the nature of "the Brook" to go on for ever, but I don't, and what's more I won't, but I'll "flow" on now till I run out. New Years day is the day of days to the rising generation of Dutson'. Other examples include *GT*, 14 March 1862, using flow metaphors to describe cricket matches between Bairnsdale and Omeo; *GT*, 29 May 1863, the cup to overflowing celebrating the marriage of the Prince of Wales; *GT*, 9 August 1865, 'no channel into which the stream of benevolence would not flow'.
13   A longer extract makes the adulation of movement clear: 'Air ceased to palpitate, and earth to quake/ The sea grew torpid as a stagnant lake./ There bloom'd no living plant on vale or hill;/ The trees stood darkly calm and deadly still;/ The laws of nature lost their vital force;/ Fair streams were palsied in their onward course,/ And stretch'd as motionless over sterile plains,/ As frozen currents in a dead man's veins./ There stirr'd upon the earth nor pulse nor breath;/ The world was wrapp'd in universal death.' Morgan, *The literature of Gippsland*.

Mitchell River published in 1865 expressed a hope for 'stately mansions' and 'palaces gorgeous' to adorn its banks and sweetly flowing waters.[14] Further examples are cited when discussing the opening of the permanent entrance later in this chapter.

The consistency of flow metaphors in religion is most striking. The *Gippsland Times* regularly printed a religious column throughout the study era. Accounts of church events described with flow metaphors are common. On 5 April 1878, the column began with 'O God, Thou faithful God/Thou fountain ever flowing'. In 1882, the sermon dwelt at length on thirst, water and flowing rivers:

> In heaven alone, the thirst of the immortal soul after happiness shall be satisfied. There the streams of Eden will flow again. They who drink of them shall forget their earthly poverty, and remember the miseries of the world no more. Some drops from the celestial cup are sufficient for a time to make us forget our sorrows even while in the midst of them. What then may we not expect from the full draughts of those pleasures.

It is this fundamental religious symbolism of water – symbolism rooted in the self evident and natural attributes of water – that permeates the Bible and the whole biblical story of creation, fall and salvation. We find water at the very beginning, in the first chapter of Genesis, where it stands for creation itself, for the 'cosmos' in which the Creator rejoices for it reflects and sings His glory. We find it as wrath, judgement and death in the stories of the Flood, and of the annihilation of the Pharoah and his chariots under the waves of the Red Sea. And we find it finally as the means of purification, repentance and forgiveness in the baptism of St John, the descent of Christ into the waters of the Jordan, and in his ultimate commandment 'go ye and baptise'.[15] Biblical flow metaphors reinforced the central role of baptism.[16] While it would be wrong to argue that religious belief created flow-altering behaviour, the Bible nevertheless was the go-to-book for how to live and think. It offered justification for looking at the colonial waterscape and finding it wanting.

---

14   *GT*, 25 February 1865.
15   A Schmemann, *Of water and the spirit: A liturgical study of baptism*, St Vladimir's Seminary Press, New York, 1974, pp. 39–40.
16   Y Feliks, *Nature and man in the Bible: Chapters in biblical ecology*, Soncino Press, London/Jerusalem/New York, 1981. In addition to the role of water in the significant biblical stories, he shows how ecology in general played an important role in biblical texts: 'The Biblical writings are the work of men who were close to nature and to agriculture, and so derived their inspiration from them'.

A second source of messages about the desirability of pure, flowing water came from the temperance movement. Temperance societies advocated the limited use of or total abstinence from alcohol, and commonly employed imagery of flowing water in their work. Without making a detailed comparative study, Gippsland does not appear to be more or less drunken than any other part of colonial Australia. Those who you would expect to complain about alcohol (the clergy) certainly did.[17] Going on 'a spree' was a regular outlet for the monotony of the laborious country life. For example, at Eagle Point there were annual regattas where 'unlimited supplies of liquor were served to lusty pioneers, who thought nothing of riding or rowing a distance of twenty miles to souse themselves with Jamaica rum or beer'.[18]

Gippsland had a variety of temperance organisations, and was regularly visited by evangelists, like Mr Horspool who spent three days in December 1888 in Warragul.[19] Non-temperance groups sometimes had temperance talks, such as the 1881 Working Men's Meeting at which the Rev. Hardy of the Baptist Church spoke:

> [He] urged upon those present the duty of wholly freeing themselves from the pernicious habits of the period, and of influencing others to abandon indulgence in ardent liquors. He mentioned some facts showing that alcohol not only injured

---

17   For example, Rev. Francis Hales was one of the early Church of England clergy, dispatched to Gippsland in July 1848. Soon after his arrival in the Port Albert region, Hales preached to about 50 'of their drunkenness and sins in plain language', and later complained that some of his morning attendees for service rode up drunk for the evening one. AE Clark, *The church of our fathers: Being the history of the Church of England in Gippsland, 1847–1947*, Diocese of Gippsland, Sale, 1947, pp. 17 and 22. Rev. Login wrote: 'The marriage ceremony was too often ignored, drunkenness was sadly prevalent – not a continuous imbibing of strong drink, but in violent outbursts of the vice after a considerable time of self restraint'. JW Leslie & HC Cowie, *The wind still blows: Extracts from the diaries of Rev WS Login, Mrs H Harrison and Mrs W Montgomery*, the authors, Sale, 1973, p. 34.
18   Recollections of Bushman (actually Rowland Bell of Metung), published in *Every Week*, 18 June 1940. Reproduced in *East Gippsland Historical Society Newsletter*, vol. 1 no. 3.
19   The index to the *Gippsland Times* records five groups: the Church of England temperance society based at Longford, the Gippsland Temperance League, the International Order of Rechabites, the Rosedale Temperance Society and the Sale Temperance League. L Kennett, *Index to the Gippsland Times 1861–1900*, Centre for Gippsland Studies, Monash University, Churchill, 1995. There were also organisations not listed in the index, such as the Independent Order of Good Templars and the branch of the Women's Christian Temperance Movement, which formed in 1887, the result of the 'crusade' by Rev. Phillip Moses. It was principally associated with the Presbyterian Church. Maffra had a temperance hall by 1884, which the Salvation Army used to hold services in. D Kemp, *Maffra: The history of the shire to 1975*, Shire of Maffra, Bairnsdale, 1975, p. 72. For Mr Horspool, see *Moe Register*, 8 December 1888. Part of the proceedings included a public march through the streets by children.

the body, but weakened and debased the mind, making men careless of their moral obligations, neglectful of their families and sometimes brutal to lower creation.[20]

There were also innumerable tea meetings recorded in the pages of all papers, at which both resident and visiting clergy would speak.

Temperance organisations all shared the metaphor of flowing water. Temperance meetings and publications regularly employed poetry, prose, song and lectures incorporating water metaphors. Sadly, records of actual meetings in Gippsland are rare. One of the few found is in the *Gippsland Mercury* on 3 February 1877, which recorded that at a temperance meeting in Bairnsdale the entertainment (apart from the lecture) was songs, one of which was called the 'River of Life'.[21] A better indication of the probable content of the lectures and sermons delivered during crusades can be derived from temperance journals. The Victorian Temperance Society published a monthly magazine commencing in 1851. Its first issue contained a dialogue between six characters representing six different virtues of drinking water instead of wine. Water was 'the draught of inspiration, the liquor of the skies, the nectar of gods, the true stream from Mount Parnassus, the best mirror of the beauty of Narcissus, and the wine of bards and patriots'. And that was only the first character speaking! Others went on to note the worship of water in other religions, its health-giving properties causing 'the blind to see, the lame to walk and the leper to be cleansed. It is the Aesculapius of the elements, the saviour of the sick and the nourisher of the hale'. Finally, it made the perfect beverage to toast political victories.[22]

Again, it is unwise to base an argument for promoting flowing waters solely on the activities of temperance campaigners. There are enough sardonic quips and outright hostile letter exchanges recorded in the papers to know that the temperance movement was not universally well regarded.[23] In a lengthy exchange in 1891, the Briagolong correspondent to the *Gippsland Times* said:

---

20   *GT*, 11 July 1881.
21   *GM*, 3 February 1877.
22   *The Victorian Temperance Pioneer, or Monthly Magazine*, vol. 1, no. 1, August 1851, pp. 10–13.
23   For example, *GM*, 19 June 1877. 'The unpropitious state of the weather prevented others from attending, for much as total abstainers may advocate the use of cold water inwardly, very few of them care about getting wet jackets'. In October and November 1891, there was a sustained exchange of letters between 'Water Drinker' and the Briagolong correspondent in the *Gippsland Times*. In early October a temperance meeting in Sale saw Rev. Roberts of Stratford express regret that the temperance cause in Sale was at such 'a low ebb'. A useful flow metaphor in itself. Immediately following on was the Briagolong correspondent writing about government support being needed to improve grape cultivation. *GT*, 2 October 1891.

> When we are becoming so sensitive to influenza, la grippe, colds, diphtheria, typhoid, it does not look as if the water regime was such a panacea for all evils physical and social as it is professed to be … I fancy the Blackall sands transformed into prosperous vineyards with a population quaffing its wines instead of the sluggish and depressing waters of that chain of waterholes would have done more to revive Stratford physically mentally and socially than the gospel of total abstinence.[24]

While completely opposing temperance, the Briagolong correspondent still uses imagery that suggests the positivity of abundant flow.

These meanings of positive flow were generally held by Gippslanders, and remain consistent across time.[25] In light of this, the flow-altering actions of nineteenth-century Gippslanders can be revisited with a more tolerant eye. The prevailing correlation between constant, moderate flow, wellbeing and progress means that such alterations were in accord with their world view, and therefore perfectly rational. The next section details the ways that the colonial settlers found waterways so in need in improvement.

Water supply

A major problem facing any town or village in the catchment was how to procure a steady supply of clean water that could be reticulated at reasonable cost. For this reason, water frontage to creeks, rivers or swamps was an important advantage that generally outweighed the risk posed by floods. Everyone had tanks to catch rainwater and a lucky few had springs. Otherwise carting was the only way to obtain water, and Frederick Gray complained of how onerous this task was.[26]

There are multiple attempts at centralised water supply in the catchment in the region. Table 5.2 gives a selected list of towns and their water supply activities.

---

24   *GT*, 9 October 1891. On 14 October 1891, Water Drinker's reply was published: 'The fact that "drunkards who take the pledge break it" does not prove the total abstinence party in the wrong, nor have the Stratford people any excuse for drinking the Avon water if that water is known to be unwholesome.' Water Drinker dislikes the excesses of his abstinence colleagues as much as the Briagolong correspondent. An example of such a person might be Rev. Matthew Barnett, who the *Gippsland Times* described in 1878 as 'mad, or injudicious'. *GT*, 18 March 1878.
25   V Strang, *The meaning of water*, Berg Publishing, Oxford and New York, 2004, p. 3.
26   Letter from Frederick Gray, 24 June 1854, written while working at Lindenow, *East Gippsland Historical Society Newsletter*, vol. 1, no. 2, p. 3.

Table 5.2: Table of water supply activities.

Date	Place	Type of work
1863	Sale	Diversion of Thomson River flow into Flooding Creek
1875	Rosedale	Pumped from the Latrobe
1880	Sale	Artesian groundwater supply
1882	Traralgon	First motion for artesian water. No construction for 25 years
1882	Maffra	Council first discusses water supply. No construction for 30 years
1892	Omeo	Livingstone Creek sourced, to reservoir 1 mile from Omeo
1894	Bairnsdale	Reticulated water supply. Caught up in rural irrigation debate
1893	Bairnsdale	Construction of Glenaladale Weir
1900	Warragul	First motion for supply, based on springs. Moved to head of Tarago River in 1906

Note: Water supplies to towns in Narracan Shire are outside the time period.

The first debates about the desirability and viability of a water supply for Sale came in the 1870s, after a period of sustained population growth from land selection and gold exploration. However, the rising population exacerbated the pollution of surface waters. As shown by Haslam in Table 5.1 above, one of the major uses of rivers in Europe was for waste disposal. Colonial Gippslanders used rivers in the same way. Many understood the difference between use of rivers as source and sink, but with the absence of enforceable regulations, water pollution was a normal feature of the catchment.

Much of Gippsland's economy depended upon the cleansing ability of flow, especially the wool and gold industries. The use of rivers and wetlands to wash sheep prior to shearing was a fundamental part in the cycle of sheep raising. Mining operations relied on river flow to dispose of their tailings. In relation to drinking water supply, the use of rivers as source and sink proved a significant problem. An excellent example comes from 'Pure Water', who wrote to the *Gippsland Mercury* in February 1877. The letter described the clash between polluting activities and water supply, observed when the writer watched a water carrier fill his barrels immediately downstream of where 40 or 50 'knights of the anvil and bellows' were washing themselves:

> I called the attention of a gentleman to the scene; he said it was a running stream, not stagnant water. True, the river is running but remarkably slow at this time of year, so much so that it has no chance of getting cleansed before it is dipped up into the casks.

'Pure Water' suggested that the water collection point should be 50 yards above or 500 yards below the washing area.[27] Significantly, there is clear acknowledgement of seasonal variations in flow, and it perfectly illustrates the perceptual differences between running and still waters.

Pure Water's letter is only one example of the acknowledgement of pollution of surface water.[28] Much of the debate about Sale's water supply focused on the split between supporters of a scheme fed from the Thomson River, and those who opposed it because of pollution concerns. Opposing a second attempt in 1885 to revive the Thomson River scheme, the health officer Dr McDonald described Stringer's Creek, a tributary to the Thomson:

> At Walhalla nearly every resident keeps pigs; there are numerous slaughteryards on the steep hills sloping towards the creek, the cesspits are constantly running over, or oozing through the surface … and all this excreta, with the filth from the slaughteryards, where scores of pigs are kept feeding on the offal, also that from the private pigstyes and debris from the crushing plant, flows into the creek which, after running three or four miles though a narrow rock bound channel, empties into the Thomson.[29]

Given this, the possibility of groundwater as a reliable, clean source became very attractive.

---

27 *GM*, 5 February 1877.
28 Diary of Rev. Bean, 14 December 1848, discussing Sale. 'I think probably, the water is bad, being either procureable in the neighbourhood out of the morass or the dirty pools in the Creek, or else from the River Thompson at 2 miles distance. These remarks are the result rather of private conversation in both neighbourhoods than my own observation, through somewhat of both.' Cited in Clark, *The church of our fathers*, p. 35. Also *East Gippsland Historical Society Newsletter*, vol. 2, no. 2, June 1982, report of a trip to Bullumwaal, noting that early Bairnsdale residents feared their water supply from the river would be polluted by stormwater from the cemetery, p. 4; *GM*, 12 April 1877, editorial; *GM*, 21 April 1877, letter from HP re diphtheria; *GM*, 3 May 1877, Maffra Council meeting of 2 May noting the police report received from the superintendent that Timothy O'Sullivan was causing a pollution of a stream at Donnelly's Creek and requesting council to have the nuisance abated. See also the *Moe Register*, 8 December 1888, for a similar point on thinking about water pollution, drainage and water supply together; *GT*, 18 March 1891, pollution on the Thomson from George Pruden's slaughter yard at Wurruk – need to shift the killing pen so that stuff can't be washed into the river; *GT*, 18 May 1896, letter about how to deal with pollution of Lake Guthridge by planting eucalypts and willows.
29 Cited in P Synan, *Precious water: 100 years of reticulated supply in Sale*, City of Sale, Sale, 1988, p. 6.

The two men who were instrumental in tapping artesian water for Sale were Samuel Lacey, an engineer, and John August Niemann, a water borer. They were working together from the late 1870s at various sites, including the Airley Run, Kilmany, the Turf Hotel and at the back of Paterson's Buildings in Raymond St.[30]

Collectively, these discoveries altered the community debate about water supply. Niemann persuaded the council to allow him to drill for a water supply bore in Macalister St, opposite the Victoria Hall. Unlike the relatively shallow depths at which water had been previously found, this location caused major problems with pipe breakages. He was eventually successful at a depth of 196 feet (59.7 metres). The resulting 10-metre-high water jet was gratifyingly dramatic.[31]

The artesian water supply was an instant hit.[32] The *Gippsland Times* described how an 'inhoate Spa' had developed around the site, with 'many persons' in town assembling to drink the 'medicinal' waters.[33] This description gives the soundest evidence for local knowledge about European spa culture and the healing tradition of 'taking the waters'.[34] The paper noted that most townspeople had switched to the artesian supply, despite its sulphurous odour. 'Medical authorities', reported the paper, 'state that the artesian water possesses eminently hygienic qualities and several citizens may be seen with their glasses in hand every morning taking some early draughts of a liquid they confidently believe to be as good as Apollinaris or Friedriebshal water.'[35] It is unclear on what basis the medical claims were made, given that in the same issue council's resolution to send water to Melbourne for analysis was recorded. Clearly though, nasally challenged residents needed no convincing by external authorities. To contain the flow, the council approved the erection of four 400-gallon tanks while the overflow would be piped to various gutters to

---

30   For locations of the boring, see Synan, *Precious water*, p. 2; *GT*, 1 April 1881, which notes that Niemann was boring at Kilmany Park; JL Cafiso, *A Niemann family history*, the author and EH Niemann, Morwell, 1986; and SL Lacey, *Laceys Of Gippsland: The history of a pioneer firm 1870–1970*, the author, Sale, n.d. There is little agreement between sources of starting dates. Synan in *Precious water* says September 1878, Cafiso says 10 June 1880, but this refers to the work for council and doesn't acknowledge the first private bore at Patersons. Lacey does not give a date.
31   Synan, *Precious water*, p. 2.
32   *GT*, 21 June 1880.
33   *GT*, 3 September 1880.
34   There are multiple articles on spa culture and bathing in CS Anderson & BH Tabb, *Water, leisure and culture: European historical perspectives*, Berg, Oxford and New York, 2002; and AL Croutier, *Taking the waters: Spirit, art, sensuality*, Abbeville Press, New York, London and Paris, 1992.
35   *GT*, 20 August 1880.

cleanse the town drains. It should come as no surprise that soon after this, residents began agitating for the draining of the eastern lagoon, where all the overflow ended up.

For all that colonial Gippslanders understood about the source versus sink aspect of flow, this did not translate into coherent behavioural or institutional change. They kept on merely shifting the problem of their own unsanitary habits, using water's ability to shift material. Their main perception was on the supposed inexhaustibility of the ground water supply, unlike the sluggish and polluted surface water streams in the summer.

Flow-dependent industries

While everyone accepts the beneficial nature of flowing waters, few devote thought to the intricate interrelationships that flow sustains. These webs of water sustained most economic activities but were only the subject of explicit discussion in the public realm when those webs were disrupted. In this section, I illustrate the economic connections that flow sustained, using the little-known tanning industry as an example. While gold mining would have also been an excellent example, its water dependencies have been well covered.[36]

The tanning trade lived off, and lived with, other parts of the rural economy. Its foundation was pastoralism, but tanning also had links with the mercantile world of tradesmen who turned leather into saleable goods like shoes and saddles. Tanners relied on tannin from bark, and bark stripping and carting formed an important source of supplementary income for many settlers, especially those eking out a living on the agriculturally marginal soils next to the Lakes. Both George Auchterlonie and Duncan Johnston recorded stripping and carting in their diaries.

---

36   For a discussion of their impact on water resources, see D Garden, 'Catalyst or cataclysm? Gold mining and the environment', *Victorian Historical Journal*, vol. 72, no. 1, 2001, pp. 28–44; and MM Tracey, 'No water, no gold: Applied hydrology in nineteenth century gold mining', *Australasian Mining History Conference 1996*, University of Melbourne, Melbourne, pp. 76–84.

Figure 5.3: Detail of *Acacia mearnsii*.
Source: Author.

The process used by all tanners (until the introduction of mineral tanning at the very end of the nineteenth century) was vegetable tanning, first developed by the Egyptians and the Hebrews around 4,000 BC.[37] Leather is created when the middle layer of the skin of the dead animal is treated to reduce its water content and strengthen the chemical bonds of the dermis.[38] In Australia, the source of tannin was principally from wattle bark, a commodity that the Gippsland Lakes was richly endowed with: *Acacia mearnsii* was a common plant around the Gippsland Lakes.

Its habitat was described in the 1878 Wattle Bark Board of Inquiry as 'growing in most parts of the colony, covering large areas in the neighbourhood of the seaboard and contiguous to rivers, creeks and marshes, also in gullies and ravines'.[39] This indicates a much wider range than Gippsland, but its preference for wet feet made Gippsland a place of reliable supply.

---

37 *The New Encyclopaedia Brittanica*, vol. 7, Micropaedia, Ready Reference, 15th edn, Chicago, p. 225.
38 'Fresh hides contain between 60 and 70 water by weight and 30 to 35 percent protein. About 85% of this protein is collagen, a fibrous protein that is held together by chemical bonds. Basically leather making is the science of using acids, bases, salts, enzymes and tannins to dissolve fats and non fibrous proteins and strengthen the bonds between collagen fibres.' *The New Encyclopaedia Brittanica*, p. 225.
39 Quoted in S Searle, *The rise and demise of the black wattle bark industry in Australia*, CSIRO Division of Forestry, Canberra, 1991, p. 5.

As early as 1814, reports were being made to the colonial administrators about the superior tanning qualities of wattle bark, and recommending its export.[40] However, it was not until much later that the properties of the bark were widely appreciated and it became a valuable commodity. Its ubiquity was a financial saviour to many selectors. As clearing was a condition of gaining title to selected land, the presence of *Acacia mearnsii* helped new arrivals by providing an unanticipated source of income. Selectors were able to get some measure of economic return for the hard work of clearing, and they could access extensive Crown lands to cut bark to supplement incomes during low times of year. It provided employment for the carters, men who had small boats that drew little water and that could access the lake and riverside places where the cutters were working.[41]

Tanneries, which processed the skins of the slaughtered animals, were absolutely water dependent. The Longford operation was on the banks of Long Waterhole.[42] Jackson's tannery at Bairnsdale dating from 1876 was located on the banks of the Mitchell, notably off the rise that the main town was situated on.[43] Paul Cansick started the Rosedale Tannery in 1868; Werner and Klux traded in skins and other related products such as fat and tallow in Sale; John Lloyd operated at Wurruk; Henry France's operations were at Heyfield; and the partnership operation between Alphonse Ducret (tanner), John Fitzpatrick (blacksmith) and James Keneally (furrier) was based in Stratford on the Avon River. There were other tanneries at Prospect, Tarraville, Traralgon, Lucknow and Port Albert.[44]

---

40 Searle, *The rise and demise of the black wattle bark industry*, p. 3.
41 Dow, 'Tantangalung Country', p. 106.
42 P Synan, *Gippsland's lucky city: A history of Sale*, City of Sale, Sale, 1994, p. 72.
43 Hal Porter made much of the town's elevation in his book about Bairnsdale, which was part history, part reminiscence, part poison pen letter about his birthplace. H Porter, *Bairnsdale: Portrait of a country town*, John Ferguson, Sydney, 1977, p. 36.
44 *Traralgon and District Historical Society Bi Monthly Bulletin*, vol. 6, no. 1, March 1975, Memoirs of W Power of Carragjung: doesn't supply a date but says early days of Carrajung was GD Clark, who had two properties called Richmond Vale and Top Camp at Carrajung and had a tannery at Richmond. *Traralgon and District Historical Society Bi Monthly Bulletin*, vol. 7, no. 1, March 1976, Walter West's history of Traralgon: At the November 1880 meeting of Council an application from Mr Whalley to erect a tannery on Traralgon Creek was approved. See also Porter, *Bairnsdale*, p. 36; Synan, *Gippsland's lucky city*, p. 72; J Adams, *Path among the years: History of Shire of Bairnsdale*, Bairnsdale Shire Council, Bairnsdale, 1987, p. 88; J Adams, *From these beginnings: History of the Shire of Alberton (Victoria)*, Alberton Shire Council, Yarram, 1990, p. 29; and M Fletcher, *Avon to the Alps: A history of the Shire of Avon*, Shire of Avon, Stratford, 1988, p. 69.

The processes involved explain the uniform choice of location besides water. The first step was to dehydrate the skin so that it would not decompose before reaching the tannery. Second, the hide would be rehydrated and dehaired. According to the *Blackwell Encyclopaedia of Industrial Archaeology*, a tannery undertook this by suspending hides in pits of lime solution. Lime was available locally in Dutson. The lime loosens the hair in the hair follicle so that a worker could scrape it off. Suppleness of the skin was achieved by massaging in infusions of animal manure. After this, the hides were suspended again in pits of tanning liquor. Hides were swapped into vats of increasing strength of tannic acid, which acts on the collagen of the hide as a sort of combined glue and water expellant.[45]

This outline of the technicalities of turning animal skin into leather illustrates why the tanning trade was both dependent upon water, and considered to be an offensive industry. Much of the processing requires soaking in various solutions of different strengths, hence a reliable water supply was critical. It also required somewhere to dump the spent process liquors. The smell, especially in summer, of rotting bits of flesh, fat and hair, would hardly have made the tanners ideal neighbours. Porter wrote of Jackson's Tannery that 'exhalations from the tannery were an addition to the town's medley of unfragrant odours – slaughter houses, drains, cesspits and hoofs singed in the smithies'.[46]

The tanning trade connected the hydrological cycle with selectors, shipping companies, exporters, pastoralists, lime workers, slaughterers and butchers, and a variety of trades that needed leather such as bootmakers. They depended on flow, from the flow of water raised by capillary action inside the stems of *Acacia mearnsii* to the flow of water beside riverside tanneries. Any diminution of that flow affected the region's economic prosperity.

---

45   *Blackwell Encyclopaedia of Industrial Archaeology*, Blackwell, Oxford, UK, and Cambridge, Mass., 1992.
46   Porter, *Bairnsdale*, p. 154.

Improving flow

Realising their dependency on surface flowing waters, and coming from a culture that continually praised permanent flow, colonial Gippslanders found themselves in a place with unparalleled opportunities to reshape local hydrology. Except for the migrants who came from relatively unchanged northern Scotland, no settler would have seen an unaltered, wild river before.

Gippsland's rivers were sinuous, high banked with thick fringing vegetation, and associated with large swamps and morasses towards their end as they emptied into the three lakes. In 1864, Rawlinson described the riparian vegetation of Gippsland's rivers as 'generally densely grown with eucalypt and mimosa, varying from the ordinary size of forest trees up to that of the largest dimension of forest and scrub'.[47] He emphasised the steepness of the banks, excepting only the small courses of the Latrobe and Thomson in the flattest parts of the Latrobe Valley.

Figure 5.4: Livingstone Creek, below Omeo, illustrating the steepness of banks.
Source: Author.

---

47   *GT*, 15 July 1864, correspondence to the editor from TE Rawlinson.

Figure 5.5: Nicholas Caire's image of the Dargo River, n.d.
Source: Pictures Collection, State Library Victoria, Accession no. H89.173/2.

Figure 5.6: A bend in the Avon River.
Source: Author.

The following section details some key ways that settlers responded to the nature of Gippsland's rivers, and attempted to improve them.

Crossing points and bridges

On 28 November 1845, the *Port Phillip Herald* optimistically announced the discovery of a track from Gippsland to Westernport 'in which there was not a single creek to cross'. John Gellion's 1844 experience, cited in the epigraph, puts the importance of such a claim into perspective. Gippsland's rivers already had a fearful reputation. Getting safely across creeks and rivers was a combination of luck, favourable conditions and skill.[48] Without all three, people and animals died.

In the absence of the kind of divine intervention offered to Moses, finding a safe crossing point was vital for the earliest arrivals. Their location represented the best amateur reckoning of factors such as current, bottom conditions and depth.

Figure 5.7: Crossing place at Wombat Creek, Thomas Henry Armstrong Bishop, n.d. Buckley recorded a log crossing on 25 February 1849.
Source: Pictures Collection, State Library Victoria, Accession no. H40967.

---

48   For another example of a near miss, David Parry-Okeden saved the life of his stockman while trying to cross the Mitchell. The stockman's horse had gotten into a hole and he had fallen off. There is no mention of whether the horse survived. D Parry-Okeden, 'The Parry Okeden family at Rosedale', *Gippsland Heritage Journal*, vol. 10, 1991, p. 46.

When money became available for bridges, the reckoning expanded to include geomorphological conditions suitable to siting good foundations. As technological capacity and capital grew with the century's age, colonial Gippslanders were increasingly able to ignore the constraints on the travel patterns placed by rivers. They began instead to change the creeks and rivers to suit themselves, rather than accommodate themselves to the rivers.

The ideal crossing point had particular characteristics. Shallowness and a gentle current was highly desirable, combined with low sloping banks and a firm, clear bottom. It was also located near a suitable place for a homestead and good fertile soil.[49] Alluvial soils dictated the spread of early pastoralists and farmers.[50] The catchment's rivers did not often produce this combination. Angus McMillan's record of exploration is marked by long rides upstream from where the rivers emptied into the lakes, to find a safe crossing. Complaints about steep banks, thick scrub and mud mark other accounts. Rev. Bean, one of the earliest Anglican clergy who spent most of his time travelling, wrote in 1849:

> After taking some refreshment at Mr Okeden's that gentlemen road with me to a ford opposite to Mr Hobson's [Fulham] where we both crossed after some difficulty; the rivers of Gippsland being by no means easy to ford or cross, the banks being general steep and the rivers deep. The Avon is the only that appears not so![51]

---

49   Diary of Patrick Coady Buckley, 4 November 1847, 'removed from the old hut near the crossing place on the south side of Merriman's Creek where I first sat down in 1843 to the new place'. See *GT*, 6 June 1881, for a reference to the shallowness of a crossing on the Tambo river and the footbridge over it, located next to Peter McDougald's inn providing a resting place for travellers.
50   For example, letter from Edward Hobson to his wife Mary, dated 7 October 1847, 'The country is by no means an interesting one, being very thickly timbered except on the banks of the rivers where the plains are extensive and exceedingly rich. On my run I could plough 1000 acres of the richest soil in the world without cutting down a tree', in J Court, *From squatters hut to city, Traralgon, 1840–1976*, City of Traralgon, Traralgon, 1976, p. 19; 'The [Maffra] district originally consisted of alluvial plains and river flats – the plains nearly treeless, and the latter have the usual fringe of timber along the Thompson and Macallister rivers, and around the edges of the numerous creeks. Most of the land surrounding being of a rich quality was taken up under the Brooks and Nicholson Land Acts …', *Middleton and Manning's Directory*, 1884; Evidence of Alfred E Otter, selector at Lake Victoria, cited in JM Powell, *Yeoman and bureaucrats: The Victorian Crown Lands Commission 1878–9*, Oxford University Press, Melbourne, 1973, p. 366. Real estate advertising also highlights alluvial soils, e.g. the opening quote of this chapter and similar, such as *GM*, 5 April 1877, re Denis Connolly's runs at Dargo; *GT*, 1 March 1870 and 1 April 1897.
51   Diary of Rev. Bean, 29 January 1849, cited in Clark, *The church of our fathers*, p. 45. See also Diary of Patrick Coady Buckley, 6 February 1844, 'Made several attempts to cross the Morwell but could not on account of the banks being so steep'.

McMillan concurred with Bean and settled on the Avon. The crossing place he chose between Boisdale and Bushy Park on the Avon met with the approval of at least two diarists. Elizabeth Montgomery described it as:

> The water was about two feet deep and the approach from our side was across a stretch of lovely flat stones, then came the clear brown water running over a bed of many coloured pebbles and the opposite side was of white sand. Both banks were shaded by wattles, tea trees and other scrubs with a dense undergrowth of tussocks and ferns.[52]

This is in stark contrast to how Alick Hunter described a crossing on the Macalister as 'beastly, 'deep' and 'muddy'.[53] One of the most graphic descriptions of a near disaster at a crossing place was given in a letter by John Gellion in 1844, author of the second epigraph. After eight days of rain, his stock refused to cross at Hobson's Crossing. They decided to attempt shifting their belongings across:

> It came on a dreadful storm of hail and wind, which raised the river to an enormous height. It was now approaching dark, and Billy Lonsdale told us that if we did not leave we should loose all our baggage, and probably our lives ... Having got the things over and placed on the opposite bank, our next business was to get them secured, but before we had time to do this the water came flowing in and in less than half an hour we were knee deep in water, and all our things covered.[54]

Possibly no one had a worse time with crossings than Joe Daniel, who was collecting statistics in the Dargo area in 1871. He had to cross the 'the Wonnangatta alone 180 times' and gave up tracking his crossings on the Crooked and Wonnangarra rivers.[55] More commonly, troubles at crossing places merely meant lost time and aggravation – for example, Duncan Johnston got stuck crossing the Tambo with a loaded dray of wood. He had to unload, release the dray, split the load and make two trips. In another accident, his horse fell over the precipice at St Patrick's Creek.[56]

---

52   Quoted in Fletcher, *Avon to the Alps*, p. 19.
53   C Daley, 'How the Hunters came to Gippsland', *Gippsland Heritage Journal*, vol. 3, no. 2, 1988, pp. 6–7.
54   Letter from John Gellion, dated 1 August 1844, cited in *Coach News*, vol. 2, no. 3, March 1975, pp. 7–9.
55   Diary of Joe Daniel, cited in *Bulletin of the Maffra and District Historical Society*, no. 49, June 1986. The Wonnangatta and Wonnangarra rivers are, confusingly, both tributaries in the Mitchell River catchment.
56   Diary of Duncan Johnston, 15 December 1882, for getting stuck and 6 August 1882, for the horse.

The characteristics of flow did more than just determine how people travelled through the landscape. Flow also influenced where they settled. The actions of Archibald McIntosh illustrate this. McIntosh, an engineer turned blacksmith, was the first settler at Flooding Creek (later renamed Sale), probably in 1844. His shop was on the banks of the creek, adjacent to the track that led to the punt over the Latrobe River. McIntosh gambled that the location would become important following the construction of the punt.[57] He was right. Bairnsdale was created similarly, around the punt that replaced the canoe used by travellers to cross the Mitchell River. Much of Adam's story of the early history of Narracan Shire, which comprised the early settlements around Moe, conforms to this pattern. A safe(ish) crossing place led to a hut or inn, which then might lead to a bridge, around which a village would grow.[58]

As the population grew, crossing points were no longer sufficient. Bridges were needed, both within Gippsland and linking Gippsland to Melbourne. The main road to Melbourne was always a subject of debate.[59] Gippslanders were perhaps no more or less argumentative than any other group in the nineteenth century; however, one cannot help but be impressed (or is that exhausted?) by the amount of arguing about bridges recorded in the papers. Fighting could take place about whether to build one, where to build one, what materials to use, what design it should be built to, the expense and how to maintain one. In the previous chapter, I used as evidence for the impact of rain a comment made by Miss Caughey in her diary. She hoped that all the hard work spent preparing for a 'picnic' would not be wasted if it rained. This picnic was much more than a light-hearted social gathering. The real reason for the picnic was that Traralgon Shire councillors and staff were on a site visit,

---

57   Synan, *Gippsland's lucky city*, pp. 23–5. Apparently McIntosh called it Flooding Creek, although Synan says that in normal rainfall it was more like a chain of billabongs. McIntosh arrived in Gippsland in 1841 with the laird Glengarry, bringing his sisters and his son with him. McIntosh bought Glengarry's plant when he failed and moved to Nuntin Station, staying approximately two years. There was the suggestion is that he left Nuntin because he was sick of the flooding on the Avon River, but removal to Flooding Creek doesn't suggest that being inundated was a issue for McIntosh.
58   J Adams, *So tall the trees: A centenary history of the southern districts of the Shire of Narracan*, Narracan Shire Council, Trafalgar, 1978. Maffra also grew from a crossing point. M Fletcher, 'The small farm ideal: Closer settlement in the Maffra district 1911–1938', MA thesis, Monash University, 1987, p. 7.
59   For example, 'Gippsland and its resources, by a correspondent', *Argus*, 3 May 1864. 'A few inquiries respecting the bridges on the main road to Gipps Land would not be out place in the Legislative Assembly … No same man could have expected that bridges built of logs simply resting upon sandy banks of mountain torrents could possibly endure for a longer period than the interval between the time of their erection and the occurrence of the first flood afterwards.'

FOLLOWING THE WATER

inspecting possible bridge sites across the Latrobe River. The Caugheys had a vested interest in the site that was closest to their holdings, and part of the purpose of the picnic was to persuade council to their own interests.

The Caugheys were by no means unusual in their attempt to influence the outcome of a bridge's location. Negotiations about bridges could take years to resolve, especially when the bridge crossed a river that formed a boundary with a neighbouring shire. This doubled the amount of confusion and wrangling. The Traralgon correspondent for the *Gippsland Times* was pleasantly understated when s/he wrote in 1881:

> Are we really at last to have bridge over the Traralgon Ck? Hurrah! Not before it is wanted certainly. There has been a little difference of opinion as to site. A public meeting was called to protest against the action taken by Council but it failed in its object, and now that everybody has had their say, and no serious objection has been raised to the proposed site it is hoped that all bickering and ill feeling will be at an end ...We ought not to put anything in the way of its being built, for the old bridge is very shakey, and may come down during any flood time, and this would cut off a large business from the township, and consequently stop money from coming in – which would be a calamity to be deplored.[60]

Rarely were the links between water flow and cash flow so plainly put.

Engineering reports to the council meetings illustrate how much of community time and money was devoted to roads, crossing places and bridges. In fact, Adams suggests that roadmaking was the main reason why the Narracan shire was formed.[61] For example, the *Gippsland Times* reported on 28 January 1875 that the Avon Shire had four water management related contracts for tender: 26 chains of draining and filling and three culverts on the road past Kee's to Nuntin Creek, 30 chains draining on the road to Bundalaguah common, a bridge across the creek between Clarkes and Clements at Upper Maffra and a bridge

---

60   *GT*, 9 February 1881.
61   Adams, *So tall the trees*, p. 85. According to Daley, the causeway was constructed at Longford in 1856, followed shortly by a timber bridge. C Daley, *The story of Gippsland*, Whitcombe & Tombs for the Gippsland Municipalities Association, Melbourne, 1960, p. 44. Construction of the first bridge over the Mitchell was started in 1870, on the site of the punt. A bridge over the 'Backwater' had been built in 1868 by John McLaren. Daley, *Story of Gippsland*, p. 51. See also Fletcher, *Avon to the Alps*, p. 58, on the long history of the Avon bridge, first built in 1854; survived floods of 1865 and 1866 but not the one in 1870; temporary punt and bridge started again until new bridge erected in 1874; extensive damage again in floods of 1878 and 1879, and the shire sought government help for repairs; narrowly survived 1891 flood but only due to efforts of volunteers.

5. 'FAIR STREAMS WERE PALSIED IN THEIR ONWARD COURSE'

and approach near Browns, on Crooked River Rd. The Rosedale Shire advertised four contracts: construction of Log Ford at Morwell Swamp on the main Melbourne road, a log bridge and approaches at Waterhole Creek (also on the main Melbourne road west of Joe Smith's), a timber bridge over the Latrobe to the north-east of Traralgon, and 60 chains of draining on the upper main side of the road at Loy Yang.[62] These tenders, and the multitude of others, are part of a collective attempt by local communities to render the hydrological features of the catchment invisible to traffic.

There were times, though, when this invisibility was an impossible fiction. The importance of bridges was highlighted by the actions undertaken by volunteers to save them during floods. The major flood that struck Sale on 26 March 1870 saw surveyor William Dawson, policeman John Sadleir and at least three other men spend hours in the dangerous work of attempting to deflect or hoist out debris pushing at the piles of the Cunninghame St bridge.[63] The regular public calls for adequate maintenance, such as the following about Flinn's Creek only a month before the above mentioned flood, show that at least some commentators appreciated that the hydrological cycle still had the upper hand:

> The bridge across Flinn's Creek is at present in a most dangerous condition. The logs of which the flooring consists are little else than a series of pitfalls through which the feet of passing horses are liable to slip at any moment, and in addition the 'structure' is so shakey that a wavy motion is distinctly perceptible when the coach crosses it. It is feared that the first flood will deprive us of this means of crossing, and steps should therefore be taken to make the bridge secure.[64]

By spanning actual flowing water, bridges facilitated the flow of people, products and cash. The Traralgon correspondent was recognising what economists call deprival value. The real value of an asset like a bridge goes beyond the value of its construction. It's what a bridge facilitates that

---

62   There is a multitude of other calls for tenders, or reports of debates in council in the paper. Other examples include the advertisement in *GT*, 11 March 1881, of a notice of motion in Maffra Shire to raise a loan for £6,000 over 15 projects, of which six were related to crossing or managing water and road making, i.e. two stone fords to be made over the Avon River, draining roads, bridges – each ford worth £500; £400 for draining main road between Stratford and Maffra; £1,000 for draining and forming roads between Maffra, upper Maffra and Newry. See *GT*, 4 February 1881, 25 April 1881 and 23 July 1886.
63   *GT*, 26 March 1870.
64   *GT*, 1 March 1870; *GT*, 4 January 1876, Rosedale correspondent.

should be valued as well, not just the structure itself.[65] The destruction of bridges caused major impacts for a council's budget. In 1891, a flood along the Avon made immediate repairs necessary to no less than 10 bridges or crossing points.[66]

Of all the bridges ever constructed in Gippsland, perhaps Sale's swing bridge encapsulates both the barrier and opportunity that flowing waters meant to colonial Gippslanders.

The swing bridge was constructed over the Latrobe River in the early 1880s and connected Sale and South Gippsland. However, the need for a bridge at this critical location posed a substantial problem. Gippslanders relied equally upon marine trade, and a bridge would cut access to any commercial-sized steamer: one kind of flow would stop another.

Unwilling to accept this, Gippslanders designed a variant of an opening bridge, of which London Bridge with its hoisting decks is perhaps the best known. In this case, the swing bridge pivots on a central pillar, which sits in the middle of the river. At the approach of a steamer, the operator detached the bridge platform from its anchors on the bank, and swivelled it 90 degrees so that the whole deck sat parallel to the river banks, allowing steamers passage on either side. The bridge opened to traffic on 22 September 1883.[67]

---

65 D Brunckhorst, 'Understanding design for planning alternative landscape futures to adapt to climate change: Learning from temporal inconsistencies in vulnerability and adaptation studies', Paper delivered at the 1st International Conference on Adapting to Climate Change: Preparing for the unavoidable impacts of climate change, Gold Coast, Australia, 29 June – 1 July 2010.
66 *GT*, 11 September 1891. At the Maffra Shire meeting of 9 September, the engineer reported: 'Since the last meeting of council the flood waters have subsided and a more careful inspection of the actual damage done has been made'; Hagan's bridge was destroyed, recommends a new site be chosen; temporary ford set up on the property of the late James Boland; at Glenmaggie the culvert washed away, recommends a pitched stone crossing instead: 'the Glenmaggie Creek at Gleeson's is now twice its original width, and the bridge is standing in mid stream, with approaches washed away', same for Manley bridge at Tinamba and Andersons Creek; Coombs bridge will need a new span and strengthening; Newry footbridge gone; crossing at Valencia Ck completely gone, Mayhews culvert gone, approaches to Avon bridge gone, temporary punt installed at Bushy Park.
67 Synan recounts how plans for the opening celebration were eclipsed by the dying wish of John Campbell. Campbell had lost the partial use of one of his hands during the flood of February 1863. Despite his disability, he continued to contribute to rescue efforts. No one therefore wanted to deny him his dying wish to be the first person to cross the Sale Swing Bridge. P Synan, *Highways of water: How shipping on the Lakes shaped Gippsland*, Landmark Press, Drouin, 1989, p. 65.

Figure 5.8: The Sale Swing Bridge, showing the central piers around which the deck pivots.
Source: Author.

Figure 5.9: Looking upstream from the Sale Swing Bridge at the new, and substantially higher, Latrobe River bridge.
Source: Author.

Removing obstructions to shipping

The Sale Swing Bridge was only one part of a much wider vision to remake the lakes and rivers of Gippsland to facilitate the flow of people and products. This vision was developed by Sir John Coode, who had been contracted by the Victorian Government to review the existing entrance works and advise on the best way to make the lakes permanently open to trade. His scheme, enthusiastically adopted by the Sale Council and a range of other supporters, included a canal to link the town to the Thomson River and a comprehensive program of desnagging rivers and dredging their mouths to facilitate access to ports at Sale and Bairnsdale, the main distribution points. The creation of the permanent entrance was merely a much more substantial application of the same principle.

The desire for permanence and order was always particularly focused around estuarine systems. Horton and Eichenbaum provide the most memorable reason why:

> The ... collision of sweet and salt – fresh river water flowing seaward and ocean pushing inward – makes what we call an estuary. The Latin verb *aestuare* – to heave, boil, be in commotion – gives fair warning that that this place is no mere river running in one direction for all time. Nor is it a lie, its waters turning over sedately once or twice a year as the surface layers cool and warm. Neither does it feature the predictable currents and constant salty chemistry of the oceans. Estuaries in their behaviour are among liveliest natural systems of the planet. They are the aquatic world's three ring circus of motion, productivity and changeableness.[68]

The reengineering of rivers and estuaries was hardly a pioneering departure.[69] Colonial Gippslanders were merely following a well-established European, and then American, precedent. British rivers have been regulated since the first century AD, when the Romans started land drainage and navigation improvements. By 1086, there were approximately 5,000 watermills, growing

---

68   T Horton, *Turning the tide: Saving the Chesapeake Bay*, Island Press, Washington DC, 1991, p. 9.
69   Substantial reengineering also took place in the gold industry. For example, *East Gippsland Historical Society Newsletter*, vol. 2, no. 4, December 1982, Trip to Deptford at Pub Gully, 'a walk through the bush along the riverbank brought them to the tunnel dug under the river bank to divert the flow of water, said to have been dug by hardworking Chinese miners', p. 8.

by 1800 to approximately 12,000.⁷⁰ Waterwheels were also employed in Gippsland – for example, in 1855 Frederick Gray noted that his employer had a watermill that was used to grind a few bags of wheat a day.⁷¹

Figure 5.10: Oriental claims area, near Omeo.
Source: Author.

---

70   S Owen, C Pooley, A Folkard, G Clark & N Watson, *Rivers and the British landscape*, Carnegie Publishing, Lancaster, 2005, p. 28. The Thames had a waterwheel on average every 3 kilometres during medieval times. Small two horsepower Saxon-style waterwheels (placed horizontally in the water) could do the work of 30 people using hand mills, and had the added advantage of not needing extra land to grow fodder for horses. Vertical wheels were a later development that could be made more powerful through constructing a weir and a flume, which allowed a more constant and steady flow of water to turn the wheel. For the 1800 figure, see pp. 104–5.
71   Letter from Frederick Gray to Mr Lewis, dated 4 August 1855, describing his life at Lindenow, *East Gippsland Historical Society Newsletter*, vol. 1, no. 3, 1980.

Snagging and dredging were the preparatory steps to facilitating major shipping trade on the lakes system. These practices ironed out, so to speak, some of the variability inherent in natural river systems, such as how Rawlinson described the upper Tambo in 1864 as alternating shallow gravels and deep pools.[72] The Tambo's upper reaches, like many of the gold-bearing alpine rivers, would be seriously altered by mining practices like sluicing. Figure 5.11 shows the visible effects of sluicing near Omeo.

Unlike gold mining activities, which leave clear evidence of human activity, snagging and dredging cannot be seen. Snagging refers to the practice of removing submerged logs and vegetation to improve navigation. For aquatic biodiversity, snagging obliterates habitat. Dredging refers to artificially deepening channels. Its ecological impacts are also destructive, both for the species in the area dredged and for those where the spoil is dumped. At the same time snagging was carried out, banks were also frequently cleared of riparian vegetation, leading to significant bank slump, and exacerbating wider landscape-scale changes in erosion and sedimentation rates.[73]

A lengthy report from May 1881 on the progress of the snagging contract in the Latrobe River highlights the changes made:

> The river has to be cleared of snags to a depth of 9 feet below summer level, and overhanging trees on both sides of the stream cut away for a distance of ten feet ... Some of the snags fished out are giants, 40 to 90 feet in length, and five or six feet in diameter.

The report noted how the snags formed habitat for eels, but highlighted the efficiency of the contractor and his block and tackle equipment, and how the works were speeding the progress of Sale.[74] The Tambo was extensively snagged, as was the Mitchell.[75]

The Avon became a textbook case in how the removal of riparian vegetation exacerbated erosion. The now bare banks of the rivers increased the susceptibility of the banks to erosion.[76]

---

72   *GT*, 15 July 1864.
73   B Abernethy & I Rutherford, 'Riverbank reinforcement by riparian roots', Paper delivered at *Second Annual Stream Management Conference*, Adelaide, 8–11 February 1999.
74   *GT*, 4 May 1881.
75   For positive reports on this, see *Middleton and Manning's Directory*, 1884, entry for Bruthen; and *Our Trip to the Gippsland Lakes*, by the Publishers, an advertorial for the Lakes Navigation Co. in *Middleton and Manning's Directory*, 1884, pp. 15–17.
76   B Abernethy & ID Rutherford, 'Where along a river's length will vegetation most effectively stabilise stream banks?', *Geomorphology*, vol. 23, 1998, pp. 55–75. doi.org/10.1016/S0169-555X(97)00089-5.

5. 'FAIR STREAMS WERE PALSIED IN THEIR ONWARD COURSE'

Figure 5.11: Bank erosion on the Avon River, 1872.
Source: Pictures Collection, State Library Victoria, Accession no. H40967.

Figure 5.12: Washaway at bridge over Avon River near Bushy Park, Victorian Railways, 1893.
Source: Pictures Collection, State Library Victoria, Accession no. H1077.

Ironically, this would lodge more sediment on the bar, which would then need dredging, thus locking settlers into a repeating cycle. By the mid-1940s, the Avon was reported as the most eroded river in Victoria.[77] The crossing that Elizabeth Montgomery once admired on the Avon was only a few feet wide. By the time Charles Daley published *The Story of Gippsland* in 1960, the bridge spanned 380 feet.[78] The change in the Avon was evident from as early as 1882, noted when a journalist visited Boisdale to write a story.[79] By 1897, the river was so changed that the railway bridge no longer spanned the river.[80] Severe erosion along the Mitchell was also reported in the 1890s. Bairnsdale Shire Council, with no apparent sense of irony, sought permission to use the dredgings from the mouth to fill the washaways.[81]

The supply of sediment from the catchment built bars at the river mouths, and this natural process was exacerbated by clearing in the catchment. A study carried out in 1998 suggested that deposition rates of sediment were approximately twice pre-European rates.[82] The need to dredge was exacerbated through low rainfall periods, as water levels dropped. In January 1881, the steamer Murray was stuck for several days on the bank at the mouth of the Mitchell. Shipping companies were refusing heavy freight as a result.[83]

The location of the dredge was yet another arena for jealousy and rivalry between the councils of Sale and Bairnsdale.[84] Ship groundings or near misses were by no means rare, and such incidences were generally used to make an argument to the government for more funding of some kind.[85]

---

77 M Fletcher & L Kennett, *Wellington landscapes: History and heritage in a Gippsland shire*, Maffra and District Historical Society, Maffra, 2005, p. 43.
78 Daley, *Story of Gippsland*, p. 48.
79 J Hales & J Little (eds), *Gippsland estates 1882: A series of articles which appeared in the Gippsland Mercury from January to August 1882*, Maffra and District Historical Society Bulletin, Supplementary Issue no. 4, Maffra, 2011, p. 23.
80 *GT*, 1 July 1897.
81 *GT*, 7 December 1896. Permission was refused by the Inspector General of Public Works on the grounds that silt was unsuitable, and that the government had no interest in protecting freehold land.
82 RB Grayson, C Kenyon, BL Finlayson & CJ Gippel, 'Bathymetric and core analysis of the Latrobe River delta to assist in catchment management', *Journal of Environmental Management*, vol. 52, 1998, p. 370. doi.org/10.1006/jema.1998.0181. The study also noted that while sedimentation was less than previously thought, the fine nature of the sediment made it easy to transport high levels of nutrients.
83 *GT*, 19 January 1881.
84 *GT*, 19 January 1881.
85 For example, *GT*, 18 April 1876 and 1 April 1891.

Figure 5.13: Flora Gregson's watercolour of the *Lady of the Lake* trying to pull a stranded vessel off the bar at Lakes Entrance, 1878.
Source: Pictures Collection, State Library Victoria, Accession no. H16561/5.

From virtually the moment white settlers arrived, they argued continuously about the best way to facilitate transport through the region. The difficulty of road travel, nicely symbolised by the three abandoned attempts of the first Crown Lands Commissioner, Charles Tyers, to reach his new post from Melbourne, fuelled discussion of the various merits of road, rail and ship. To date, these have been understood by many historians to have been in opposition to each other. However, both the lobbyists for the permanent entrance and the lobbyists for rail were in perfect agreement on the fundamentals. Underlying both campaigns was an argument based on metaphorical flow. No one in Gippsland disagreed with the idea that people and goods should be able to flow swiftly and unimpeded to Melbourne, and then outwards into the imperial economy. Their disagreement revolved around the mode of conquering the vagaries of flowing water. J Rodgers of Greendale in North Gippsland argued for both, in a hyperbolic poem published by the *Gippsland Times* in 1881. The first three stanzas elaborate on the silent, primitive wilderness, complete with unclothed 'tawny native', before launching into a rallying cry for progress via rail and ship:

> Be ours, a free unfettered Press, the Plough, the Rail.
> Ah yes! The Rail! Prepare the iron way,
> Tear up the rock, prostrate the wood,
> Drive through the mountain, and make no stay
> For Thomson, M'Allister or Latrobe.
> Lay the long rail across the verdant green
> And wake to roar and echo of whirling wheel and steam.
>
> Remove the sand bar! Open Gippsland to the world!
>
> …
>
> Gippsland will yet arouse from its slumbers
> And become the most flourishing spot of Victoria
> We have the land, the beautiful sparkling streams
> Coupled with magnificent climate.
> Its with God's blessing &c. What can we want more?[86]

Rodgers was not the first or last to write truly awful poetry about Gippsland, and nor was s/he the first to advocate for opening the entrance. From as early as 1846, colonial settlers were trying to regularise the estuarine entrance. Mary Cunninghame wrote about how flow variability was impacting on her life, demonstrating how, after only eight years of colonisation, the preference for moderate flow was set:

> Owing to very heave rains great parts of Gipps Land is flooded this spring and it has been alarming increased by the outlet of the Lakes into the sea being closed by a large sand bank, after waiting and hoping that the water would force its own way out Boyd and my brother determined to go down to the Entrance to see if they could force a passage, fortunately a few days before Boyd had exchanged a couple of horses for a good sized boat which is indispensable where there is so much water carriage – Accordingly they were able to go without danger the distance being nearly fifty miles – They cut a passage through the bank of twenty feet long and about ten deep – and now we are anxiously waiting for the waters to subside, as they are doing much injury to the grass, the country being for miles under water.[87]

---

86   *GT*, 7 September 1881.
87   M Fletcher, 'The Cunninghame letters', *Gippsland Heritage Journal*, vol. 16, June 1994, p. 48. Patrick Coady Buckley also hand-cut entrances at smaller creeks that emptied out into the sea on his various runs, and there are occasional references to cutting bars at Prospect in the late 1890s, when it became a favourite camping and holiday place. *GT*, 14 January 1897 and 4 November 1897.

5. 'FAIR STREAMS WERE PALSIED IN THEIR ONWARD COURSE'

The previous quote illustrates precisely what the authors of the Gippsland Lakes Environmental Audit described in more measured academic language in October 1998:

> Before the opening of the Entrance in 1889, the Lakes were a series of coastal lagoons that only opened to the sea after heavy rainfall and runoff from the major catchments … The level of the lakes would have fluctuated markedly after rainfall, with the level rising and inundating the freshwater marshes around the open water areas before the flow through the opening scoured out enough sand to lower the level again.[88]

The authors went on to note that it was an almost entirely freshwater system, and that the country that Mary Cunninghame thought was being 'injured' was actually dependent upon such periods of inundation.

Investigations into the nature of the bar at the entrance commenced early. Kirsopp, on a coastal charting expedition in 1841, reported that there was no permanent entrance. Squatter John Reeves visited the entrance in 1842 and, shortly after, he took a larger party including the Port Albert harbour master for a closer inspection. They concluded that the entrance was unsafe.[89]

The year 1844 was dry. With insufficient flow, the entrance did not open at all. It reopened by the time Smythe made a survey in 1849, but rarely opened during the 1850s. After noting the damage that floods were causing, *a la* Mary Cunninghame, surveyor WT Dawson proposed a scheme to cut a 60-mile-long canal to Port Albert.[90] The real turn in fortune for lakes transit advocates came in 1858. First, Phillip McArdell took advantage of a flood to float the *Enterprise*, his purpose-built lakes trader boat, across inundated flats, around the low bridge at Longford and into the Lakes.[91] Second, on 23 April 1858 Malcolm Campbell successfully navigated the *Georgina Smith*, his schooner with a draught of 7 feet 6 inches, through the entrance and almost to Bruthen with

---

88 G Harris, G Batley, I Webster, R Molloy & D Fox, *Gippsland Lakes environmental audit: Review of water quality and status of aquatic ecosystems of the Gippsland Lakes*, prepared for the Gippsland Coastal Board by CSIRO Environmental Projects Office, Melbourne, October 1998, p. 4.
89 ECF Bird & J Lennon, *Making an entrance: The story of the artificial entrance to the Gippsland Lakes*, James Yeates & Sons, Bairnsdale, 1989, p. 5.
90 Bird & Lennon, *Making an entrance*, p. 8.
91 Morgan, *The literature of Gippsland*, p. 89.

cargo for the Omeo goldfields. Support for a lakes entrance scheme was instantly won from all the gold districts in the catchment, suffering under the crippling costs of road freight.[92]

Campbell's success fostered much more interest in the possibility of a permanent entrance. It also coincided with a run of wet years, which kept the entrance open and relatively navigable. With a growing population in the gold districts, and the arrival of selectors, the potency of the idea took hold.

Given that just about every history of Gippsland recounts this story, it is not the intent here to recount the minutiae of the claims, counter claims, petitions, tenders or works. Rather, I highlight the intent to create permanence, order and moderation in nature. This was not an option available when it came to rainfall, as Chapter 4 demonstrated. Weather vagaries were always commented upon, and sometimes even with a flow metaphor:

> The old saying 'As changeable as the weather' has been exemplified here last week. The heat of Thursday and Friday was such as Queensland could not excel. It was so intense on Friday that it roasted the ferns at Sandy Creek. A welcome change set in on Saturday evening; the rain came down in torrents, and it really looked at one time as if the flood gates of heaven had opened up. Let us hope that they will continue to flow, for never was Gippsland, at least this part of it, more in need of rain than at present.[93]

In the face of such changeableness, regularising flow meant exerting control. The manipulation of surface water was increasingly an option for nineteenth-century settlers. The construction of a permanent entrance is the apotheosis of this attitude that nature could be corrected.

---

92   KMcD Fairweather, *Time to remember: The history of gold mining on the Tambo and its tributaries*, the author, Doctors Flat, 1975, p. 22. The high cost affected the affordability of food – Fairweather says that Omeo district miners paid the exorbitant sum of 2 shillings and 6 pence for blackberry roots for the pleasure of fresh berries in season, contributing to the widespread weed problem of the next century. Steenhuis suggests that miners around Donnelly's Creek relied on shooting native fauna such as the lyrebird and wallaby. L Steenhuis, *Donnelly's Creek: From rush to ruin of a Gippsland mountain goldfield*, Paoletti's Maps and Videos PL, Lagwarrin, Vic., 2001, p. 15.
93   *GT*, 24 February 1886, Stratford correspondent.

Figure 5.14: The permanent entrance from Jemmy's Point.
Source: Author.

Very few gave thought to the possible knock-on effects of the change. Only 'Tom Cringle' and John Coode noted that water levels would drop and that it would become more marine.[94] It is doubtful that either of them appreciated the full ramifications. Creating the permanent entrance changed the fundamental nature of the system. The 2003 Ramsar plan for the Lakes lists altered water regimes, salinity, pollution, pest species, resource utilisation, dredging, recreation and tourism, fire and erosion as main risks, although obviously this reflects the full sweep of time rather than just the impacts created by actions in the nineteenth century.[95] As an intermittently closed and open lagoon (ICOL), the Lakes were generally a freshwater system, only becoming salty during periods of breach.[96] They supported a wide variety of plant and animal species that

---

94   ECF Bird, *A geomorphological study of the Gippsland Lakes*, Research School of Pacific Studies, Department of Geography Publication G/1, The Australian National University, Canberra, 1965, p. 83; Synan, *Highway of water*, p. 80. Cringle was a pseudonym for William Walker.
95   Department of Sustainability and Environment, *Gippsland Lakes Ramsar site: Strategic management plan*, the department, East Melbourne, July 2003, p. 24.
96   ICOLs can be fresh, brackish or salty. There would also be changes in salinity during drought due to evaporative processes and a reduction of groundwater discharge. Sara Beavis, pers. comm., 30 September 2011.

were adapted to this regime of primarily fresh water. When the system became predominantly marine, all those species slowly died, leading to widespread shoreline erosion.

As Bird notes, it was this gradual change that misled most for decades into thinking that the construction of the entrance was wholly benign. A few years of high rainfall following the opening buffered the freshwater levels. Changes only became obvious just before World War I and became pronounced in the 1920s, both after long periods of drought. Changes included the death of *Melaleuca ericifolia* stands on the lake fringes, the proliferation of eelgrass (*Zostera* spp.) and invasions of marine crabs.[97] Severe eutrophication problems came later in the twentieth century, when farmers began to apply cheap, artificial fertilisers.

Ironically, the attempts to regularise flow only created long-term change and uncertainty for settlers' descendants. According to Harris et al.:

> The only long term solution is to markedly reduce the nutrient loads from the Latrobe system by both restoring the catchment to a more sustainable land use and by replacing riparian vegetation and reducing erosion. The importance of wetlands, in reducing nutrient loads to the lakes should not be underestimated.[98]

From having energetically tried to remake the catchment, current Gippsland citizens are now attempting to restore the catchment to something more like its situation in the nineteenth century.

Conclusion

This chapter has demonstrated that the ideal of permanent, moderate flowing waters was valued across many different aspects of colonial life. Initially, colonial settlers were forced to accommodate themselves to the power of flowing water. However, as wealth and technical capacity grew, they were able to effect significant changes to the hydrological cycle. The ability to make hydrological changes reflects a mathematical and quantitative understanding of hydrology; the common paradigm of the

---

97 Bird, *A geomorphological study of the Gippsland Lakes*, p. 83.
98 Harris et al., *Gippsland lakes environmental audit*, p. i.

nineteenth century. The changes, which included permanent piped water supplies, altering river morphology and the creation of the permanent entrance, were universally regarded as desirable and progressive. The limitations to both a quantitative vision of hydrology and the notion of progress would only become clear generations later.

CHAPTER 6

'A useless weight of water':[1] Responding to stagnancy, mud and morasses

> Mud mud mud; nothing but mud, watery mud, creamy mud, treacly mud, mud like school boys' 'stick jaw', mud which wont be passed by, mud which clings like Potiphar's wife, and mud like unto Dante's sea of defilement, for seven long and weary miles![2]

> Whatever may be the means employed by the landholders to carry out the work of the reclamation of the swamp, there can only be one opinion as to its value … Their success would ensure immediately increased value to the land so created, and would add materially to the progress and prosperity of the town.[3]

Introduction

The nature of the hydrological cycle is one of constant, varied movement. Sometimes that movement is barely perceptible, and with groundwater, it is invisible. This chapter turns the focus to places in the landscape where the movement of water is slowing down, where the force of the current is dissipating.

---

1   *GT*, 16 April 1886.
2   Anon., *Guide for Excursionists from Melbourne 1868*, H Thomas, Melbourne, p. 155 quoted in SM Legg, 'Arcadia or abandonment; The evolution of the rural landscape in South Gippsland, 1870–1947', MA thesis, Dept Geography, Monash University, 1984, p. 171.
3   The editor of the *Gippsland Times* commenting on a cooperative proposal to drain the Heart Morass, 18 January 1886.

According to Giblett, 'the distinction … between flowing and stagnant waters is one of the fundamental organising principles of the dominant western cultural construction of nature'.[4] The previous chapter argued that colonial Gippslanders had a preference for moderate amounts of permanent, flowing, channelised water. This chapter continues the exploration of this ideal by examining its opposite; those parts of the catchment that were soggy, muddy and still.

In contemporary environmental parlance, places in the landscape where water is slow moving or still are called wetlands.[5] They are recognised as being absolutely critical for a wide variety of important ecological functions, including habitat for fish and fowl, providing breeding grounds, retaining floodwaters, preventing erosion, and water filtration and cleansing.[6] In 1982, parts of the Gippsland Lakes catchment (the coastal brackish/saline lagoons, permanent saline/brackish ponds and permanent fresh water marshes) were nominated by the Australian Government as internationally significant wetlands under the Ramsar convention. They were listed for their ability to support waterbirds, and for their representation of 'natural or near natural wetland characteristic of the appropriate biogeographical region'.[7]

---

4   R Giblett, *Postmodern wetlands: Culture, history, ecology*, Edinburgh University Press, Edinburgh, 1996, p. 23.
5   Definition of 'wetland' in M Allaby (ed.), *A dictionary of ecology*, Oxford University Press, 2006. *Oxford Reference Online*. Oxford University Press, www.oxfordreference.com/views/ENTRY.html?subview=Main&entry=t14.e5956, accessed 16 July 2008. The word 'wetland' is an invention of the mid-twentieth century, which passed into common use when it was defined at the world's first international convention on wetlands in 1971 (the Ramsar Convention) as 'all areas of marsh, fen, peatland, or water, whether natural or artificial, permanent or temporary with water that is static or flowing, fresh, brackish, or salt, including areas of marine water the depth of which at low tide does not exceed six meters'.
6   Millennium Ecosystem Assessment, *Ecosystems and human wellbeing wetlands and water synthesis*, World Resources Institute, Washington DC, 2005; G Harris, 'Inland waters', theme commentary prepared for the 2006 Australian State of the Environment Committee, Department of the Environment and Heritage, Canberra, 2006, www.environment.gov.au/soe/2006/publications/commentaries/water/index.html, accessed 20 August 2008; G Aplin, *Australians and their environment: An introduction to environmental studies*, Oxford University Press, Melbourne, 1988, reprinted 1999, p. 45; AH Arthington and BJ Pusey, 'Flow restoration and protection in Australian rivers', *River Research and Application*, no. 19, nos 1–3, 2003, p. 379; KH Taffs, 'The role of surface water drainage in environmental change: A case example of the Upper South East of South Australia; an historical review', *Australian Geographical Studies*, vol. 39, no. 3, 2001, pp. 279–301. doi.org/10.1111/1467-8470.00147.
7   Department of Sustainability and Environment, *Gippsland Lakes Ramsar site: Strategic management plan*, July 2003, East Melbourne, p. 3. The listing includes Lake Wellington (18,000 ha), Lake Victoria (10,850 ha), Lake King (7,100 ha), Lake Bunga (460 ha), Lake Tyers (1,186 ha), Macleod's Morass (520 ha), and Lake Reeve (5,158 ha), of a total plan area of 58,824 ha.

Such an action could scarcely have been dreamt of by the catchment's residents 100 years earlier. In fact, their agitation to drain wetlands was fierce. To them, the best thing to do with a swamp, bog or morass was to drain it and make it 'productive'. This perception permeated the highest levels of society. Reporting on their trip through Gippsland in 1874, the Surveyor General and the Secretary for Mines remarked several times on the extensive swamps and the need to drain them for agricultural use. This attitude was so common that the 2006 Australian *State of the Environment Report* concluded that 'unfortunately, wetlands have frequently been regarded as useless swamps to be drained, farmed or otherwise "reclaimed" for agricultural and other uses'.[8]

This chapter explores the multifaceted reasons why colonial Gippslanders had such a dislike for wetlands, and what they did about it. In the first part, I discuss how Gippslanders perceived still waters. The second part examines the range of responses made to them, including legislation, policy and physical changes, and discusses the long-term environmental implications of the changes made to wetlands in the nineteenth century.

Naming names

The main words used by Gippslanders to denote places of slow or still water were lake, morass, swamp, waterhole and backwater. The *Oxford English Dictionary* (*OED*) defines a lake as 'a large body of water entirely surrounded by land; properly, one sufficiently large to form a geographical feature, but in recent use often applied to an ornamental water in a park, etc.'[9] The key feature of this definition is the size, distinguishing it as permanent landscape feature. Within the GLC, the term 'lake' was applied to Lakes Wellington, King and Victoria, which have no hint of ephemerality, and Lake Guthridge in Sale which became an ornamental lake. Originally part of the extensive low-lying floodplains around Sale, Lake Guthridge was often disparaged in the press as being unworthy of the name, as it was used as a sewerage and drainage dumping ground.

---

8   Harris, 'Inland waters'. This report noted that the Murray-Darling Basin has a 90 per cent rate of alteration to wetlands, while Western Australia's Swan Coastal Plain has a 75 per cent loss.
9   The *Oxford English Dictionary*, revised edition, C Soanes & A Stevenson (eds), Oxford University Press, 2005, online, accessed 22 July 2008.

'Morass' was an important word in the lexicon of nineteenth-century Gippslanders. It was commonly used by the Scots migrants, and passed into general use with the naming of key landscape features, such as Clifton's Morass, MacLeod's Morass and Dowd's Morass. It derives from Middle Low German and Middle Dutch through the Old French noun *marais*, or marsh.[10] Morasses are land and water together, places of mud and marsh with low-lying vegetation adapted to permanent or periodical inundation. Gippslanders gave the term morass to areas that were predominantly fresh water, like MacLeod's Morass on the fringe of Lake King. They tend to be intermittently open areas of reeds and small *Melaleuca* trees. Areas of saline water such as around Maringa Creek at Nyermilang tend to have the characteristically red-tinged samphire plant.

'Morass' has another, less neutral usage. It is used to denote 'a complicated or confused situation which it is difficult to escape from or make progress through' and reflects the physical nature of this type of landscape. The examples given in the *OED* link morass with vice, doubt, addiction, outdated scholarship and politics. More often than not, the complicated or confused situation that Gippslanders found themselves in was the actual physical morass, which impeded their progress. However, they were also using the word in its figurative sense. For example, Sir Charles Gavan Duffy speaking on the Bill to connect Melbourne and Sale via rail described parliament as one of the most odious morasses in the world.[11]

Other terms common in England did not make their way into place names, although sometimes they were used adjectivally. The word 'fen' is occasionally used to denote a place of ill health and bad air, reflecting contemporary beliefs about miasma. 'The bleak misty air generated by the outlying swamps and morasses, indeed, reminds us of the fens of Lincolnshire.'[12] Where 'mere' might have been used, swamp tends to be substituted. The *OED* is unclear on the origins of 'mere', but notes that its first recorded usage is in Virginia. The dictionary suggests that the origin of the word may be Germanic, and may relate to 'sponge' or 'fungus', both organisms preferring watery surroundings. In the American usage, it denoted a tract of rich soil having a growth of trees and other vegetation, but too moist for cultivation. Waterhole is another term defined as being of colonial origin, and means a being a permanent pool of water in a river

---

10   *Oxford Dictionary of English* online, accessed 22 July 2008.
11   Cited in H Copeland, *The path of progress: From the forests of yesterday to the homes of today*, Shire of Warragul, Warragul, 1934, p. 125.
12   *GT*, 23 April 1870.

course. Gippslanders use the term somewhat more freely than this, but nor are they specific as to which waterhole they could be driving their stock to. It may be a farm dam; however, as these were generally very small and ephemeral, it is more likely to be part of a fluvial system.[13] Waterhole simply seems to refer to any place where stock can drink.

Finally, 'backwater' can have three meanings: 'a part of a river not reached by the current, where the water is stagnant'; 'an isolated or peaceful place'; or 'a place or situation in which no development or progress is taking place'.[14] Colonial Gippslanders usage of the term tends towards the first two senses. Backwaters were a common feature of the landscape, and they were used by squatters and selectors. For example, Isabella and Dalmahoy MacLeod selected near the backwater of Lake King, and Auchterlonie regularly talks in his diary about the backwater paddock, from which he regularly retrieved straying stock.[15] In private writings, references to these places are generally neutral, although Auchterlonie recorded his fury when he found Bessie Auchterlonie in a compromising position with the hired help up in the backwater.[16]

The definitions discussed above already indicate a negative viewpoint. The most used word in relation to wetlands is the word 'stagnant'. If there was anything water should not be, it was stagnant. Stagnant means foul, unwholesome and sluggish. Nor is the figurative use of the word flattering.[17] Gippslanders rarely described wetlands as still, even though they were,

---

13   Pers. comm. Sara Beavis, 30 September 2011. 'Farm dams at that stage were extremely limited in size because were constructed by horse and rudimentary scourers. These ended up being so shallow – all the ones I have looked at tend to be about 20–30 square metres and no more than a metre deep … so they would have been extremely ephemeral.'
14   Definition of 'backwater *noun*', *Oxford Dictionary of English* online, accessed 22 July 2008.
15   Diary of George Glen Auchterlonie, CGS, 4060, 12 March 1869, 'Found the steer loose this morning. Had an hour's work hunting out of the backwater scrub', also 25 September 1870; P Macleod, *From Bernisdale to Bairnsdale: The story of Archibald and Colina Macleod and their descendants in Australia, 1821–1994*, the author, Nar Nar Goon North, 1994, pp. 223–5.
16   Diary of George Glen Auchterlonie, 15 October 1871.
17   *GT*, 31 January 1862, editorial berating stagnancy and lassitude in effecting infrastructure improvements; *GT*, 29 March 1865, Town Talk on the 'unsightly repository of filthy water, dead dogs and other fruitful generators of nuisance'; *GT*, 19 April 1866, letter from Samuel Toynbee to the Editor on draining stagnant swamps; *GT*, 22 June 1869, advertising for stomach bitters by Hostetter 'malaria arising from unhealthy soils and stagnant water'; *GT*, 23 November 1872, Rosedale correspondent; *GT*, 4 November 1873, Walhalla correspondent on the 'somewhat stagnant state of [affairs]'; *GT*, 18 October 1876, arguing from protection from imported books to stop local trade from stagnating; *GT*, 4 December 1889, 'Nothing in the colony could remain stagnant'; *GT*, 2 July 1890, Briagolong correspondent wishing for a flood to clear out stagnant waterholes; *GT*, 25 June 1896, Portfolio Sermon, 'Idleness is the stagnant marsh which exhales the pestilence …'; *GT*, 13 January 1898, Melbourne Letter: 'Politics is still very stagnant'.

and regularly emphasised the aspects of unwholesomeness. This comes out most clearly in newspaper reports from local correspondents, who were attempting to get some kind of public response to a perceived problem. Concepts of heath and disease, and the kinds of topography that were mentally linked with those concepts drive much of the action in relation to wetlands.

What stillness helps

Wetlands are particularly productive and useful places. Because they are the margin between fully terrestrial and fully aquatic ecosystems, wetlands have greater biological diversity than many other parts of the landscape. Wetlands provided protein in the form of birds, eels, fish, crustaceans and eggs; important building products such as reeds for thatching; a supply of peat for heating and cooking; and summer grazing for stock. Coles's survey of wetland archaeology in Europe serves to illustrate that the advantages of wetland life often outweighed the disadvantages for our distant ancestors.[18] There were settlements at Glastonbury and Meare in the Iron Age whose economies were built around wetland resources, without resorting to the need to drain them to make firm land for cultivation. Michael Williams's work on the Somerset Levels and Darby's multiple volumes on the Fens provide ample evidence of how wetlands were a focus of settlement, detailing both the pre- and post-drainage histories of the areas.[19]

Prior to the large-scale modern drainage projects that changed the face of English wetlands, the resources they supported were so valuable that strict rules evolved about their use. These could become hotly disputed. In mid-1300s Somerset, disputes escalated to sabotage, arson and threats of excommunication and in the Lincolnshire Fens there was an arbitration court, the Court of Sewers, which kept track of who had rights to what and dealt with the frequent and varied transgressions.[20]

---

18 B Coles and J Coles, *People of the wetlands: Bogs, bodies and lake dwellers; a world survey*, Thames and Hudson, London, 1989.
19 M Williams, *The draining of the Somerset Levels*, Cambridge University Press, Cambridge, 1970, p. 17; HC Darby, *The changing fenland*, Cambridge University Press, Cambridge, 1983; HC Darby, *The draining of the Fens*, Cambridge University Press, Cambridge, 1956.
20 Williams, *Draining of the Somerset Levels*, p. 28.

When the enclosure and draining movement gained pace from around the 1500s, it was these wetland areas rich in resources that were the most hotly contested. There were repeated sabotage attempts of newly constructed drainage channels in the Fens by dispossessed peasants. As Mingay noted:

> It is clear that when large areas of widely used common or fenland, marsh, moor or other extensive wastes were threatened by enclosure, the local inhabitants were very likely to protest … The enclosures were regarded as depriving the poor of what had always been theirs, and claims to exclusive ownership by wealthy landowners were rejected as unjust and treated with scorn.

Despite protests, in some counties as much as 50 per cent of the available land, including wetlands, was enclosed.[21]

In England, there was a strong relationship between class and the kind of rural activity pursued or proposed. In Victoria, this class divide tended to be reversed. Generally in England, it was working-class labourers whose subsistence and seasonal lifestyle depended upon the wetlands, while educated, wealthy and often titled men with an interest in agricultural improvement advocated drainage, frequently employing a suite of negative imagery to bolster their case against the traditional users.[22] This use of moral geography will be discussed more fully in a latter section of this chapter. In Victoria, however, it was the wealthy pastoralist who strove to retain wetlands undrained, while poorer farmers agitated for drainage.[23] As Harris notes, mobility was an important strategy for pastoralists attempting to live with the variability of the climate. Large runs with a variety of water sources that they could circulate stock around increased their chances for survival and profit.[24] This mobility is abundantly clear in Patrick Coady Buckley's diary of his life as a pastoralist, covering his

---

21 GE Mingay, *Parliamentary enclosure in England: An introduction to its causes, incidence and impact 1750–1850*, Addison Wesley and Longman, London, 1997, p. 133.

22   F Willmoth, 'Dugdale's "History of Imbanking and Drayning": A Royalist Antiquarian in the 1630s', *Historical Research*, vol. 71, no. 176, 1998, pp. 281–302. For an account of the men involved in the early nineteenth-century innovations such as James Smith and Josiah Parkes, see GE Fussell, 'The dawn of high farming in England: Land reclamation in early Victorian days', *Agricultural History*, vol. 22, April 1948, pp. 83–95. Engineers, mill managers, bankers and lords are some of the occupations of the different men discussed in this article.

23 To the best of my knowledge, no publication has observed this contrast. It is my observation.

24   E Harris, 'Development and damage: Water and landscape evolution in Victoria, Australia', *Landscape Research*, vol. 31, no. 2, 2006, p. 171. doi.org/10.1080/01426390600638687.

arrival in Gippsland from the Monaro in 1844 up to his death in 1873. He was constantly out rounding up and shifting cattle from one place to another in search of pasture and water.

Many Gippslanders followed in this tradition of utilising wetlands. The diversity of uses is impressive. Some towns established grazing commons on low-lying lands, such as the one in Sale, which has survived.[25]

Attempts to privatise and subdivide the common were regularly opposed.[26] Other commons were proclaimed in Bairnsdale and Bundalaguah, to name only a few.

Figure 6.1: Sale Common, now a wildlife refuge.
Source: Author.

---

25 Letter from CJ Tyers, dated 9 April 1856, referring to a petition from residents of Sale to establish a common between the river, the punt and Flooding Creek.
26 For instances of debate about the use of Sale and other commons, see *GT*, 19 December 1874, 6 May 1876 and 23 May 1876; *GM*, 18 and 25 January 1877; *GT*, 11 October 1876, 25 and 30 November 1881, 18 August 1886 and 10 August 1891. See also *GT*, 16 November 1881, for attempts by Longford residents to prevent part of the morass (acting like a common) around Long Waterhole from being selected.

Settlers recorded in their diaries how they harvested various food and building resources from wetland areas. Annie Prout at Flaggy Creek near Bairnsdale collected cress while Broome, on the Nicholson River, would collect swan eggs from the morass.[27] Dow has discussed the use of reeds and ti-tree as building products, especially by settlers adjacent to Lakes Wellington, King and Victoria, as well as extensive recreational hunting by holiday-makers and residents.[28]

Broome also made an interesting reference to using mud from a morass to amend soil on his property abutting the Nicholson River. Clay or mud is an important way of building up organic matter, which in sandier soils assists in retaining water in the root zone.[29] The Sale Council in 1874 debated the establishment of a loam reserve, which indicates that many more settlers used loam for property improvement reasons.[30] Thus, Gippsland's colonial settlers continued to use wetland resources in ways very similar to their English ancestors.

---

27   Diary of Annie Prout, 15 November 1885, SLV, MS 12306, box 3054/5; Diary of Charles Alfred [Alf] Broome, 25 June 1881, SLV, MS 10774, box 1542.
28   C Dow, 'Tantungalung Country: An environmental history of the Gippsland Lakes', PhD thesis, Monash University, Melbourne, 2004, ch. 5. See also M Watson, 'William Odell Raymond', *Gippsland Heritage Journal*, vol. 2, no. 2, 1987, p. 35, for Raymond's bricks being made from clay from Lake Wellington; Diary of Patrick Coady Buckley, 19 November 1856, CGS, 2806, using ti-tree to make a crossing place across to an island in Lake Reeve; Diary of Alf Broome, between 9 and 14 July 1883; *East Gippsland Historical Society Newsletter*, vol. 1, no. 11, Colin McLaren who lived on Rotumah Island in the late 1880s had a jetty built of ti-tree. Extracts from the Barton Family diary in *East Gippsland Historical Society Newsletter*, vol. 3, no. 1, March 1985, pp. 5–7, discuss the family using ti-tree to make their camp 'dining room' and for the framework of their house. Indigenous people continued to use wetland resources throughout the period, e.g. M Drysdale, 'Survival of culture: Alice Thorpe and her basket', *Gippsland Heritage Journal*, no. 15, December 1993, where Alice tells that green *Lepidosperma* was used for bodies of baskets and red grass *Carex tereticaulis* was used to weave patterns in. For examples of hunting, see Diary of Aleck McMillan, 10 October 1872, 20 November 1872 and 26 December 1872, SLV MS 12545; Francis Barfus, 'A visit to the mission station Ramahyuck, at Lake Wellington Gippsland (Victoria), January 1882', [manuscript], SLV MS 12645, p. 16; R Jones, 'An ideal holiday experience: Guesthouses of the Gippsland Lakes', *Gippsland Heritage Journal*, no. 17, pp. 25–30; 'My first and I trust my last New Year's Day at Omeo', D LaTrobe Leopold from the 1863 diary of WH Foster, police magistrate and goldfields warden, *Gippsland Heritage Journal*, no. 8, 1990, pp. 23–5.
29   Diary of Alf Broome, 14 June 1899, 'carted some mud from the morass for cultivation ground'.
30   *GT*, 19 December 1874. 'Cr Guthridge could not understand about the loam reserve. Loam was intended to be used for building purposes, and it was never contemplated that the reserves should be stripped by persons wanting the earth to make up their gardens or apply it in other ways to the improvement of their properties.'

Figure 6.2: Shooting swans by moonlight, first published in *The Australasian Sketcher with Pen and Pencil*, 22 May 1880.
Source: Pictures Collection, State Library Victoria, Accession no. A/S22/05/80/101.

The quality of stillness contributes to one of the wetland's key ecological functions. Previous chapters discussed how rainfall transports organic and inorganic material, which is then washed into rivers and swept downstream. As water slows down, this sediment drops out. Wetlands therefore perform a crucial role in maintaining water quality. Wetlands are also sites of significant geochemical processes, including the filtering of elements such as iron and manganese, and the cycling of sulphur, carbon, nitrogen and phosphorus.[31] They are often compared to the kidneys in the human body, which filter and purify the blood of waste products. Their capacity to retain water, nutrients and silt creates one of the highest values of wetlands. They are drought refuges. In highly modified landscapes, they act as places of retreat.

31   Pers. comm., Sara Beavis, 30 September 2011.

Figure 6.3: Macalister Swamp, on the outskirts of Maffra.
Source: Author.

Squatters regularly used wetlands for summer grazing of cattle. In 1897, a drought year before the massive bushfires, the *Gippsland Times* reported that 'the Sale Common, small in area as it is, has furnished evidence of its fattening abilities when the grass in the paddocks of the surrounding district is burnt up'.[32] When grass was scarce after the fires, surviving stock were sent to graze on the unburnt portions of Moe Swamp.[33]

---

32 *GT*, 16 December 1897. See also as examples, Diary of Patrick Coady Buckley, 25 January 1858, 'Collecting [stock] on my run about Fidler's Flat and Farrell's swamp'; *GM*, 17 March 1877, editorial: 'The weather has recently caused grave fears to be entertained as regards the future. We have had a long continuance of dry and hot weather which has had the effect of scorching the grass in all places except swampy land, and the plains of Gippsland, so green and refreshing to the gaze, present now a desert like appearance.'

33 *Coach News*, vol. 3, no. 3, March 1976, pp. 6–7. It should be noted that this pattern of using the drier lands of swamps was also practised in the United States, see H Prince, 'A marshland chronicle, 1830–1960: From artificial drainage to outdoor recreation in Wisconsin', *Journal of Historical Geography*, vol. 21, 1995, pp. 3–22. doi.org/10.1016/0305-7488(95)90003-9. Wisconsin farmers survived in drought years through sowing wheat on drier wetlands, p. 6. DC Smith, 'Salt marshes as a factor in the agriculture of north eastern America', *Agricultural History*, vol. 63, no. 2, Spring 1989, pp. 270–94, for uses of coastal swamps in the US.

Morasses and swamps also acted as places of shelter and refuge for other animals. The Gippsland Lakes are internationally recognised as habitat for waterbirds. The waterbirds of the lakes helped feed settlers during leaner times and there are many reports of hunting as a pastime. Many of the hotels dotting the lakes in later years offered hunting equipment to guests.[34] These hunting parties were capitalising on the still and protected qualities of wetlands, which provide safe nesting spots for the many different types of birds that live on or visit the Lakes.

Stillness is also a precondition for the growth of certain bacteria. These can affect ecological processes and have a negative impact on human health. Dow has pointed out one nineteenth-century occurrence that sounds suspiciously like an algal bloom:

> Fisherman Jock Carstairs described what must have been an algal bloom in the 1880s. After a few dry years, he wrote, the entrance became shallow and the 'lake grass' and weed rotted. 'The water up the lakes became very stagnant, and the lakes became full of rotten green sediment. The water turned green like a 'green field', 'tonnes of fish' died and fisherman Charlie O'Neill, who fell in and swallowed some water, died.[35]

This is the only evidence to date of nineteenth-century toxic blooms in the lakes. Stillness in the water column is needed for bacterial pollution and algal blooms to occur. The population explodes, then collapses, leaving a mass of stinking and decomposing weed. They also require an excess of nutrients, which is why serious algal bloom problems in highly developed catchments occur mainly in the decades after the introduction of artificial fertilisers.[36]

For the nineteenth century, evidence abounds of complaints about foul-smelling waters in towns. The 'pong factor' is important in understanding responses to wetlands. The bad smells associated with bogs and marshes were taken as a sign of unwholesomeness, and in some English localities

---

34 For example, *East Gippsland Historical Society Newsletter*, vol. 1 no. 3, letter from Frederick Gray to Mr Lewis, dated 4 August 1855, describing his life at Lindenow, 'there are plenty of wild ducks and pigeons, one or two of us go shooting if we have nothing to do just a little amusement'.
35 Dow, 'Tantungalung Country', p. 246.
36 Pers. comm., Sara Beavis, 30 September 2011. Beavis has advised that blooms can occur naturally in catchments where there is basalt parent material, and it is the delivery of sediments with adsorbed phosphorus that provide the source nutrients for algal blooms to develop.

were associated with supernatural spirits.[37] This folk belief reinforced wetlands' bad reputation. The generally inadequate nature of stormwater and sewerage disposal in all the settlements in the GLC meant that most rubbish ended up in the nearest low-lying swamp. For example, Mirams of Sale pleaded for some of the water liberated by the artesian well to be used to cleanse the gutters of York St, described as 'the foulest open sewer in Sale'.[38] Settlers generally neglected to think about where the contents of the gutters ended up. This excess of nutrients was far more than what the natural system could possibly decompose, and so Gippslanders reinforced their learnt attitudes about wetlands being dangerous by their own behaviour. They then followed this with regular bouts of lobbying to drain the, by now, stinking wetland, or planting eucalypts to freshen the air.[39]

Places with still waters also have a dark side, being associated with murder, militarism, destruction, punishment and criminal behaviour. Whether or not this is a good thing depends on whether you are the hunter or the hunted.

For the hunted, the morasses of Gippsland were particularly important as places of hiding. Stock regularly escaped or strayed into morasses. On his journey into Gippsland, Buckley shot a cow that had gotten stuck in Clifton Morass.[40] John King had trouble with five bullocks who escaped their handlers by bolting into the morass.[41] In 1882, Mr Jenner's

---

37 Marsh gas is produced by anaerobic decomposition of organic matter in swamps and may contain any of several odoriferous gases, but its principal component is usually methane, which is odourless and flammable. It is lighter than air, and floats in a pale, eerie haze until blown away by a strong wind. It also burns with a mysterious blue flame when ignited, thus accounting for popular belief that such places are the province of ghosts and evil spirits. Definition of 'marsh gas', JM Last (ed.), *A Dictionary of Public Health*, online, Oxford University Press, 2007, www.oxfordreference.com/views/ENTRY.html?subview=Main&entry=t235.e2711, accessed 23 July 2008.
38 *GT*, 3 September 1880. For other examples or urban water pollution, see *GM*, 13 March 1877, Walhalla correspondent reporting a small flood that 'washed away a good deal of accumulated tailings'; *GT*, 20 November 1876, Sale Council meeting report of a debate on appropriate location for a bonemill, morass site approved; *GT*, 25 December 1876, 'Complaints are still rife at Traralgon respecting the sanatory arrangements of that township'; *GT*, 11 November 1881, 'Dr McDonald drew attention the existence of stagnant water under some houses in Raymond street, in one instance to the depth of 12 inches. When the hot weather sets in these stagnant pools are likely to be productive of much mischief'; *GT*, 28 May 1896, 'Since the village settlers have been established along Flooding Creek there have been complaints of rubbish and filthy being deposited in the bed of the creek, which in some classes is the sole supply of drinking water for the settlers'.
39 For example, *GT*, 18 May 1896, letter about how to deal with Lake Guthridge to make it healthy by planting eucalypts and willows. Eucalypts in particular were thought to be effective air cleansers. *GT*, 3 September 1870, for a discussion about how to deal with Lake Guthridge.
40 Diary of Patrick Coady Buckley, 4 January 1844.
41 King Station Day Book, 30 October 1844, SLV, MS 11396, MSB 404.

horse was grazing in a swamp when it took fright. In its flight, it staked itself and died. Mr Glassford's prize bull, called Royal Frederick and worth £300, slipped into a waterhole while trying to drink and had to be towed to the bridge half a mile downstream before he could be got out.[42] Buckley also regularly recorded either retrieving his own stock from morasses, or assisting neighbours or business partners to do the same. For anyone involved in pastoralism, keeping stock out of the morasses in good times was a regular part of working life.[43] The presence of morasses in the landscape was a central feature to their life and a matter of strategic advantage during drought.

Settlers in later years recorded how they used purposely created pools of still water to trap and drown native animals. In one family, the first attempt at killing wallabies in hand-dug water traps actually caught the family cat.[44] The Rosedale correspondent wrote in 1874:

> The caterpillar has invaded the crops of some of the farmers in the Rosedale district. The crop in many instances is too far on to be much injured; in others trenches and pit falls have been constructed to snare unwanted visitors.[45]

The use of trenches must have been effective because in 1882, the *Maffra Spectator* recorded that settlers were digging trenches to drown the caterpillars that were attacking crops.[46]

Dow devotes a whole chapter to the relationship between white squatters, the morasses and colonial violence. All the documented massacres that took place were in wetland areas. The infamous Campbell incident, where John Campbell fired a cannon loaded with shrapnel at the advancing Kurnai, involved a morass on Campbell's Glencoe Station.[47] Buckley records in his diary several punitive expeditions, many of which involved attempting to track Aboriginals through morasses surrounding the lakes,

---

42 Both from the *Maffra Spectator*, 9 March 1882.
43 Diary of Patrick Coady Buckley, 4 December 1844, 6 December 1844, 9 and 10 February 1848; extracts of Charles Macleod's diary from 1876 in Macleod, *From Bernisdale to Bairnsdale*, p. 75; Diary of Duncan Johnston, 21 November 1882. 'Started with Evans and Gove to assist them to the Tambo, two horses out of Cooper's paddock when we got there, Joe and I went to look for them, I found tracks. Tracked them to the morass could not find them. Heavy rain in the evening and night.'
44 J Adams, *So tall the trees: A centenary history of the southern districts of the Narracan Shire*, Narracan Shire Council, Narracan, 1978, p. 20.
45 *GT*, 22 December 1874.
46 *Maffra Spectator*, 14 December 1882.
47 P Morgan, 'The Campbell incident', *Victorian Historical Journal*, vol. 68, no. 1, April 1997, p. 22. According to Morgan, an unspecified number of blacks were killed.

for example between 8 and 11 April 1844. He and his party found it very hard going. His entry for 10 April stated: 'If it had not been for this water I do not think Marshall would have got out of the morass as it was very difficult to travel through'.[48] He was even more explicit in his entry of 10 April 1847:

> Sowed some wheat then pursued some blacks who came to kill my cattle. With William Scott and Dan Bloore, after a good deal of tracking and wading through water we found their camp. They saw us before we got near enough to shoot any of them, however we managed to get nearly all their spears. They were camped on an island in Lake Reeve.

King recorded hunting Aboriginal people around the lakes in mid-September 1845.[49] Colin McLaren, one of Angus McMillan's workers, kept a skull with a 'suggestive' hole in it.[50]

There are also references to the military or punishment uses of wetlands in a more metaphoric sense. On 21 January 1888, the *Morwell Advertiser* told its readers of a dispute between the self-dubbed 'holly boys' and the Mechanic's Institute:

> Closely copying the example of the Russians in dealing with the Bulgarian Prince, the Secretary, who stands in the way of the further advancement of the 'new church' is to be kidnapped, and placed on an island in the Moe Swamp and there detained till the election is over.[51]

Clearly, factional feeling was running high in Morwell that summer. Suspected arsonists in 1898 were threatened publicly in the papers of being drowned.[52] This suggests that such criminal behaviour as kidnapping and drowning were associated with wetlands. Second, Sale residents during the Franco-Prussian War were wont to joke that they would be perfectly safe from any potential invasion, given that they were protected by the aptly named Gluepot to the west and Punt Lane to the south, both

---

48   Marshall was suffering from dehydration, and Buckley was forced to leave him alone for a period while he searched for water to bring back to him. Diary of Patrick Coady Buckley, 8–11 April 1844.
49   King Station Day Book, 11–13 September 1845.
50   *East Gippsland Historical Society*, vol. 1 no. 11.
51   *Morwell Advertiser*, 21 January 1888.
52   *GT*, 20 January 1898, Briagolong correspondent.

swamps notorious for causing grief to travellers.⁵³ This echoed knowledge of English military history, such as when King Alfred eluded capture by taking refuge in the swamps of southern England, knowing full well that his enemies would not find their way through the marshes. During the early years of the Roman invasion, rebel Britons behaved exactly as the Kurnai were doing, protecting their land in any way they knew how and seeking safety in marshes.⁵⁴

In summary, wetlands provided drought refuge, grazing fodder, building products, food, and waste processing. They were the scene of actual violence between settlers and the Kurnai, and imaginary violence between different groups of settlers.

What stillness hinders

By virtue of their still qualities, the swamps and morasses of Gippsland played an important, if underappreciated, part in the economy of colonial Gippsland. However, these advantages were not enough to outweigh what Gippslanders saw as the profound disadvantages of places of still water.

Gippslanders disliked wet and boggy areas, and reserved a special kind of loathing for 'stagnant' waters. In this, they were following a path well-trodden by their European ancestors, explored in depth by Giblett. He concluded, from a survey across literature, painting, philosophy, medicine, religion and psychology, that an early modern European could hardly have escaped the message that swamps were places of moral and physical danger.⁵⁵ The same perception is found among the GLC settlers, reworked through their own lens as the insecure colonisers of an unfamiliar and highly variable landscape.

---

53   *GT*, 28 May 1870, Letter from T Booth; *GT*, 5 November 1870, report of Mayor Guthridge's speech to the anniversary dinner of the Latrobe Oddfellow's Lodge.
54   Giblett, *Postmodern wetlands*, pp. 206–7. There was considerable concerns about national security at this time as a result of the various wars occurring in Europe. See *GT*, 23 March 1873, editor calling for compulsory military training in state schools to resist any invasion.
55   Giblett, *Postmodern wetlands*.

Figure 6.4: A part of MacLeod's Morass, Bairnsdale. The low featureless expanse of reeds made perfect hiding grounds, but would exacerbate the troubles of the genuinely lost.
Source: Author.

Generally, broad open lakes were regarded positively, as indeed Lakes Victoria, King and Wellington were. The popularity of summertime lake cruises attests to this. The very first description of the lakes provided by Angus McMillan emphasises their size and openness. In contrast, the lack of visibility in morasses caused much grief to settlers: McMillan was by no means as complimentary about the boggy fringes of his admired lakes. Clifton's Morass is named after the horse he nearly had to abandon there when they got stuck and lost. His descriptions of the surrounding country also place close attention on the closed or open nature of the vegetation, such as on 21 January when he noted that the Avon, where he would establish his run, has 'high banks' and flows through a 'fine country of fine open forest'.[56] Rev. Frances Hales 'got into forest and scrub and waterholes that I almost got bewildered' on his way to the Snake's Ridge run.[57]

---

56 TF Bride, *Letters from Victorian pioneers: a series of papers on the early occupation of the colony, the aborigines, etc. / addressed by Victorian pioneers to Charles Joseph La Trobe*, CE Sayers (ed.), Lloyd O'Neill for Currey O'Neill, Sth Yarra, Melbourne, 1983, pp. 205–6.
57 Journal of the Rev. Francis Hales (edited by Michael H Wilson), 5 May 1848, SLV, MS 12950, box 1716/14.

Figure 6.5: 'Cleared' land near Mirboo, Thomas Henry Armstrong Bishop, n.d. (c. 1894 - c. 1909).
Source: Pictures Collection, State Library Victoria, Accession no. H40967.

McMillan is not the only one to emphasise this closed/open dichotomy. Most of the histories that valorise the clearing efforts of selectors describe those efforts as precisely so valorous because the forest was dense, closed and trackless.[58]

There is no doubt that clearing was phenomenally hard, dangerous work.[59] This basic structural difference of the catchment's vegetation, between the dense forest and the lightly wooded central grasslands, had a substantial impact on the processes of settlement in South Gippsland:

---

58   The two best examples of valorising are, from my perspective, the following two early twentieth-century histories: H Copeland, *The path of progress: From the forests of yesterday to the homes of today*, Shire of Warragul, Warragul, 1934; and Committee of the South Gippsland Pioneer's Association, *The land of the lyre bird: A story of early settlement in the great forest of South Gippsland*, Shire of Korumburra for the South Gippsland Development League, Clayton, Vic., 1966.

59   For an example, diary of George Glen Auchterlonie, 31 August 1874, 'Went and began cutting the hazel scrub below the hut but met an accident about eleven o'clock, a piece of stick fell from the top of a spar I cut my cheek just below my eye the cut is horizontal and about an inch long and very deep'. Also 'Memories of the Early Settlement of Narracan' by Lucy Bell, in R Murray Savige, *History of the Savige family*, Frankston, the author, 1966, p. 115, 'Tom did a lot of his clearing single handed. Annie told later on, how she would hear his axe all the morning, then the crash of the forest giant and how anxiously she would await the sound of the axe and renewed chopping'. Lucy also remembered the death of a man while rolling logs.

> It is clear from the pattern of alienation … that every attempt was being made to select flat, open, lightly timbered, land with permanent water but well drained, unlike the southern lowlands along the coast to the west of Wilson's Prom, and ready access to market. Only the outstanding soils of the Brandy Creek district disrupted this pattern …[60]

Once the open flats were taken up, newcomers had no alternative but to select heavily forested land. While parts of these forests were admired for the magnificence of their trees and being picturesque and beautiful, few redeeming qualities could be found in the morasses.[61]

Wetlands combined both of the colonists' least favourite characteristics, being both closed and low. The copy of real estate agents illustrates this preference for high and dry land. McLennan described a lot at Shaving Point as having a 'most elevated and salubrious position', while a farm at Paynesville had 'an unprecedented situation commanding a magnificent view of Jones Bay and the distant mountains. The healthy situation and other advantages of this farm can only be realised by personal inspection'.[62] Descriptions published in various papers on walking trips in the mountains also linked height with health, and legal disputes about high land, lower lands and drainage also build up this picture.[63] In his work on Gippsland, Don Watson noted that 'it is the habit of explorers to compose their finest thoughts on hill tops, or at least to locate them there in their memoirs. Gullies do not encourage the sense of man's domain, or favour a friendly communion with God'.[64]

---

60   Legg, 'Arcadia or abandonment', p. 180.
61   See T Griffiths, *Forests of ash: An environmental history*, Cambridge University Press, Port Melbourne, 2001.
62   Both in *Bairnsdale Advertiser*, 14 February 1882. See also, *GT*, 5 October 1885, McLean and Co. auction notice for Mount View Estate at Tralargon; *GT*, 15 July 1871, Victoria Hotel in Bourke St West Melbourne, advertising its 'elevated' position; *GT*, 11 March 1873, Hawthorn Grammar School in Melbourne advertising the school as being in the healthiest suburb in Melbourne and 'elevated and pleasant'.
63   *GT*, 27 February 1891, 'An Australian Midsummer Holiday by a Country Parson'; *GT*, 25 May 1872, Land Board proceedings noting a dispute between Emson and Creaton over who had the high ground; *GT*, 13 September 1873, Legal Intelligence, Traill vs Rosedale Shire; *GT*, 9 May 1876, reprinting a letter from Sale Mayor to the District Surveyor opposing selection of high lands on the Sale Common.
64   D Watson, *Caledonia Australis: Scottish highlanders on the frontier of Australia*, Random House, Milsons Point, NSW, 1997, p. 144.

The high/low dichotomy has been previously discussed in Chapter 4 in relation to floods, and the moral symbology of popular music and hymns that associated goodness with height, light and the sun. This high and low contrast is a regular motif in Christianity. The geographic structure of hell, purgatory and heaven in Dante's *Divine Comedy* is but one example of the association of low places with undesirable emotions. It is no accident that the Fifth Circle of Hell is a 'dreary swampland, vaporous and malignant' abutting the River Styx.[65] The first epigraph referencing Dante shows that Victorian popular imagery was influenced by classical depictions of low muddy and undesirable places.[66]

Low-lying, still waters were further reviled because of concerns that they could produce ill-health in humans. Before the acceptance of microbes, stagnant and polluted water was widely associated with the concept of miasma. While the actual mechanism by which disease was contracted was mistaken, the stink of stagnant water was the best evidence a person had to avoid illness. According to the *Water Atlas*, one litre of contaminated water infects the surrounding 10 litres, very similar to the theory of miasma so beloved by nineteenth-century scientists and reformers.[67] So colonial Gippslanders were right to be worried, both in their own terms and on the evidence of modern science. Mortality statistics indicated that one quarter of Victorians could expect to die from miasmatic disease, including smallpox, scarlatina, diphtheria, typhus, cholera, influenza, dysentery and ague.[68] None of these would be a pleasant way to die. Schmitt's thesis on the Gippsland Hospital and approaches to typhoid, diphtheria and tuberculosis gives a picture of a community worried about the environmental contributors to disease but without the resources to manage them.[69] Frederick Hagenauer, in charge of the Ramahyuck Aboriginal Mission, expressed the kind of apprehension about illness that would have been common:

---

65 Dante, *Inferno*, www.online-literature.com/dante/Inferno, accessed 6 May 2008. Longfellow edition, canto 7, lines 103–30.
66 Giblett also discusses the use of swamp imagery by popular Victorian writers like Charles Dickens. For example, Pip from *Great Expectations* was born in a swamp.
67 R Clarke and J King, *The water atlas: A unique visual analysis of the world's most critical resource*, The New Press, New York and London, 2004, p. 11.
68 This figure is derived from Statistics of the Colony of Victoria for the year 1864, vital Statistics etc, in Victoria, Legislative Assembly, *Votes and Proceedings of the Legislative Assembly and Papers*, First Session 1866. It gave deaths from all sources from independence up to 1864. The yearly average of deaths from miasmatic causes varied from a low of 22 per cent to a high of 37 per cent.
69 DJ Schmitt, 'Provincial pestilence: A study of the impact and response to typhoid, tuberculosis and diphtheria in the Gippsland Hospital catchment, 1866–1931', MA thesis, Monash University, March 1993, p. 62 for the statistics and p. 70 for the comment about resourcing.

Another reason for gratitude has been, and still is, that whilst almost every part of Victoria and especially Gippsland has been visited by illness of different kinds, no disease has come near the station and all are enjoying good health. This may be partly accounted for by the healthy situation of the place, but perhaps also to the protection and mercy of God whose <u>loving kindness has no end</u>.[70]

The phrase 'healthy situation' is indicative of the belief that the environment impacts upon health in important ways. To a modern reader this sounds quaint, used as we are to the powerful effects of antibiotics. David Servan-Schreiber suggests that the twentieth-century revolution in healthcare obliterated other aspects of healthcare that had been important in medical practice for centuries, such as considerations of diet, location and the doctor/patient relationship.[71] The consensus was that a healthy place was high, dry and well ventilated. The local and regional papers in Gippsland regularly reported on polluted water and their attendant health problems.[72] This example from the *Morwell Advertiser* of 7 February 1890 is typical:

> In a little town like Boolarra (where, by the way, Health officer visits are like Angels, few and far between) a stagnant pool is allowed to remain unattended to, and the inhabitants are compelled to 'breathe contagion to the world' as Hamlet puts it. Directly behind Clarke's Hall, a place of public amusement, this dirty pond is allowed to remain. What wonder that Mr. Biles has had the misfortune to have to send his daughter to Melbourne, suffering from typhoid in its most malignant form.

There is a small counter strand in European history that has valued stillness. For example, the monks that settled at Glastonbury in the fourth century were attracted to the wetlands there because they provided a place of retreat and contemplation. Well-known European myths and legends engage the use of a reflective pool as part of the storyline, with the most

---

70 Underlining in original. Annual Report, 1874, Letterbooks of FA Hagenauer, vol. 2, NLA, MS 3343.
71 D Servan-Schreiber, *Anti-cancer: A way of life*, Scribe Publications, Carlton, Vic., 2008, p. 56.
72 *GT*, 13 September 1870, report of the Avon Shire Council meeting of Sept 12; *Castner's Rural Australian*, March 1876, p. 17; *GT*, 1 January 1876, Bairnsdale correspondent; *GT*, 11 September 1876, Toongabbie correspondent; *GT*, 11 October 1876, Letter from 'Public Improvements'; *GT*, 11 November 1881, health report to Sale Council; *GDN*, 4 April 1890, Bairnsdale Inspector of Nuisances; *GT*, 25 February 1887; *GT*, 9 October 1891, Sale Council health officer reports; *GT*, 5 March 1896, Sale Council correspondence from F Chaafe; *GT*, 13 January 1898, report by Dr Chapman to Avon Shire Council.

famous being the tale of Narcissus and Echo. The Book of Common Prayer has the famous saying 'be still and know that I am God', while in the Bible, God is again described as 'a small still voice'.[73] Yet, Gippsland's colonisers were anything but still. As Chapter 5 demonstrated, their wholehearted faith was in movement and progress.

The overwhelming impression from diaries and day books is one of constant movement and activity.[74] Settlers were constantly on the move, getting their goods flowing into the colonial economy and maintaining their social relationships. Gippsland's muddy roads presented a major impediment. Writing about Warragul, Francis Barfus said:

> After rain the roads are practically impassable. A friend of mine, a German clergyman in Melbourne gave up his intention of visiting several friends who are settled in this district on one occasion when he beheld the roads leading to the forest deep in mud and mire. He probably remembered the words of Schuller 'Let man not be tempted by the Gods nor wish to behold what they have kindly clothed in darkness and terror.'[75]

Their preachers were equally as mobile. Rev. CJ Chambers, reflecting on his time as the vicar at Yarragon wrote:

> Today, theorists, learning by experience have reclaimed and transformed the water [of Moe Swamp] into a glorious blessing. No need now for the Minister of religion to seek the privilege of keeping his appointments by riding along the railway track.[76]

Francis Hales's extended trip through Gippsland made him saddle sore and bone weary. Later clergymen with smaller areas still travelled regularly, alternating services throughout their district.[77]

---

73 Book of Common Prayer; King James Bible, 1 Kings, 19:12.
74 S Legg, 'Farm abandonment in South Gippsland's Strzelecki Ranges, 1870–1925: Challenge or tragedy?', *Gippsland Heritage Journal*, vol. 1, no. 1, 1986, p. 17, referring to 'enormous' levels of geographic mobility.
75 Francis Barfus, A Visit to the Mission Station Ramahyuck, SLV, MS 12645, box 3486/3, p. 4.
76 'A short history of the Church of England at Moe', *Coach News*, vol. 12, no. 3, March 1985.
77 For example, Diary of Rev. Thomas Moorhouse quoted in *Traralgon and District Historical Society Bi Monthly Bulletin*, vol. 1, no. 1, March 1970. Extract of his diary from Sunday 17 August 1879: 'Service this morning was at 10 o'clock. The weather was very rough. It rained nearly all the morning but we had a good attendance not withstanding. My horse being ready I rode off immediately to Flynn's Creek. Called at Mr Sykes for a cup of tea, and he rode on with me to the Church The storm kept people away, and only four of us were present … I came on to Traralgon through mud and water to the Presbyterian church, being very few present'.

The King family papers are a testimony to the constant travel that went with the squatting occupation. In 1849, an economically challenging year, King and his men were regularly travelling to Port Albert taking skins, tallow and live animals to port and taking delivery of supplies. Buckley's life is a constant round of movement, shifting stock on his run or moving stock across the entire region. Even selectors, who signed on for a settled agricultural life, never stayed still. This was especially the case for men who had to take supplementary work cutting bark, or transporting goods, until they could get a return from their farms. Men of other professions were also on the move, especially government staff like surveyors or land commissioners. The only pool of stillness in all this relentless activity was the Sabbath, regularly observed by the majority.

It is possible to link this constant movement with the dislike for swamps at both a practical and metaphorical level.[78] Spurning stillness outside suggests spurned stillness inside. Slowing down and being still creates space for questioning, for peering into the reflective mirror of the catchment's waters. But Gippslanders show little evidence of self reflection about the process of colonising a foreign country. As the definition of ecological perception in Chapter 1 showed, the capacity to reflect is critical to understanding ecological processes. Having committed to the colonial endeavour, it was important to retain the certainty that they had 'permission' from God to subdue and remake the catchment and civilise the natives. Of all the farms created in the catchment, perhaps the Ramahyuck mission holds up the most revealing mirror. Just as Gippslanders tried to channel and tame the waters of the catchment and make them conform to preset ideas of usefulness, regularity and predictability, so it was with the surviving Kurnai. In his annual report in 1877, Rev. Hagenauer took pride in quoting the findings of the royal commission, which reported:

> Everything in and about the Ramahyuck Mission was found in a faultless state of order. The children were cleanly and well clad, and many of them educated up to a standard that would compare favourably with schools frequented by white children. The adults were also found to have acquired industrious and well regulated habits. Not only are the Blacks on this station well cared for and ably instructed in those arts that pertain to industrious rural life, but they are taught to be contented and happy.[79]

---

78   For an example of metaphorical stagnation, see *GT*, 30 December 1895, editorial: 'The true cause of stagnation which has hung over this portion of Gippsland during the past year or two is that the people drifted into easy and idle ways'.
79   Letter from Rev. F Hagenauer to Mr Hamilton, 5 October 1877, Letterbooks of FA Hagenauer, vol. 2.

FOLLOWING THE WATER

The Ramahyuck mission, with its neat cottages and grid layout, was the antithesis of the supposedly random and meaningless life the Kurnai had pursued amongst the wetlands.

Given that activity and movement was the unspoken creed of settlers, it is no wonder they reserved such loathing for the morasses, swamps and bogs that caused prolonged trouble for travellers. Copeland said that 'the bush tracks would bog a duck'.[80] Nearly every reminiscence of pioneering days contains an account about mud and morasses and the difficulty of travelling through them. 'Instead of trying to find a suitable mode of transport for the wetland,' Giblett writes, 'the western response has been largely the attempt to cross the marsh by building roads through it or around it or to canalise it in order to provide a means of communication through it.'[81]

Government officers discussed this 'problem' in their reports. Skene and Smythe wrote in 1874 that:

> For a distance of eight miles from the part just described [Melbourne to Brandy Creek] the road over which the rich chocolate soil showed marks everywhere of difficulties met and overcome by travellers in wet weather – holes where wheels had been buried to the axle, and deep ruts partly filled with dust.[82]

In the wetter western parts of the catchments, complaints about bad roads and poor drainage abound. Two examples are characteristic of this tone. In May 1897, the Sale Borough Council received a petition from 19 ratepayers, requesting repairs to culverts and a footpath to be gravelled 'so that their children could get to town dry-footed which is impossible during the greater part of the year'. Councillor Coverdale spoke in support. Since the village settlers had fenced off their allotments, pedestrians were forced to keep to the road 'which was little better than a swamp'.[83] In 1886 the *Morwell Advertiser* commented on the state of the road outside Mr Date's blacksmith shop:

---

80   Copeland, *Path of progress*, p. 332.
81   Giblett, *Postmodern wetlands*, p. 18.
82   *Report on the physical character and resources of Gippsland*, by the Surveyor General and the Secretary for Mines, with a Map and Geological Section, 1874, John Ferres, Govt Printer, p. 19.
83   *GT*, 13 May 1897. For an earlier example of a similar problem of fencing and bogs, see *GT*, 11 October 1876, Rosedale correspondent: 'People are never satisfied. Complaints were made some time since of a large landholder in this vicinity having unfenced roads running through his property, and now that he has commenced to fence the roads off, the gentle public are wroth because they cant get comfortably along them. Where, for instance, a swampy place existed, and it was possible until lately to make a short detour and avoid the difficulty, fences compel the traveller to "keep to the road" and suffer the bogging or whatever else may be attendant thereon.'

It would be difficult to find a greater quagmire in all the country around. What is extremely nice about it is that the public generally have no idea of the spongy nature of the 'earthworks' and so they are taken by surprise when they reach the centre and realize with the utmost felicity that they are gradually sinking a few feet below the surface. We have gone through the experience and enjoyed it exceedingly.[84]

Gippsland's muddy roads stood in the way of settlers' desire for firm and reliable roads of communication and commerce. Mud was also a domestic nightmare for women, trying to keep clothes and houses clean. Maggie Lamb wrote in her diary with satisfaction how she was able to get all the family out of the house, so that she was 'able to scrub out in safety as no one was about to keep walking in and out'.[85]

Behind these complaints was a deeper level of worry, one that reflects a metaphorical understanding of the problem. No one was immune from the mud. Schoolchildren, farmers carrying produce to markets, dairy farmers getting to the creamery, the shire engineer; all faced this obstacle to their daily work. Mud was nothing if not democratic! Mud interfered with the orderly, planned and defined round of life that colonial Gippslanders were trying to build. When your horse fell into a bog on the way to the railway to send butter to Melbourne, you not only faced inconvenience, and probably expense, it was also a reminder that you were not really the one in command. Mud destabilised more than the material body. It also destabilised one's ambitions and goals. The sticky, oozing, soft, unpredictable and unformed character of mud unsettled the settlers, especially those living in the heavily forested western portions of the catchment. Mud prevents a purposeful making of one's way. Instead, it produces floundering, an anathema to these progress-oriented pioneers.

One of the world's most famous texts, and one highly important to devoted Christians, is called *Pilgrim's Progress*. This text bequeathed the phrase 'slough of despond', which equates the physical floundering caused by mud with a symbolic floundering. The book is a parable for finding Christian faith in God. The slough of despond is the first obstacle that

---

84   *Morwell Advertiser*, 27 November 1886. Other examples of complaints about stagnant waters include *GT*, 23 August 1870, report of the Avon Shire Council meeting of Monday 22 August: correspondence from JM Clark complaining of stagnant water opposite his hotel, granted 1,000 yards of earth. See also fn. 59.
85   Diary of Maggie Lamb, Saturday 15 January 1910, CGS, 05317.

Christian meets after having abandoned his wife, children and home. He struggles to extricate himself from it because he is burdened down with sin. There are a couple of uses of the term 'slough' or the phrase 'slough of despond' which lend credence to this connection between mud and stagnation in nineteenth century Gippsland. Most often they were in connection with Punt Lane in Sale, and generally after a flood. The year 1870, already discussed in Chapter 5, yielded three references by the *Gippsland Times*.[86] One of these made a direct comparison to Bunyan's character, Christian:

> We have been told by a gentleman who has lately crossed from Longford to this town that, like Christian in the Slough of Despond, he walled for a time in the Punt Lane, being grievously bedaubed with dirt, and only succeeded in getting out of his difficulty by the help of a good Samaritan who was journeying by the same road. Anyone who from choice would live at Latrobe bridge must have extraordinary amphibious instincts.[87]

The phrase 'grievously bedaubed with dirt' is a direct quote from the text. One Gippslander recorded in his diary reading *Pilgrim's Progress* with his future wife. Less than a month later, he resolved to give up dancing because he believed it sinful, indicating that Christian's tale is taken seriously by this couple at least.[88] As a place in the landscape where water apparently ceases to move, swamps and morasses were both physical impediments to the flow of people, goods and stock, and they represented a fear of spiritual stagnation.

Swamps, bogs and drains could indeed be dangerous places. Without clear paths, it was easy to get lost. Children were particularly vulnerable. They were often sent out to find straying animals, and as feed was normally good in such places, this combination was too often lethal. Kate Robinson drowned in a waterhole after being chased by an angry goat, and a young boy (only identified in the paper as Jones), who was searching for a horse

---

86   *GT*, 28 May 1870, a horse got stuck in the 'slough' and while trying to extricate itself fell over the embankment. The report did not state whether the horse sustained injuries. *GT*, 12 July 1870, the Rosedale Road Board received a letter from Mr MacMahon who said his address was the 'Slough of Despond'. He was requesting drainage works, which were approved.
87   *GT*, 28 June 1870.
88   Diary of Alf Broome, 4 September 1887, for the reference to *Pilgrim's Progress*, and 4 August 1887 for the dancing. For another example of the use of the term slough of despond in relation to crime, see *GT*, 4 October 1861.

in the Cowarr Backwater, also drowned.[89] One of Catherine Currie's children, Katie, saved the family's black mare, Bess, when she got trapped in a waterhole.[90] Later, the drowning of Catherine's daughter ultimately sent her mother to the Yarra Bend Asylum.

Stock often met their deaths in the morasses. George Auchterlonie recorded in his diary how one of his bullocks had been so weakened by its struggle to extricate itself from the mud that he offered the skin to his neighbour 'for the trouble of killing him'. The fact that the backwaters and morasses acted as drought refuges, and that many farmers tried to increase their usefulness for grazing by sowing seed, contributed to this problem.[91]

Another reason why wetlands caused such ire is that they are changeable, with water levels fluctuating seasonally. This is particularly a problem when it comes to systems of property. Western systems of property assume a uniformity that is highly at odds with how the hydrological system actually works. Vileisis identifies this misfit between property and ecology as a driving theme in her work on wetland destruction in the United States.[92] Some Gippslanders recognised this, particularly those who opposed the cutting up and selling of Sale Common. In essence, they asked how was it possible to expect selectors to establish a farm based on the European style of agriculture, when at any time their land might change from dry to wet? In other words, how do you impose a fixed property and farming system on something that fluctuates? The extremely wet year of 1870 illuminated this problem. In one of the earlier floods of that year, the *Gippsland Times* commented:

> The flooded state of these localities for the past two or three days must have convinced all those who had a doubt on the subject that three fourths of the area are wholly unsuited to the purposes of settlement. People may select these lands – *waters* just now would be the true descriptive word – if they please …[93]

---

89 For Kate Robinson, see *GT*, 23 January 1869; for Jones see *GT*, 29 April 1870. See also *GFJ*, 2 August 1887; *GT*, 8 March 1897, Charles Smith, age 21 months, drowns in a drain opposite the Port Albert PO.
90 Diary of Catherine Curry, 8 October 1882.
91 Diary of George Glen Auchterlonie, 16 September 1871, 'Sowing grass in the back waters', and 21 July 1873.
92 A Vileisis, *Discovering the unknown landscape: A history of America's wetlands*, IslandPress, Washington DC, 1997, p. 5.
93 *GT*, 26 March 1870, emphasis in original.

When the next flood came, the paper wrote:

> Of course the Chinese gardeners in the north west backwater were obliged to leave their dwellings, which became surrounded, and many of the residents on the Common had to clear out – a forcible illustration of the utility of the common for agricultural or residential purposes. It is unlikely after this that the land will ever be coveted for purposes other than what it now fulfils – those of grazing.[94]

The colonial project had as one of its cores the extension of private property systems to the Australian landscape. As Sheryl Breen notes, 'from Plato to Locke and Marx, conceptions of property have been central to our visions of political liberty, equality and order'.[95] The idea of the possession of one's private and productive estate has been cherished for generations and was an especially strong motivator of the selection era.

The imposition of the Western ideal of property and ownership over the catchment caused a number of problems for selectors, because it attempted to place a uniform grid and uniform rules over a geographic space that was, manifestly, not uniform. And it was the areas of wetness where it was the most liable to change. Wetlands shrink in summer and expand in winter, while creeks and rivers change course. Floodplains flood. The residents of Bundalaguah understood this in 1870 when they opposed their common being thrown open for selection. It was a 'dangerous locality', not providing the kind of certainty needed for success and profit.[96]

---

94 *GT*, 19 April 1870. The topic of a definition of a morass arose again in *GT*, 7 May 1870, when the newspaper reported on a deputation made to the minister about which morass lands and which high lands were included in lands thrown open for selection.
95 SD Breen, 'Ecocentrism, weighted interests and property theory', in M Humphrey (ed.), *Political theory and the environment: A reassessment*, F Cass, London, 2001, pp. 36–51.
96 *GT*, 30 April 1870.

The final reason for the dislike of still waters was their negative impact on agriculture. Rich, productive lands meant dry lands.[97] The *Morwell Advertiser* could not have put it plainer than this:

> Fixed Facts in Agriculture: One: No lands can be preserved in a high state of fertility unless clover and grass are cultivated [text illegible]. Two: Deep ploughing greatly improves the productive powers of a variety of soils that is not wet. Three: Subsoiling sound lands, that, land that is not wet, is eminently conducive to increased production. Four: All wet land should be drained. Five: Draining of wet lands and marshes adds to their value by making them produce more and better crops and producing them earlier. Six: To manure or lime wet lands is to throw, manure, lime and labour away.[98]

This was only echoing sentiments expressed nearly two decades previously by the *Gippsland Times* on behalf of those objecting to the selection of morass lands. 'There can be neither wisdom nor kindness in encouraging agriculturalists to take up land for cultivation that is neither fit for ploughing and seed sowing or safe as a place of residence.'[99] At the same time, settlers knew from long experience in their home countries that drained swamps provided highly desirable soil for intensive horticultural enterprises. There is a good reason why the Rhine has lost 90 per cent of its floodplain, largely to agriculture.[100] One of the additional benefits of the creation of the permanent entrance to the Lakes was the transformation of morass lands abutting the Latrobe and Avon rivers into 'rich summer and autumn pasturages', thanks to a 21-inch drop in the level of Lake Wellington.[101]

In this section, I have discussed how Gippslanders found places of slow and still waters to be an impediment. While they had some distinct advantages and were regularly used, on balance Gippslanders found swamps, morasses and boggy land to be a great deal of trouble. There were multiple reasons for this intense dislike. Physically, they impeded movement and sight, and therefore increased danger to travellers. They could harbour vermin and criminals. They were thought to be unhealthy places of bad air and tainted water, endangering life and health.

---

97   *GT*, 18 January 1886, editorial on draining the Heart morass.
98   *Morwell Advertiser*, 5 February 1887.
99   *GT*, 23 April 1870.
100 M Cioc, *The Rhine: An eco-biography*, University of Washington Press, Seattle, 2002, p. 16.
101 *GT*, 13 August 1896.

Taken collectively, these physical characteristics were endowed with symbolic distaste. They were low in the landscape, symbolising the inability to see forward. Their ability to slow people down became symbolic for emotional, economic and spiritual stagnation. Finally, by being neither land nor water, they refuted the boundaries that colonial settlers attempted to impose on them.

Responding to mud

Civilisation, as we fondly imagine it, and drainage, the conscious art of alteration to the hydrological features of a place, are closely related.[102] The practice of drainage, to facilitate agriculture and the creation of urban settlements, was pivotal to the development of ancient civilisations.[103] Drainage is also usually regarded as a good thing, such as Nace's assertion that 'the Romans emerged from barbarism and achieved civilization under the influence of the Etruscans, who were masters of the arts of swamp drainage and irrigation'.[104] It is only in recent decades, under the influence of the fifth version of the hydrological cycle, when the cumulative environmental impacts of wetland loss has become more clear, that this assumption has been questioned.

---

102 Williams, *Draining of the Somerset Levels*, p. 3.
103 In the *Handbook of ancient water technology*, Wilson provides a brief summary of lake, marsh and land drainage up to the Roman period, while noting that pre-Roman drainage techniques are poorly studied. Drainage of lakes can be dated as early as the Bronze Age in the Lake Kopais Basin in central Greece, when an 11 km by 18 km lake was drained for agriculture. In the second century AD, Roman soldiers in North Africa were set to work to drain marshes to settle colonists and make money, p. 309. Numbers of private profit schemes of marsh drainage have been identified throughout Roman Europe, while those undertaken in England appear to have been state led. Land drainage was of a more medium scale and mostly appears to have been undertaken by individual owners. Works to improve drainage of vineyards and orchards have been identified during archaeological works and various agricultural writers of the time give detailed descriptions of different types. See Wilson in O Wikander (ed.), *Handbook of ancient water technology*, Technology and Change in History, vol. 2, Brill, Leiden, 2000, pp. 304–11.
104 R Nace, 'UNESCO History of Hydrology', in *General evolution of the concept of the hydrological cycle*, Three Centuries of Scientific Hydrology: Key papers submitted on the occasion of the celebration of the Tercentenary of Scientific Hydrology, 9–12 September 1974, UNESCO, Paris, 1974, p. 44; also JR Ravensdale, *Liable to floods: Village landscape on the edge of the Fens AD 450 to 1850*, Cambridge University Press, Cambridge, 1974, p. 1.

Emigrants to Gippsland from Europe or America were certainly familiar with drainage. In Europe, drainage of wetland areas increased from around the 1500s, with examples in many countries.[105] Virtually every country in Europe had at least one iconic drainage project, and many hundreds of smaller ones. While discussing draining at The Heart run near Sale, the *Gippsland Times* compared the catchment to Holland. Using drainage work in Lake Haarlem completed in the late 1850s as the comparison, drainage saw a 'dangerous lake' transformed into an ordered and productive landscape of orchards and gardens.[106]

Colonial Gippslanders wanted prosperity and progress, and transport, water supply and an expanding agricultural sector were the yardsticks of their progress. Practically, this meant firm, dry roads, well-made bridges and ground that would not rot their crops:

> We must make the lands more profitable than by their existence in their pristine covering, this is the age of science and advancement and we must progress with both, or forever remain in the same state we are now in. We well know we cannot all be agriculturalists, but at the same time we know that agriculture is the foundation of progressiveness.[107]

Above all they feared stagnation, and wetlands symbolised this fear. Their collective actions were largely about changing the hydrological cycle: to rechannel water where it spread over the landscape, to create permanent, moderate flow in defined channels wherever possible. In short, they drained and reclaimed. That drainage equalled improvement was an automatic assumption which colonial Gippslanders supported, and regularly put into practice.

---

105 Vileisis demonstrates clearly in *Discovering the unknown landscape*, her comprehensive history of wetland destruction in America from the Puritan era onwards, how drainage facilitated the expansion of settlers and intensified agriculture and settlement in waves. Ciriacono discusses parts of northern Italy, with a strong emphasis upon Venice, while Cioc and Blackbourn tackles the destruction of the floodplains of the Rhine: S Ciriacono, *Building on water: Venice, Holland and the construction of the European landscape in early modern times*, Berghahn Books, New York and Oxford, 2006; Cioc, *The Rhine: An eco-biography*; D Blackbourn, *The conquest of nature: Water, landscape and the making of modern Germany*, WW Norton and Co., New York and London, 2005. See also P van Dam, 'Sinking peat bogs: Environmental change in Holland 1350–1550', *Environmental History*, vol. 6, no. 1, 2001, pp. 32–46; D Speich, 'Draining the marshlands, disciplining the masses: The Linth Valley hydro engineering scheme (1807–1823) and the genesis of Swiss national unity', *Environment and History*, vol. 8, 2002, pp. 429–47. doi.org/10.3197/096734002129342729; T Glick, *Irrigation and society in Medieval Valencia*, Belknap Press of Harvard University Press, Cambridge, Mass., 1970. doi.org/10.4159/harvard.9780674281806; Darby, *The changing fenland*; S Halliday, *Water: A turbulent history*, Sutton Publishing, Phoenix Mill, Gloucestershire, 2004.
106 *GT*, 18 January 1886.
107 *GT*, 21 August 1861.

As demonstrated above, wetlands held a mixture of positive and negative qualities. Which of these prevailed for individuals was largely a function of their particular circumstances. Collectively in Gippsland, the negative qualities held sway. This was not necessarily the case for other parts of Victoria. The deliberations of the government's Professional Board on Swamp Reclamation in 1865 gives an illustration of this geographically modified approach.[108] The board reported in 1866, rejecting a number of applications in other regions of Victoria but approving all those in Gippsland.[109] The board based their rejection on climatic considerations. They concluded that the generally drier climate of the western and north-western regions of Victoria meant that the wetlands performed greater service to a larger number of people through their ability to supply water and moderate the climate around them:

> They arrest the rapid conduction of that water to the sea, and render less unequal the summer and winter discharges of the streams which are fed by these swamps … In nearly every case the general interests of the public would be more promoted by raising the levels of the water in these swamps by dams than by lowering those levels by drains.[110]

This public benefit was considered more significant than the benefit that would accrue to the aspiring lessees. Gippsland's higher rainfall thus worked in squatter's interests. Dow notes that the Gippsland applicants were primarily using the application process to both frustrate the selection process, and retain the valuable drought refuges under their own control.[111] They had no real intention of draining. William Pearson

---

108 Section 38 of the 1865 Land Act allowed people to lease swamps for a period of 21 years in order to facilitate drainage. The board was composed of Charles Ligar, Clement Hodgkinson and Robert Brough Smyth. Of the three, Hodgkinson had the most experience with water, having been involved with water supply for Melbourne, by being on the Board of Health and giving advice to Sir John Coode. HW Nunn, 'Hodgkinson, Clement (1818–1893)', *Australian Dictionary of Biography*, vol. 4, Melbourne University Press, Melbourne, 1972, pp. 403–4.
109 *Report of the Professional Board on Reclamation of Swamps*, in Votes and Proceedings of the Legislative Assembly and Papers, First Session, 1866.
110 *Report of the Professional Board on Reclamation of Swamps*, p. 4.
111 Reminiscences of William Blennerhasset, about the Bengworden and Meerlieu district from 1876. 'When the selectors took up land they curtailed the squatters' runs. They were often pushed into the sandy scrub. Surveyors were influenced to survey the land into the 3 cornered blocks so the cockies would have too much fencing to do. A junction of 6 roads at the Meerlieu school corner was one example.' *East Gippsland Historical Society Newsletter*, vol. 1 no. 11. For other examples of squatters attempting to frustrate the selection process, see V Wardy, *Beneath blue hills: A history of Mewburn Park, Tinamba and Riverslea*, Kapana Press, Bairnsdale, 1994, p. 7; Diary of Patrick Coady Buckley, 25 May 1869, 13 and 14 July 1871 and 1 October 1871, CGS, 2806.

of Kilmany Park became embroiled in a parliamentary enquiry over his actions in relation to the morass lands in the Wurruk district. Despite allegations of corruption and dummying, Pearson's wealth and social influence eventually held sway after four years. He retained the morass.[112]

For agriculturalists, a different attitude prevailed. Wet lands were undesirable for agriculture, and drainage was regularly discussed and undertaken. Like European nations, Gippslanders had their iconic projects, as well as many smaller ones. The two key projects that required massive investment by the state were the drainage of Koo-Wee-Rup Swamp and the Moe Swamp. Koo-Wee-Rup lies in Gippsland, but, as it is not part of the GLC, it is not considered here.[113]

The Moe Swamp area was universally decried for causing trouble to travellers. It covered the area from Yarragon to the banks of the Latrobe River. Adams described the Moe River as being a series of lagoons spreading up to 5 kilometres in width.[114] With its permanent pools, soft soils and thick vegetation of paperbark, ti-tree and sword grass, it is not hard to understand the special loathing that travellers reserved for it. In 1849 Tyers wrote, at length, to his superiors in Melbourne about the best way to spend his limited road-making funds:

> It seems to me that the Moe Swamp cannot be avoided except by crossing the Narracan Rivulet near the present bridge – that is in between the junction of the Moe with the Narracan and that of the Rivulet with the La Trobe (a distance perhaps of no less than a hundred yards) and that this [word illegible] is the key between Melbourne and Gippsland – its improvement should be first effected.[115]

---

112 J Power, 'Squatters and selectors: An historical geography of the central Gippsland Plains 1840–1880', BA Hons (Geography) thesis, University of Melbourne, 1979, p. 53; and VPP, 1876, vol. 25, p. 1873. The furore about Kilmany Park is exhaustively discussed in the local papers. *GT*, 6 and 27 June 1876, 20 and 27 September 1876 and 18 and 22 December 1876.
113 It lies in the Lang Lang catchment, which drains into Westernport. Because it was closer to Melbourne, this area became more densely settled earlier, which prompted calls for drainage earlier than at Moe. The first attempts to drain Koo-Wee-Rup date from 1857 and were by private individuals, which failed because of the size of the area, approximately 4,000 ha. K Sher, 'Catchment management and littoral consequences: A case study approach from Victoria, Australia', BA Hons (Geography) thesis, Monash University, 1986, p. 23.
114 Adams, *So tall the trees*, p. 21.
115 Letter from CJ Tyers to David Lennox, Superintendent of Bridges, 9 July 1849. SLV, MSM 157–8.

Figure 6.6: Selections in the parishes of Warragul, Moe, Yarragon and Narracan in 1883. Moe Swamp is not delineated except on the south by the main road.

Source: National Library of Australia, Accession no. 3408361.

6. 'A USELESS WEIGHT OF WATER'

Figure 6.7: Parish map of Yarragon, 1908.
The bottom section of the map depicts the now drained and regularised landscape of the Moe Swamp. The bottom, relatively straight line is the main road, while the next is the main drain. Sandwiched in between is the contour drain.

Source: National Library of Australia, Accession no. 3359879.

He went into detail describing how soft the bottom of the creek was, how low the surrounding floodplain was, the shape and strength of the banks, and known flood heights, from which he concluded that to 'render this road available throughout the year', three quarters of a mile of road and bridges would have to be elevated above the bank level of Narracan by 4–5 feet. This work alone, he said, would take the entire budget, and he did not bother to list other places that were troublesome to travellers.

Tyers wrote in 1849, well before the gold rushes and the Land Selection Acts brought an influx of new migrants. Moe Swamp continued to be a source of aggravation for all travellers. The first formal moves to drain Moe Swamp date from 1881, when the Narracan Shire president called a meeting to approach the government for support. This was prompted by the influx of new residents brought by the railway, and by a perceived shortage of land, girded by the longstanding knowledge that drained swamps usually have fertile soils.[116] The precedent of Koo-Wee-Rup encouraged the lobbying process, but it took until 1888 for the first contracts to be let. The intervening seven years were taken up with surveying and design work.[117]

By the time the work commenced, the colony was in the grip of a major depression. The reclamation works became a way of addressing high unemployment, but even with extra labour available, progress was slow. As late as 1897, election candidate AC Groom told a meeting in Moe 'that it was scandalous that a vast tract of country like the Moe Swamp was lying unoccupied' and that 'in its present state it was a detriment to the country'.[118]

---

116 Moe and District Historical Society, *A Pictorial History of Moe and District*, Moe City Council, Moe, 1988, p. 112. According to Hans Faubel, George Auchterlonie was present at this meeting, *Coach News*, vol. 4, no. 2, December 1986, p. 3. For an example of general discussion on fertility of wetland soils, see *GM*, 7 April 1877, 'The soil on the bank of the swamp near Foster St bridge must be very fertile judging by the sample of fine potatoes sent for the inspection of the Borough councillors on Thursday evening. They were grown in Mr Morris' garden and the second crop of the same patch of land this year'.
117 Faubel, *Coach News*, vol. 4, no. 2, December 1986, p. 5.
118 *Narracan Advocate*, 9 October 1897, cited in Faubel, *Coach News*, vol. 4, no. 2, December 1986, p. 6.

6. 'A USELESS WEIGHT OF WATER'

Figure 6.8: The former Moe Swamp, now the Moe Flats.
This photograph illustrates the lowness of the former swamp in comparison to the surrounding elevated lands.
Source: Author.

Groom failed to realise that the Moe Swamp was a valuable ecological asset. The interest in converting swamps and morasses into agricultural lands indicates a basic understanding of hydrology, certainly of the movement of organic matter in rivers and its removal in wetlands. Beyond this, there is no evidence of any interest, appreciation or understanding of the other functions that wetlands perform.[119] They filter out sediment and recycle nutrients, thus providing food for a wide range of flora and fauna. They also mitigate floods, and prevent erosion, by providing a space where water can spread out and slow down. When the Moe Main Drain was constructed from the raw material of the Moe Swamp, it removed approximately 5,000 ha of space where these functions could be performed.

In 1898, some of the drained lands were released. The general mood seemed to one of anticipating a boom, but it would not be long before the forces of hydrology would show themselves to be greater than the forces of men.

---

119 A Bullock & M Acreman, 'The role of wetlands in the hydrological cycle', *Hydrology and Earth System Sciences*, vol. 7, no. 3, 2003, pp. 358–89. doi.org/10.5194/hess-7-358-2003.

As with the Koo-Wee-Rup Swamp, the initial drainage designs were soon found to be inadequate. When a large flood hit in April 1900, the swamp was back to its old self. The engineering works had proved completely incapable of restraining the hydrological cycle. Having sold the lots, the government was now forced to commit to another major round of construction to solve the problem it had just created. The bigger, better system simply shifted the problem downstream. In the following years, farmers along the Latrobe River began to complain that floods were worse, because of the greater amount of water being discharged at greater velocity.

There is also substantial evidence for smaller-scale drainage works across the GLC, beginning soon after settlement. The most voluminous evidence comes from newspaper reports of council meetings or the popular country tours of journalists and visitors. 'Clodhopper' from the *Morwell Advertiser* said of land around Traralgon:

> This soil, although poor in quality in its natural state, may be made into really good arable land, that would increase its value to a wonderful extent for cultivation or for grass, and this may be done almost wholly by draining.[120]

In his lengthy discourse, he explicitly linked life with flowing water and death with stagnant water. Local papers included a wide range of information about agricultural matters designed to help selectors. This often included advice on draining. Agricultural shows performed a similar function. 'Progress' wrote to the *Gippsland Times* pleading for the show organisers to expand the display of drainage materials, because 'I know of scarcely any district where there is so much land requiring thorough drainage as Gippsland, and yet there was nothing in the shape of drainage pipes to be found at the late show'.[121] At least two popular almanacs included drainage advice. In 1864 *Dr LL Smith's Medical Almanac* advised readers that:

> Next to the character of soil, stands in importance the facility for easy drainage, and no matter how tempting the position or quality of the soil is, let this be regarded as an essential requirement. If practicable, select only land the condition of which you have had the opportunity of observing in winter. In the absence of this

---

120 *Morwell Advertiser*, 25 June 1887.
121 *GT*, 9 April 1870.

knowledge, be careful to note if cattle have left the marks of their feet visible in deep indentations, colonially termed 'glue pots'. If so it has arisen from the water lodging on the ground during winter; and such land where no natural inclination for drainage exists, should be avoided as totally unfit for horticulture.

In his 1895 edition, Dr LL Smith included a helpful table to calculate the pipes required to drain an acre of land at various widths of trenches.[122] *Stevens and Bartholomew's* almanac for 1867 gave detailed instructions for drainage, covering location, direction of feeder channels, depth, width and materials.[123]

While the majority saw the benefits to be gained from drainage, it was also expensive. Some squatters had undertaken works, such as Lemuel Bolden at Lake Wellington. Mary Ann Barton's memory of Bolden on their arrival on the selection was that he was sarcastic, probably angry that others were to benefit from his outlay.[124] Around Gormandale, the cashed up Horsley brothers employed men and bullock teams to drain their swamp-dominated property and set about breeding horses for the Indian Army. Adjacent landholders followed their example.[125]

In the series by the *Gippsland Mercury* of prominent farms in 1882, all important improvements including drainage were noted, such as at Park House, which had been flood prone until the works were completed.[126] Hezekiah Harrison supervised 20 Chinese workers on the Airlie estate, cutting drains and undertaking other work like cutting thistles, but this scale of employment tended to be the exception. Large families could achieve something similar.[127] Goulding notes in her article on the

---

122 *Dr LL Smith's Almanac*, 1864 edition and 1895 edition.
123 *Stevens and Bartholomew's Sandhurst, Castlemaine, Echuca, Maldon, Dunolly, Maryborough, Back Creek, and Avoca district directory for 1867*, p. 63.
124 M Fletcher, *Avon to the Alps: A history of the Shire of Avon*, Shire of Avon, Stratford, 1988, p. 42.
125 K Huffer, 'Dedicated to the pioneers of Gormandale: To those who succeeded and those who failed', Typescript, NLA Npf994.56 H889, p. 6.
126 J Hales & J Little (eds), *Gippsland estates 1882: A series of articles which appeared in the Gippsland Mercury from January to August 1882*, Maffra and District Historical Society Bulletin, Supplementary Issue no. 4, Maffra, p. 7.
127 JW Leslie & HC Cowie, *The wind still blows: Extracts from the diaries of Rev WS Login, Mrs H Harrison and Mrs W Montgomery*, the authors, Sale, p. 90; *East Gippsland Historical Society Newsletter*, 1980, vol. 1, no. 7, p. 8. William and Jane Jennings of Estella Park, Broadlands, initially selected 80 acres in 1867, then a second 97 acres. 'They worked the land as a family group, clearing, draining cultivating and improving'; they bought out failed neighbours.

Aboriginal people of east Gippsland (which includes Swan Reach) that Aboriginals were often employed to drain, in addition to a wide range of other farm tasks.[128]

Some farmers cooperated in ventures to share the costs involved. The *Gippsland Farmer's Journal* observed that at Tyers:

> An effort is to be made to drain a further proportion of the swamp and bring it under cultivation. Three or four of the selectors, whose holdings abut the flooded land, talk of combining to cut another large drain, shall carry off the surplus water into the Latrobe.[129]

Pooling of resources and labours was a logical way of addressing the problem that wetlands and wet lands posed for selectors. The second epigraph commends another joint approach for the drainage the Heart Morass. Given that many settlers were inexperienced and had little money, a joint approach would have been beneficial. While the extant diaries don't provide evidence of cooperative draining, they do provide plenty of evidence of cooperation on nearly every other aspect of farm life so it seems likely that settlers worked together to drain land.

The *Morwell Advertiser* told its readers that:

> Of course, drainage is an expensive operation, but it can be effectively done at a much less cost were other materials than pipes used. At this very time there are under drains, freely running, which were sunk twenty years since and filled with tea scrub, stones being laid upon a layer of scrub to within eight inches of the surface. Let stock owners bear this in mind, that ten acres of artificial grasses upon well drained land will yield more and better food than fifty acres of undrained, uncared for grass land.[130]

Diary evidence tends to support the less expensive version of draining. Buckley employed small-scale, hand-dug drainage on his run, draining water away from his house and from key grain paddocks. John Disher wrote on 11 October 1869 that he was 'clearing drains outside the garden and clearing around cottage'.[131] It is reasonable to assume that Aleck

---

128 M Goulding, 'Placing the past: Aboriginal historical places in East Gippsland, Victoria', *Historic Environment*, vol. 13, no. 2, 1992, p. 48.
129 *GFJ*, 27 January 1887.
130 *Morwell Advertiser*, 5 February 1887.
131 J Hibben, 'The Disher family in the nineteenth century', BA Hons thesis, University of Melbourne, 1978, p. 22.

McMillan's shorthand 'at home all day working about the garden', which crops up regularly in his diary in 1870, would have included draining off the accumulated waters from that very wet year. He also discusses forming paths in the garden and around the house, in order to allow family and servants to get from garden to house with dry feet. Alf Broome at the Nicholson makes three references to digging drains through the run of his diary, mostly to drain water from his orchard.[132]

The patchiness of references to drainage in surviving diaries must be interpreted in light of what stage each farmer was at, where the farm was and the type of weather at the time of recording. Most of the diaries that have survived are generally of a brief nature, a paragraph per day at most, and only in exceptional circumstances do they say more. This is understandable given how physical their work was and how tired they must have been. For instance, Margaret McCann kept a diary beginning in 1894 and her entries rarely went over two or three sentences. She lived at Stradbroke, where the soil becomes more sandy, free draining and coastal in character. Her common refrain is a plea for rain, hardly a situation when drainage would be required. For selectors just starting out, it would have been a matter of first things first. Providing a watertight house and barn, followed by clearing were the essential tasks. Draining would have come later, but as the diaries do not generally have long runs, it is not possible to say how much time farmers devoted to forming draining systems.

Hence, it is the unending regularity of references to drainage in council reports that is the most reliable guide to breadth of local drainage practices. Councils engage in local drainage construction and maintenance works continuously throughout the time period.[133] The decision to drain commits residents to an unending expenditure. This is hardly new. Histories of Dutch drainage constantly emphasise the maintenance duties required

---

132 Diary of Patrick Coady Buckley, 20 October 1870, 21 December 1870, 19 July 1871 and 17 November 1971; Diary of Aleck McMillan, 19 August, 3 and 27 September 1870; Diary of Charles Alfred Broome, 14 November 1882, dug round the fruit trees in the orchard, 17 May 1884, dug a drain by the orchard, 5 June 1899, digging drains at the barn – wet weather.

133 *GT*, 23 January 1869, Shire of Avon has four tenders advertised: one for draining and repairing the road to Clydebank and the other three are for culvert construction and maintenance on other roads. *GT*, 1 June 1869, Shire of Bairnsdale lets eight tenders for work, six of which involve an element of water management, either drainage of a road or construction of culverts.

to keep dikes in good repair.[134] The desire to remake local hydrology by keeping water moving did, however, have the effect of keeping money circulating throughout the local economy. Local contractors and labourers profited from the community's dislike of still and stagnant waters.

Appropriate drainage in the catchment's urban areas was seen as a sign of progress and achievement. As early as 1861 the *Gippsland Times* was calling for it during election campaigns: 'Our other requirements are paltry in comparison to this [opening the entrance]; nevertheless they require attention. We require safe and commodious public buildings, we require our roads clearing and our streets draining'.[135] The *Gippsland Farmer's Journal* disparaged the state of roads in Traralgon in 1887, and called for street lighting to prevent residents falling into the numerous holes and bogs. It reported the findings of the inspector of the Central Board of Health in 1887, which criticised both Traralgon's and Morwell's drainage. In fact, the streets of Traralgon were so bad that the Glenmaggie correspondent to the paper dubbed it Mudditown.[136] In 1897 the Maffra Shire Engineer complained that:

> For its size and its prospects of expansion in the near future, Maffra is one of the worst drained towns in Victoria. Leaving out all minor considerations such as ease of travelling and comfort to pedestrians, some change must be urged in the interests of health.[137]

---

134 PJEM van Dam, 'Ecological challenges, technological innovations: The modernisation of sluice building in Holland 1300–1600', *Technology and Culture*, vol. 43, no. 3, July 2002, pp. 500–20. doi.org/10.1353/tech.2002.0144; A Kaijser, 'System building from below: Institutional change in Dutch water control systems', *Technology and Culture*, vol. 43, no. 3, July 2002, pp. 521–48. doi.org/10.1353/tech.2002.0120; L Roberts, 'An Arcadian apparatus: The introduction of the steam engine into the Dutch landscape', *Technology and Culture*, vol. 45, no. 2, April 2004, pp. 251–76. doi.org/10.1353/tech.2004.0089; M Ruess, 'Learning from the Dutch: Technology, management, and water resources development', *Technology and Culture*, vol. 43, no. 3, July 2002, pp. 465–72. doi.org/10.1353/tech.2002.0135.
135 *GT*, 14 August 1861.
136 *GFJ*, 23 June 1887, and 1 July 1887 for the Glenmaggie comment.
137 *GT*, 18 February 1897.

Figure 6.9: Township of Warragul, by Nicholas Caire, 1886.
Note the newly metalled road, and the deep drainage ditches on either sides with planks to help pedestrians cross the ditches.
Source: National Library of Australia, Accession no. 3550774.

Examples of the dominance of water management in local government budgets are easy to find. In January 1869, the Avon Shire's four advertised tenders were all to drain and maintain roads. In June 1869, the Shire of Bairnsdale advertised eight tenders, of which six were for road draining and culvert construction. The 1870 budget for the Sale Borough Council included works for 'repairs to streets and rectifying water tables at 10d per chain'.[138] Under the Health Acts, councils employed inspectors of nuisances whose job was to maintain public health by serving notices on residents who were causing a problem. Mr French of the Club Hotel in Sale received one such notice, and took such exception to it that he wrote a lengthy objection to it to the paper, detailing the cleanliness of his operations. He blamed residents further up Raymond St who, he claimed, had not installed drainage and the council for getting the levels wrong.[139]

---

138 *GT*, 23 January 1869, 1 June 1869 and 22 February 1870.
139 *GT*, 31 May 1870.

It is reasonable to expect numerous complaints from Sale given its low-lying topography, but mountainous towns also had their share of urban drainage headaches. The *Walhalla Chronicle* described its streets as a 'sea of mud' and requested the young men to ride slowly, so that ladies' attire did not become splattered.[140] Drainage works drained council resources, with the purse never being equal to the need. The *Walhalla Chronicle* reported the 'gruesome' state of roads leading to the town from Moe and Toongabbie. Using an appropriate metaphor, the chronicle claimed that the special grant of £1,300 received for both was 'a mere drop in the ocean and would be swamped on a few chains of corduroy'. The editor took the opportunity again to call for a rail line, which were not subject to the bogs and swamps that bedevilled road travellers.[141]

Conclusion

In the whole study period, the only voices in the GLC raised against rural drainage were those who wanted to keep the morasses for grazing. Mostly, people thought that agriculture was a better use for such fertile soils, and the need for town drainage was never questioned. Settlers had a largely self-serving interest in using or eradicating wetlands: What can I get from here? How can this place serve my ends? Given the culturally and religiously entrenched perceptions about swamps as places of darkness, danger and illness, this utilitarian and individualistic approach was the norm.

To convert a swamp into farmland was economically beneficial to individuals in the short term, and was widely supported in the community as a public service. With so little incentive to look at the morasses in a different way, they could not possibly know that they started in train a series of environmental changes that would jeopardise the Lakes system. The interest in converting swamps and morasses into agricultural lands indicates a simplistic understanding of hydrology, certainly of the movement of organic matter, nutrients and sediments in rivers and their removal in wetlands. Beyond this, there is no evidence at all of any interest, appreciation or understanding of the many ecological functions that wetlands perform.[142] That is, until a natural disaster struck.

---

140 *Walhalla Chronicle*, 19 November 1897.
141 *Walhalla Chronicle*, 16 June 1899.
142 Bullock & Acreman, 'The role of wetlands in the hydrological cycle', pp. 358–89.

During drought, fire and flood, Gippsland's swamps and morasses suddenly seemed valuable. Sadly, this appreciation faded with the return of moderate conditions. Ecological knowledge gained in hard times didn't survive good times.

By supporting and encouraging the destruction of Moe Swamp and the myriad of other small swamps, colonial Gippslanders were proclaiming that they knew better than nature. The still, reflective surface of the swamp showed a people who were vain and arrogant, but with little reason to justify that position. It also shows a people who were trying to do the best for themselves with the knowledge they had. Reality, no doubt, lies somewhere between these two extremes.

CHAPTER 7

Between 'the water famine and the fire demon':[1] Drying up the catchment

The lower portion of Gippsland, sheltered, as it is, to the northward and westward by the dividing range, and watered by five fine rivers, may be rendered, by irrigation, a most flourishing portion of the colony.

Strzelecki, 1845[2]

I have seen cattle die upon the common for want of water – die in dry water-holes, trying to lick the mud. I have seen the grass so burnt up that a beast could not get a mouthful, and when a little rain came, and the fresh grass came up, they were dropping down in all directions.

Michael Landy, 1885[3]

---

1   *GT*, 10 March 1882.
2   EP Strzelecki, *Physical description of New South Wales and Van Diemen's Land*, Longman, Brown, Green and Longmans, London, 1845, p. 446.
3   Victoria, Royal Commission on Water Supply, *Further Progress Report with appendices thereto, Extracts from Minutes of Committee: Together with Minutes of Evidence, etc, (August 31st, 1885.)*, no. 53, Government Printer, Melbourne (hereafter *Royal Commission on Water Supply*), Q6930. Evidence of Michael Landy, hop grower, p. 249.

Introduction

Colonial Gippslanders preferred their hydrological conditions on the wet and permanent side, echoing the conditions in their European birthplaces. They were most happy when rain was regular and well spaced, and when it flowed in defined river channels. This chapter tells the opposite story; what happened when conditions were dry, irregular and undependable.

All species are a form of embodied water. Without at least one litre of water a day, human beings perish. Second only to oxygen, water is our most essential need. Water hydrates, cleanses and keeps our organs functioning. That vital few kilograms of grey matter, the brain, is 85 per cent water, the body 66 per cent. Losing 10 per cent of your embodied water paralyses, losing 20 per cent spells death.[4] An absence of water is therefore a profound physical and psychological threat. While the evaporative part of the hydrological cycle returns water to the sky to make life-giving rain, and thus is an essential part of the cycle, it also ensures that it is no longer available for human use.[5]

The counterbalance to precipitation, evaporation is defined as 'the transferal of liquid water into a gaseous state and diffusion into the atmosphere'. For evaporation to occur, there must be water to evaporate, energy to do it and somewhere for it to go, that is a dry atmosphere.[6] Its rate is affected by the action of winds, and its obvious impact is on surface water availability. With high evaporation, the water levels of rivers, lakes and dams can drop markedly. Australian average evaporation rates are high, at 87 per cent, compared to the approximate average of 60 per cent for Europe and North America.[7]

The hydrological reality of evaporation is an important constraint in the success of agricultural societies, and it has, for millennia, been entirely out of human control. In large portions of Europe, however, this was not a major problem. In the nineteenth century, as noted in Chapter 2, Europeans lived in one of the most temperate of climates Earth has to offer. Coming

---

4   M Newson, *Hydrology and the river environment*, Clarendon Press, Oxford, 1994, p. 5. Without oxygen for three minutes, you will likely suffer brain damage.
5   W Brutsaert, *Evaporation into the atmosphere*, D Reidel Publishing, Dordrecht, 1982, p. 2. doi.org/10.1007/978-94-017-1497-6.
6   T Davie, *Fundamentals of hydrology*, Routledge Fundamentals of Physical Geography, Routledge, London, 2003, pp. 30–1.
7   JJ Pigram, *Issues in the management of Australia's water resources*, Longman Cheshire, Melbourne, 1986, p. 18.

from a part of the planet where droughts were measured in mere weeks, colonial settlers were in for an unpleasant surprise. They had emigrated from one of the most constant hydrological regimes to the least constant.[8] For a people relying on the success of an agricultural model that required high precipitation and low evaporation, trouble and grief were inevitable.

This chapter provides a brief exploration of the role of evaporation in the hydrological cycle, explores what settlers understood about evaporation and dryness, and the positive and negative aspects to it in daily life. It pays particular attention to the experience of fire and drought in the 1880s, then discusses settlers' principal responses to ameliorating the threat, particularly focusing on the increasing availability of irrigation technology.

The evolving science of evaporation

The science of evaporation follows a similar path to that laid out in previous chapters; that is, a period of theorising in ancient Greece was followed by a long period of inactivity dominated by Christian theology. After this, the emergence of science saw improved understanding of evaporation based upon quantitative experiment.

In ancient Greece various theories were put forward.[9] As demonstrated in Chapter 2, Anaxagoras of Clazomenae had the closest conception of the hydrological cycle. His grasp of the evaporative part of the cycle was strong and succinctly expressed:

---

8   G Aplin, *Australians and their environment: An introduction to environmental studies*, Oxford University Press, Melbourne, 1988, reprinted 1999, pp. 41–9; J Tibby, 'Explaining lake and catchment change using sediment derived and written histories: An Australian perspective', *Science of the Total Environment*, vol. 310, nos 1–3, July 2003, p. 61. doi.org/10.1016/S0048-9697(02)00623-X; Pigram, *Issues in the management of Australia's water resources*, p. 15; K Johnson, 'Creating place and landscape', in S Dovers (ed.), *Australian environmental history: Essays and cases*, Oxford University Press, Melbourne, 1994, pp. 43–4.

9   Due to his belief that water was the basis of life, Thales of Miletos is supposed to have thought that evaporation is the driving force behind the cosmos. Anaximander of Miletos subscribed to the so-called dual exhalation theory, which combined evaporation from the sea with an exhalation of vapours from the earth together to form rain. It is thought that Herakleios of Ephesos (c. 500 BC) was the principal writer on the dual exhalation theory, for Aristotle derived inspiration for his work on meteorology from him. He described an upward and downward path in the atmosphere. This strongly echoes the version of the hydrological cycle that was based on a hierarchical transmutation of the elements discussed in Chapter 2. This model involved the transformation of the elements, laying a heavy emphasis upon evaporation as the mechanism of this transformation. Dual exhalation theory included more than just the transformation of water; it included the change of seasons and night and day. All the power for this transformation derived from the sun. Brutsaert, *Evaporation into the atmosphere*, pp. 13–14.

> What happens in the sky is caused by the heat of the sun; for when the moisture is drawn up out of the sea, the sweet part, which is distinguished by its fine texture, forms a cloud, and drips out as rain by compression like that of felt, and the winds spread it around.[10]

Anaxagoras's correct interpretation was again eclipsed by Aristotle, who subscribed to the dual exhalation theory. Dual exhalation theory confused all attempts to understand evaporation because it discounted wind, one of the controlling factors in the rate of evaporation. Instead, it privileged temperature and the role of the sun. To fully understand evaporation, equal emphasis on both is necessary.

In 1637, Descartes mounted the first serious challenge to Aristotlean views. He said that wind is air in motion and that it is the result of evaporation, not the cause. He also described evaporation in terms of the action of particles, thus reinvigorating the notion of atoms first proposed by Leukippos and Demokritos. Data from Perrault, based on an experiment conducted between 1669 and 1670, and by Halley, whose experiments ran between 1687 and 1694, confirmed Descartes's theory.[11] The results of these experiments finally debunked the ancient argument that water had to come from Tartarus (the lowest region of the underworld), because evaporation could not account for river flow.[12]

Despite Dalton's significant 1802 publication, which linked wind speed and the dryness of the air to the evaporation rate, there were still many gaps in understanding.[13] Stephen Hales was an early investigator of water transpiration from plants.[14] It took until 1862 for Tate to prove that the rate of evaporation is nearly proportional to the velocity of the wind. In addition, the nineteenth century was full of theory and counter theory about the nature of heat and energy, factors fundamental to understanding the processes of evaporation.[15]

---

10   Brutsaert, *Evaporation into the atmosphere*, p. 13.
11   Brutsaert, *Evaporation into the atmosphere*, pp. 26–7.
12   JR Phillip, 'Water on the earth', in AK McIntyre (ed.), *Water: Planets, plants and people*, Australian Academy of Science, Canberra, 1978, p. 41.
13   Davie, *Fundamentals of hydrology*, p. 31.
14   IR Cowan, 'Water use in higher plants', in AK McIntyre (ed.), *Water: Planets, plants and people*, Australian Academy of Science, Canberra, 1978, pp. 71–107.
15   Brutsaert, *Evaporation into the atmosphere*, p. 33.

Evaporation is also hard to measure accurately. This uncertainty around the science is the reason why the only place where it was officially measured in the whole colony of Victoria was in Melbourne. The earliest evaporation records for Gippsland appear to date from 1957, at the Yallourn State Electricity Commission.

Settlers' understanding of evaporation

So if the best scientists of Europe were bamboozled by evaporation, how would colonists grapple with it? This returns to the question of ecological perception by people who, by modern standards, were uneducated. Only a few settlers with an advanced education and an interest in science would have known about the likes of Dalton and Tate, let alone been interested in them.[16] And yet, there they all were, living and working in a catchment that would in time display a previously unthinkable level of evaporation.

Colonial settlers had three kinds of knowledge about evaporation: embodied, cultural and formal. Formal learning was the least influential, because the majority of settlers had little formal education to a standard where precepts of hydrology would be taught. The settlers' primary knowledge would have come from their direct, embodied experience; for example:

- thirst being a sign of internal dehydration
- seeing water evaporate in a boiling kettle or pot
- drying clothes, or
- witnessing plants and animals wither.

Personal embodied knowledge was available to all, regardless of class or gender. This physical knowledge about water's absence and the threat to survival was augmented and framed by cultural knowledge.

Previous chapters have detailed the use of biblical imagery on shaping perceptions of the hydrological cycle, where, crudely, precipitation and flow were good and everything else was not. Chapter 4 discussed the likening of rain to the blessing of God's love, Chapter 5 explored the importance of springs and rivers in various biblical stories that are fundamental to

---

16   Men who might fit this category include Robert Thomson, a lawyer who was the first representative for Gippsland; Crown Commissioner Charles Tyers, whose almanac held hydraulic equations; Dr Alexander Arbuckle, the Sale coroner for many years; William Dawson, chief surveyor; and any of the clergy.

the faith, while Chapter 6 looked at the negative portrayal of swamps and bogs. The biblical portrayal of dry areas is similarly negative, and this negativity is displayed in much of the remaining chapter.

It should not be forgotten that the three monotheistic religions – Judaism, Christianity and Islam – originated in the deserts. '[Israel] is surrounded by deserts, and in their visions, the prophets saw fountains causing the deserts to bloom' (Joel 4:18, Zech 14:8, Ezek 47:1–2). In Isaiah's prophecies of consolation, he describes the blossoming of the desert and the wilderness as follows: 'I will open rivers on the high hills, and fountains in the midst of the valleys' (Isiah 41:18). When the ways of man please God, 'He will make the wilderness a pool of water, and the dry lands springs of water' but sin causes Him to turn 'rivers into a wilderness, and water-springs to a thirsty ground' (Psalms 107:33, 35).[17]

Certainly, colonial settlers in Victoria as a whole used the metaphor of deserts blooming when they debated water policy, irrigation and infrastructure.[18] However, descriptions of blighted landscapes common to

---

17   Other desert/wilderness references from the Bible are discussed by Y Feliks, *Nature and man in the Bible: Chapters in biblical ecology*, Soncino Press, London, 1981. Chapter 14 of Jeremiah gives a description of drought. According to the prophet, this was punishment for the people's sins against God. It describes shepherds trying to find water for their stock and how the wild animals stagger and suffer. In Chapter 17, Jeremiah raises several central themes in his prophecies: the sins of Judah, their persecution and mockery of him, and a call for repentance, coupled with an account of the glorious future that would follow. In his admonition, the prophet employs two parables, one from the plant world, and the other from the animal world. In his parable drawn from plant life, the righteous man is compared to a tree planted by the water, while the wicked is compared to the tamarisk in the desert (p. 35). The prophet Ezekiel makes seven prophecies about the future of Egypt, which includes punishment for the Egyptians through the drying up of the Nile (pp. 141–3). The last nine chapters of Ezekiel describe the future church revealed to Ezekiel by the angels. A key element is a fountain of fresh water, enough to 'swim in'. This river will restore life to the Judean desert, cause an abundance of nurturing plant life for humans, and finally will flow into the Dead Sea and restore it to life as well (pp. 148–9). Armstrong notes that Jacob's sons leave Canaan to escape a great famine and go to Egypt, where they are enslaved. K Armstrong, *A history of god: From Abraham to the present, the 4000 year quest for god*, Vintage, London, 1999, p. 12.
18   *Argus* (Melbourne), 7 April 1870, p. 5, comparing desert-like South Australia to garden-like Western Victoria; *Argus*, 25 April 1876, p. 6, SA correspondent describing the capacities of Mennonite settlers to turn deserts into gardens; *Argus*, 6 October 1880, p. 14s, describing the SA exhibit to the world fair as showing how an oasis can be made in a desert; *Argus*, 1 November 1882, p. 6, calling it a scandal that the Mallee should be left a desert. See also J Powell, 'Environment and institutions: Three episodes in Australian water management, 1880–2000', *Journal of Historical Geography*, vol. 28, no. 1, 2002, pp. 102–3. doi.org/10.1006/jhge.2001.0376, noting the importance of the garden rhetoric; PG Sinclair, 'Making the deserts bloom: Attitudes towards water and nature in the Victorian irrigation debate, 1880–1890', MA thesis, University of Melbourne, 1994. Nor was the idea confined to Australia, see for example MJ Heffeman, 'Bringing the desert to bloom: French ambitions in the Sahara Desert during the late nineteenth century – the strange case of "La Mer Interieure"', in D Cosgrove & G Petts (eds), *Water, engineering and landscape: Water control and landscape transformation in the modern period*, Belhaven Press, London, 1990.

drought-ridden northern Victoria are rarer in Gippsland's papers, simply because of the difference in annual rainfall. But while the average rainfall was higher in Gippsland, they still experienced drought.[19]

Given the difficulties in measuring evaporation, the next best available measurement is temperature. However, temperatures were only recorded at Sale from 1892, starting 22 years after rainfall.[20] Rainfall was measured at a total of 77 stations, including Sale, Stratford, Rosedale, Omeo, Bruthen, Clydebank, Sale Mercury Office, Toongabbie, Warragul, Bairnsdale and Maffra. Dates offered by the Bureau of Meteorology for temperature records at other sites conflict, but it would appear that Omeo commenced in 1879. The next two stations were Bairnsdale Post Office, which began recording temperature in 1896, and Stratford followed suit in 1903. Four temperature stations compares poorly to 77 rainfall stations.

The lack of temperature measurement is a conundrum, given the impact of dryness on an agricultural economy. Is it possible to assume that the absence of measurement meant lack of interest? There are a number of probable and complementary reasons to this apparent disinterest. The first lay in the practical costs of establishing a meteorological network. A rainfall gauge is cheaper and simpler to supply and use than a temperature gauge. The second relates to the prevalence of the popular theories about rainfall either being linked to the extent of tree cover, or to the spread of agriculture. It is possible that the strong emphasis on the measurement of rainfall reflected the 'rain follows the plough' debate, which reached

---

19   *GT*, 26 January 1875, Rosedale correspondent: 'The excessive heat which prevailed during the week was agreeably tempered by a very acceptable fall of rain on Friday evening, which both cooled the atmosphere and cleared it of the smoky pall that hung over the earth's surface'; *GT*, 8 February 1876, 'The exceedingly warm weather of the last few says has been very severely felt, and yesterday we were favoured with a genuine hot wind, which, though according to authorities is healthy, has a most enervating effect. The glass in the sun yesterday was as high at 136 and from 90 to 96 in the shade. The country presents a very dried up appearance, and bush fires will have to be guarded against should the present weather continue'; *GT*, 25 March 1881, 'Last Friday was one of the most disagreeably hot days which have been experienced here during this season … The oppressiveness of the day was much intensified by a very hot wind which blew for hours'; and *GT*, 22 February 1886, 'The solar heat for the past four days has been overpowering, but a welcome change in the weather occurred yesterday afternoon and rain commenced to fall in the afternoon'; *MR*, 22 December 1888, General News: 'The face of nature will again, be draped in green. In town the tanks are overflowing and the dust "fiend" has disappeared'.

20   *Record of results of observations in meteorology and terrestrial magnetism made at the Melbourne Observatory, and at other localities in the Colony of Victoria, Australia*, prepared by Pietro Barrachi, Govt Astronomer, Govt Printer, Melbourne, 1919.

its apotheosis in the nineteenth century.[21] When settlers realised that an Australian drought meant months or even years without rain, it focused attention on rain even more.[22] As the Rosedale correspondent said in 1886:

> A south easterly wind and a very cloudy sky on Tuesday gave us the impression that we were in for a share of the good thing in the shape of rain ... But alas! It turned out otherwise, a sprinkling of rain fell, the dark clouds sailed away, and the dreary dryness resumed its sway. People, in another season or so, will come to believe that brown is the normal colour of grass, and a whitey grey that of soil.[23]

A season or so, compared to England where a drought lasted weeks. The strength of their experience of rainfall in England combined with the cultural precepts about the undesirability of deserts produced a perception filter. It acted like a filter on a camera lens, which alters the contrast in the image, but the person viewing the image is unaware that a filter has been used.[24] Green and blue were good, brown and grey were not. Something the Rosedale correspondent wholeheartedly agreed with.

The pros of dryness

Gippsland was settled in response to drought in New South Wales, and was generally considered to be a pleasant temperate place. George Augustus Robinson described it as sheltered from hot winds and well watered.[25] He is almost word for word with Strzelecki, author of the first epigraph. Both emphasised abundant surface water and a mild climate, suggesting knowledge of extreme conditions elsewhere on the continent at the time. Too often histories focus upon the extremes, and neglect the more subtle

---

21   For a succinct discussion of the tree–rainfall theory, see M Williams, *Deforesting the earth, from prehistory to global crisis*, University of Chicago Press, Chicago, 2003, pp. 430–2. For newspaper discussion of forest conservation issues, see *GT*, 9 July 1872, 2 November 1872, 23 July 1874, 20 February 1875, 3 June 1876 and 24 January 1879.
22   Contemporary definition of drought in England is 15 days with less than one-eighth of an inch of rain, www.weatheronline.co.uk.
23   *GT*, 16 April 1886, Rosedale correspondent.
24   For example, a K2 yellow filter increases the contrast between black and white and reduces the amount of grey. It is used in primarily in black-and-white photography. Most people are more familiar with the use of soft focus filters; however, it's possible to see that it has been used.
25   P Brightling, 'A sketch on the Avon; Tuesday October 3, 1843', *Gippsland Heritage Journal*, no. 7, December 1989, p. 38.

middle ground.²⁶ Before moving on to discuss the impacts of drought, it is worth highlighting the benefits that many settlers found in Gippsland's generally mild, dry weather.

Dryness has a wide range of advantages, many of which were crucially important to the success of agricultural economies. Broadly, dryness was beneficial for travelling, crop ripening, some types of food preservation or processing, carrying out drainage works, some aspects of health, drying clothes, fire starting and most recreational activities.²⁷ Chapters 4 and 6 have respectively covered the role of wet weather and mud in drainage and disruption of travel patterns and many social events, so it only remains to observe that dry weather encouraged Gippsland's social and mercantile life. The optimistic report from the Trafalgar correspondent to the *Moe Register* in October 1888 epitomises what warm, clear weather did for settlers:

> The month of October throughout has proved exceedingly fine, and very warm, and bids fair to continue. The roads are drying up allowing vehicles of all sorts to pass along them, and the farmers seem to be making good use of their time; grass is becoming plentiful, and cattle seem to be thriving, and our gardens flourishing.²⁸

The amount and seasonality of dryness was important in the growth cycle of crops. A middling amount of rain was needed to start ploughing, but only enough to moisten the soil and make germination likely. Following germination, gentle showers interspersed amongst days of warm, fine weather (so called 'genial' weather) provided ideal growing conditions.²⁹ Damp weather encouraged fungal diseases like rust, which farmers dreaded. As the Lindenow correspondent remarked in early December 1876: 'The steady and continuous rain which is now falling will, it is feared, prove detrimental to the crops; rust and caterpillars will be almost certain to appear if we are not shortly favoured with dry settled weather'. This was a common refrain.³⁰

---

26 For example, consider the title of Jenny Keating's book on Victoria: *'And the drought walked through': A history of water shortage in Victoria*, Dept of Water Resources, Melbourne, 1992.
27 *Morwell Advertiser*, 13 April 1889, in relation to draining Moe Swamp.
28 *Moe Register*, 27 October 1888.
29 *Moe Register*, 11 August 1888. The return of dry weather enables ploughing to recommence.
30 *GT*, 1 December 1876, Lindenow correspondent. See also, *GT*, 15 August 1881, Cowwarr correspondent; *GT*, 10 February 1886, on rotting crops in Port Albert district; *GT*, 3 March 1886, Narracan correspondent.

Figure 7.1: AW Howitt's hop kiln, Frederick Cornell, 1872. Hops were dried with the use of specially constructed hop kilns.

Source: Pictures Collection, State Library Victoria, Accession no. H87.16/75.

Dry weather was essential for harvesting and processing of grain crops. In January 1853, a rain storm forced Buckley to re-dry and restack his crop. Later attempts to thresh the wheat were constantly interrupted by drizzling weather.[31] Once a crop was in, some required further processing. This, too, was affected by levels of humidity. Warmth and damp encourages the proliferation of organisms, which attack foodstuff and other organic goods. When survival meant achieving rapid self-sufficiency, loss of food was a major problem.

The problem lay in trying to achieve the ideal conditions for the variety of foodstuffs that the settlers raised. Warm and dry conditions are ideal for drying fruits. There was a large fruit-growing industry around Bairnsdale, in which Annie Prout and the Macleods were engaged. While the bulk of product went to Melbourne as fresh fruit, most settlers would have preserved portions of their harvest for domestic consumption.[32] Some crops had to be dried before they could be used. A few settlers grew tobacco but by far the most significant crop in this category was hops, grown around Bairnsdale.[33] Large growers erected purpose-built drying kilns, such as in Figure 7.1, which shows AW Howitt's kilns.

Mild and dry weather helps to process animals and animal products. There was an extensive dried fish industry operating around the Lakes prior to the introduction of refrigerated railway trucks.[34] Tanning animal hides (discussed in Chapter 5 as an example of a flow-dependent industry) is also a preserving operation, based upon removal of moisture from the skin. Humid weather affected other preserving activities. Margaret McCann put off killing her pig because the weather was 'muggie', and had trouble with making butter.[35]

---

31 Diary of Patrick Coady Buckley, January 1853 and between 9 May and 27 June 1853, CGS, 2806. Also, 13 April to 9 May 1855 with his own threshing machine.
32 Agricultural Jottings by Ceres, *Bairnsdale Advertiser and Omeo Chronicle*, 1 May 1890, for a discussion on different methods of drying raisins. Auchterlonie sometimes bought dried fruit to sustain him when travelling, e.g. 13 February 1869.
33 *GT*, 4 May 1896.
34 Walter Spurrell, 'History of Nicholson, 1876–1976', SLV, MS 10774, box 1542.
35 Diary of Margaret McCann, 17 February and 7 March 1899, SLV, MS 9632, MSB 480. The *Moe Register* also discusses problems of hot weather and butter making. 8 September 1888.

The dairy industry presents the best evidence of how settlers tried to enhance moderate, dry conditions to maintain the quality of their product. One way was by underground storage, where temperatures are less likely to fluctuate. Buckley wrote that he 'dug a hole in the bank for dairy at the new place', and stored his potatoes in pits.[36] In 1890, the *Morwell Advertiser* reported that the government was running a design competition for dairies constructed of wood, which would keep temperatures at 50°F while outside temperatures were at 120°F.[37]

Moderate, dry weather was needed to undertake that most iconic of colonial activities, land clearing. Stephen Pyne has noted that by using fire to clear colonial settlers were imitating their ancient European forebears, but few, if any of them, would have realised this.[38] In the heavily forested western portions of the catchment, fire was the essential step in the establishment of a farm.[39] The importance of a 'good burn' crops up continuously. *The Land of the Lyre Bird*, produced in the 1920s by settlers who cleared the forests of South Gippsland, devotes two detailed chapters to the use of fire to establish grass and crops. Seriously hard and dangerous work, the book takes a valorising tone, written by a group who were looking back and finding meaning in their life's work.[40]

A 'good burn' had important economic advantages for a selector. It opened up ground, provided instant fertiliser and enabled a return on all that sweat. The Narracan correspondent complained in 1886 that poorly timed rains in summer had made 'many of the burns but indifferent'.[41] The sooner grass could be established, the sooner one could meet repayments. This was mournfully noted by Caleb Burchett, who said in his memoirs: 'It was my misfortune to have almost every time a "Bad Burn". My first picking up cost me nearly 2 pounds to make the land fit for sowing the grass seed!!'[42]

---

36 Diary of Patrick Buckley, 3 August 1846 and 28 to 31 May 1852.
37 This was also reported in *GT*, 12 February 1890.
38 S Pyne, *Burning bush: A fire history of Australia*, Allen and Unwin, North Sydney, 1992, p. 157.
39 Diary of George Glen Auchterlonie, 29 April 1869, CGS, 4060.
40 Committee of the South Gippsland Pioneers Association, *The land of the lyre bird: A story of early settlement in the great forest of South Gippsland*, Shire of Korumburra on behalf of the South Gippsland Development League, Clayton, Vic., 1966.
41 *GT*, 3 March 1886.
42 Reminiscences of Caleb Burchett, SLV, MS 8814, MSB 436, unpaginated; see also *Morwell Advertiser*, 12 February 1887 and 25 February 1888 for reports of 'good burns'.

Warm, dry weather made a difference to attendance rates at various events, which were often recorded in diaries.[43] The races are one example, with the *Bairnsdale Advertiser* commenting that the prolonged dry would pull a big crowd for a race meeting at Lindenow.[44] Agricultural shows also depended on fine dry weather for their success. One of the favoured forms of group recreation was picnicking, which relied on clear and still weather. Generally, the papers report the weather associated with these events, noting either too cold and wet or too hot and dry.[45]

As long as dryness was moderate and agriculturally well timed, it was deemed positive. It encouraged people to socialise, building relationships and communities through a wide range of activities. Pleasant, dry weather also kept the economy ticking along, because people could travel with more ease and conduct their business. It allowed farms to grow by facilitating clearing. In contrast, higher temperatures, hot winds and high evaporation rates combined with a below average rainfall were conditions which settlers dreaded.

Negative dryness

Fire, heatwave and drought are common experiences in Australia, and are frequently described as natural disasters. They have an extensive literature, especially with the looming threat of climate change.[46] There have been more deaths from heatwave in Australia than from any other natural hazard.[47] Vulnerability to heatwave is increased by a range of

---

43 Mark Daniel went to the New Year's Day picnic at the Knob Reserve in Stratford in 1885, and his lack of weather observations suggests a pleasant summer's day. While she didn't attend, having just given birth to her son, Margaret McCann noted that the weather for the school picnic cleared up by the afternoon. Diary of Mark Daniel, 1 January 1885, Daniel Family Papers, SLV, MS 10222; Diary of Margaret McCann, 3 November 1898.
44 *Bairnsdale Advertiser*, 30 March 1882.
45 *GT*, 15 November 1897. The holiday declared for the Prince of Wales's birthday in 1897 was marred 'by fierce hot winds', but nevertheless got a good attendance.
46 Natural hazards include cyclones, flooding, landslides, tsunamis, earthquakes, floods, fires, heatwave and drought. M Hulme, S Dessai, I Lorenzoni & DR Nelson, 'Unstable climates: Exploring the statistical and social constructions of "normal" climate', *Geoforum*, vol. 40, 2009, pp. 197–206. doi.org/10.1016/j.geoforum.2008.09.010; DM Liverman, 'Conventions of climate change: Constructions of danger and the dispossession of the atmosphere', *Journal of Historical Geography*, vol. 35, 2009, pp. 279–96. doi.org/10.1016/j.jhg.2008.08.008; M Parry, 'Copenhagen number crunch', *Nature Reports Climate Change*, 14 January 2010, doi.org/10.1038/climate.2010.01.
47 Bureau of Meteorology data cited in E Hanna, T Kjellstrom, C Bennett & K Dear, 'Climate change and rising heat: Population health implications for working people in Australia', *Asia Pacific Journal of Public Health*, Supp to vol. 23, no. 2, March 2011, pp. 14s–26s.

factors, including access to water, the extent and type of shading available (constructed or vegetative), and economic pressures. For example, a study of economic aspects of heatwave found that people working to a quota may ignore warning signals such as fatigue, thirst, tiredness, mental confusion, poor decision-making and visual disturbances. It also identified that 'workplaces with risks of extreme heat exposure include outdoor maintenance work, mining, shearing, farmwork, firefighting and other emergency and essential services'.[48] All of these identified places were the normal workplaces for the majority of colonial settlers.

Settlers developed a set of expectations about when dryness formed a natural part of the cycle and how much drought was tolerable. This was because of their ongoing attempts to establish a European style of agriculture, and their cultural beliefs about dry lands. The title of this chapter sums up their perception, the water famine and the fire demon. Unlike the Kurnai before them, who knew what a vast living larder they moved through, European settlers chose to reproduce an economy based upon a few staple species. Furthermore, the ecological changes produced by European agriculture increased settlers vulnerability to drought.[49] As Oelschlaeger notes, it is unlikely that hunter-gatherer societies starved. While there most certainly would have been lean times, it seems unlikely that all the ecosystems comprising a tribe's whole geographical range would have collapsed in a drought. Their populations were also much smaller, placing less pressure on their range. In contrast, poverty is more common with more populous sedentary, agricultural societies, which also developed class and wealth distinctions that can render some parts of the population more vulnerable than others.[50] With bigger populations relying on a much smaller number of food species, colonial Gippslanders were thus more vulnerable to fluctuations in the hydrological cycle than the Kurnai peoples they dispossessed.

---

48   Hanna et al., 'Climate change and rising heat', p. 17s. Vulnerability is defined by the IPCC as the combination of risk, exposure and sensitivity. Risk refers to the actual physical conditions or prevalence, e.g. if you live in an earthquake-prone zone. Exposure refers to how exposed an individual is to the risk. A person living on the edge of state forest has a higher degree of exposure to fire risk than a person living in an inner city area. Sensitivity acknowledges that different demographic types will have different reactions. For example, people with kidney and heart disease are much more sensitive to high temperatures than a fit, healthy 20-year-old.
49   D Garden, *Droughts, floods and cyclones: El Ninos that shaped our colonial past*, Australian Scholarly Publishing, North Melbourne, 2009, pp. 142–4.
50   M Oelschlaeger, *The idea of wilderness: From prehistory to the age of ecology*, Yale University Press, New Haven, 1991, p. 14.

The myth of Tiddalik the frog, first recorded in Gippsland, indicates that drought was a feature of Indigenous life.[51] But drought is a relative experience. Glantz noted that it could 'be defined meteorologically, hydrologically, agriculturally, or otherwise'.[52] Sherratt notes in *A Change in the Weather* that there are 150 definitions of drought worldwide, reflecting its social and cultural dimensions.[53] Agricultural drought was what terrified colonial Gippslanders. Without the embodied water used to produce foods and objects common to an agriculturally founded society, daily life in the catchment could not be sustained. They were forced to simply watch, wait and hope. The Swan Reach correspondent for the *Bairnsdale Advertiser* captured this sense of powerlessness in 1882:

> There is very little stirring here worth mentioning, and it seems to me that a correspondent's duty here is merely to dilate upon the harvest prospects, the want of rain and a few matters requiring the attention of our shire councillors, or to report the proceedings of tea meetings and other social gatherings, of which we have our share.[54]

'To dilate' carries overtones of passivity and helplessness. Who can be effective against the sun and wind? A knowledge of the El Niño/La Niña cycles they were experiencing might have helped, but that knowledge was 40 years away, at least.[55]

High evaporation rates affected many aspects of colonial life. The *Morwell Advertiser* complained how evaporation affected the farming cycle. 'The surface of the ground dries up so quickly now, giving out by evaporation the moisture underneath, that occasional gentle showers would be acceptable in many parts.'[56] George Auchterlonie noted how the

---

51   J Morton, 'Tiddalik's travels: The making and remaking of an Aboriginal flood myth', *Advances in Ecological Research*, vol. 39, 2006, pp. 39–158. doi.org/10.1016/S0065-2504(06)39008-3. Curr recorded this myth in the Port Albert district and published it in 1887. The drought is broken when one of the animals 'tricks' Tiddalik into laughing, thereby releasing the waters. I can't help but notice the similarity between this myth and the Japanese myth of Amaterseru. The difference is that Amaterseru is goddess of the sun, who turns the world to darkness when retreating into a cave. She, too, is encouraged to reverse through the use of laughter.
52   M Glantz, *Drought follows the plough*, Cambridge University Press, Cambridge, 1994, p. 10.
53   T Sherratt, L Robin & T Griffiths, *A change in the weather: Climate and culture in Australia*, National Museum of Australia Press, Canberra, 2005, p. 6. See also Garden, *Droughts, floods and cyclones*, p. 12.
54   *Bairnsdale Advertiser*, 23 November 1882, Swan Reach correspondent.
55   According to Garden, the first correct formulation of the El Niño phenomenon was put forward by an Indian colonist in the 1920s. Garden, *Droughts, floods and cyclones*, p. 2.
56   *Morwell Advertiser*, 12 October 1889, emphasis added.

dry winter of 1868 made ploughing very difficult: 'land part next lagoon kindly and somewhat moist, far end rather dry and snarly'.[57] While he succeeded in ploughing, the resulting crop struggled. 'The barley seem almost done through drought and heat. Oats very much in same state becoming quite withered.'[58]

The weight of expectation of how it *should be* is littered through the papers. A good example comes from February 1890:

> The remarkable heat and dryness of the Autumn is causing heavy loss in those parts of the colony where a fair amount of moisture is *customary* at this time of year, and is a necessity for crops produced in them. In North Gippsland, the maise and hops crops are pretty nigh ruined, especially the former, while the potato crops in other districts are suffering severely for want of moisture. Bushfires are commencing to be prevalent, though as yet there have been no extensive conflagrations. Altogether the autumn is reckoned as the hottest and driest experienced for many years past.[59]

In autumns like this, death and destruction seemed far too close for comfort. The tenor of the days echoed and reinforced familiar biblical stories of desolation and thirst.

The experience of drought was different depending on what kind of agricultural enterprise each settler was involved in and where they were located. Maffra and Bairnsdale suffered more than many other areas because of their location in rainshadow areas. Michael Landy from Briagolong described it as 'the fitful nature of rainfall. It came at the wrong time so far as hops were concerned, it came when it was no use to them'.[60] In contrast, areas in the west of the Latrobe valley were rarely affected. However, when drought did come to those areas, its impacts were more severe because they were unexpected. 'At South Warragul the complete reversal of the usual order of things has quite astonished the farmers', wrote the *Moe Register* in 1888.[61]

---

57 Diary of George Auchterlonie, 23 September 1868; see also *GT*, 1 August 1876, Glenmaggie correspondent: 'There will not be so much grain grown on Glenmaggie this season, as in former years, as the farmers could not get their land ploughed in time, for want of rain'.
58 Diary of George Auchterlonie, 31 October 1868.
59 *Morwell Advertiser*, 7 February 1890.
60 *Royal Commission on Water Supply*, Landy, p. 249.
61 *Moe Register*, 8 December 1888.

Figure 7.2: Bed of the Tambo River, Thomas Henry Armstrong Bishop, n.d.
Source: Pictures Collection, State Library Victoria, Accession no. H40967.

The failure of spring rains in 1888 tracked through the *Moe Register* show the variation of experience in drought. In early November, hot winds desiccated the pasture grasses. The crops were holding out better, but without a good interim fall of rain, they would not be worth harvesting. The Clydebank cheese factory was due to start operations, but its prospects hardly seemed promising under the conditions. A week later, those moister areas of Gippsland were predicting a good year if the potato crop survived because prices were being driven upwards.[62] By mid-December, most of the crops in North Gippsland were ruined, and the cattle benefited, being turned out to graze on whatever was left.[63] Heavy rains fell in the week before Christmas, which, although too late for many crops, had revived many orchards and home gardens. In the annual pre-Christmas yearly review, the editor noted that compared to experience in northern Victoria, the conditions had only been a partial drought, a 'favourably dry season' around Moe. Those who had sown early and prepared their grounds well had escaped the worst impacts.[64]

---

62   *Moe Register*, 24 November 1888 and 22 December 1888.
63   *Moe Register*, 15 December 1888.
64   *Moe Register*, 22 December 1888.

The twin concerns for pastoralists under drought conditions were pasture growth and stock watering points. In particular, the further one had to drive stock to water the more condition they lost, and therefore the less return a grazier would make at the sale yards.[65] On 13 February 1877, the *Gippsland Mercury* reported:

> The drought continues to increase in its intensity, and cattle in many parts of North Gippsland are dying for want of food and water. We hear that a good heavy shower fell at Morwell on Saturday afternoon, and a very slight one fell in Sale on Sunday evening; but if we are not blessed with a good heavy fall before long, the result will be most disastrous.

In very poor years, there was the extra expense of agisting stock in different districts. Many upper Maffra stock spent the summer of 1896 holidaying in Morwell and Warragul.[66] Prolonged dryness affected viability of milk production. The quality of pasture was critical to how much milk each cow could produce. If pasture was scorched by dry winds or not enough sprouted, daily supply to the creameries and butter factories would drop, also affecting their viability.[67]

Town dwellers were also affected. Carting water added extra expense to household budgets. In July 1882, the Maffra correspondent noted 'such a thing was never before seen viz. the carting of water for domestic purposes through the month of July'.[68] High evaporation rates compromised both the quality and quantity of drinking water. Commenting on drought in Bairnsdale, the *Moe Register* remarked:

---

65   In the *GM*, 10 February 1877, the Rosedale correspondent laments lack of rain and says that selectors will have to travel for water and feed. Previously, Sale stock agents Messrs English and Peck reported that due to the drought they did not anticipate any transactions for store cattle, *GM*, 6 February 1877. Later in March, the Flynn's Creek correspondent noted that many cattle had died or were suffering from ophthalmic disease, and it was thought locally that only a good rainfall would cure the survivors. *GM*, 22 March 1877.
66   *GT*, 27 February 1896, Upper Maffra correspondent.
67   *Gippsland Mercury*, 17 February 1877, noting impact of drought on cheese production; *GM*, 15 March 1877, the Upper Maffra factory had been forced to close; *GT*, 27 February 1896, Upper Maffra correspondent, supply of milk to the Newry creamery has decreased to 200 gallons daily; *GT*, 16 July 1896, 'There is already an increase in the milk supply at the Sale butter factory as a result of the fine rains of last month, and there is every reason to anticipate a good season'.
68   *GT*, 26 July 1882.

> The last place one would expect to hear about drought is surely in Gippsland, at a spot, moreover, where you can see snow on the mountains, and what is more, feel the cold wind that blows over their bleak summits. Such, however, is the case in the neighbourhood of Bairnsdale, where residents this winter have not been able to catch sufficient rainwater from their houses for domestic purposes.[69]

When rains returned, it was no guarantee of an instant supply either. The Heyfield correspondent complained in mid-May 1877, after a good series of showers generally through the district, that the Thomson River was still as dry as his evaporated ink bottle.[70] Rain that fell was sucked up by parched soil and plants before there could be enough runoff to raise river levels. High evaporation also affected water quality. Water levels could drop markedly, making water collected from rivers sludgy and polluted. In a classic understatement, the editor of the *Gippsland Mercury* said: 'The excessively dry season has rendered water very scarce, and that obtainable has not been of the purest quality'.[71]

High evaporation also highlighted poor drainage practices in Gippsland's towns. The habit of sheeting off waste to the nearest waterbody relied on water's capacity to dilute pollutants and flush them away, but less water meant less dilution. Walhalla residents in February 1877 probably cheered when a hailstorm arrived, as they believed that incidences of low fever and scarlatina would decline after the streets had a good wash down. The small matter of employment loomed too. The Long Tunnel Company could return to work.[72] The emotional impact of drought was clear in the newspapers. Words like 'despair', injurious' and 'apprehension' were

---

69 *Moe Register*, 8 September 1888.
70 *GM*, 12 May 1877.
71 *GM*, 13 January 1877, petition to Sale Council asking for a small dam to be constructed across Flooding Creek to: 1) provide a pedestrian link and 2) raise the summer water levels to help with domestic supplies; *GM*, 12 April 1877, editorial: 'There has recently been a good deal of illness in our district, and as dysentery is at the present time very prevalent, it is absolutely necessary that inspectors of nuisances be vigilant in looking after the cleanliness of localities inspected by them. We also see it as an opportune suggestion on the part of the Avon Shire health officer that all water being used for culinary and drinking purposes be filtered. The excessively dry season has rendered water very scarce, and that obtainable has not been of the purest quality'.
72 *GM*, 20 February 1877.

not uncommon. Further understanding of the impact of drought comes from the jubilant tones and sense of relief when rain did arrive in the nick of time.[73]

As if drought conditions were not a big enough threat to survival and prosperity, things could get so much worse. As early as 1875, some recognised that the spread of settlement exacerbated the suffering caused by the fire demon.[74] With the arrival of white Europeans, there were so many more opportunities for fire to take hold, like bored children playing with matches or careless travellers who failed to extinguish their campfires properly.[75] Extremely hot and dry conditions are the perfect conditions for bushfires. This example from George Auchterlonie's diary in 1868 is fairly typical:

> Reaping till about 10 o'clock had to knock of with the heat which was something extraordinary – hot wind blew hotter than I've ever felt before. In the afternoon a bushfire came roaring down upon us threatening to sweep all before it but the wind providentially reared round to the south west which stoped its progress just as it had reached within 200 yards of the house. I was most of the aft. helping to save Warner's hut, the fire was blazing fiercely about 20 yards from it but we succeeded in keeping it from getting any nearer. It was suffocating work.[76]

---

73   *GT*, 28 March 1862; *GT*, 8 February 1866, advocating agricultural insurance as an antidote to drought induced 'despair'; *GT*, 12 April 1866, describing apprehension of drought; *GT*, 6 April 1869, *GT*, 19 December 1877, Rosedale correspondent, the farmers therefore are not in the best of humours; *GT*, 19 April 1882, 'the usual growl of discontent' about the drought.

For examples of rain-related relief and happiness, see *GT*, 7 March 1862, editorial, 'whilst we are preparing for press, a fine soft rain is falling, which will be gladly welcomed by both agriculturists and squatters'; *GT*, 29 May 1866, editorial: 'It would not be easy to compute the amount of good which North Gippsland derived from the copious rainfall we had a few days ago. Previously, the country looked parched and uninteresting, and vegetation of every kind appeared to languish from the effects of the long-continued drought. Now, however, the aspect is altogether changed for the better – the grass and newly sown crops especially presenting a very refreshing greenness to the eye, after the burnt-up appearance that the country has uninterruptedly worn for so many months in succession'; *GT*, 1 February 1878, Traralgon storm initially welcomed as a drought breaker; *GT*, 8 March 1882, blessed with a day's rain.

74   *GT*, 28 January 1875.

75   *GT*, 29 February 1884. In another appalling incident of playing with matches, there was the death of a little girl, caused by her brother playing with matches in dry grass. Being hot and windy, the girl's clothes caught alight and she was burned to death. *Morwell Advertiser*, 3 March 1888.

76   Diary of George Auchterlonie, 24 December 1868.

Clearing by fire was a double-edged sword. Selectors were acutely tuned to the degree of dryness. Burns were usually in January and February, and they preferred the hottest and windiest day. This had the greatest effect on the area to be burned, but could also lead to serious repercussions. Some people lost everything because a neighbour's burn went out of control, for example Mr Hall of Leongatha lost his entire dairy herd.[77] One man even burned to death in his own clearing fire.[78] Good neighbours gave sufficient notice of their intention to burn, and had enough people to watch and help.[79] Catherine and John Currie nailed up notices when they intended to burn on 25 January 1881. An entry in her diary for the following month shows just how easily a fire might turn into a disaster:

> John lit an old tree … the flames Leaped High and catched the bark and was to the top of the tree in no time. We were very frightened as it was so near the Barn and Stock. The wind was blowing the spark right over this way it catched on other two or three and John chopped them down but the worst were such big ones.[80]

In *Forests of Ash*, Tom Griffiths notes that there are three fires in Victorian history that stand out for their ferocity and destruction. All of them took place in Gippsland and two of them took place within the study period. The first, Black Thursday, happened on 6 February 1851. The second was between 10 and 12 January 1898, although fires were active before and after those dates. The next major conflagration was the 1939 fires, by which all fire indexes were calibrated until recently. Since *Forests of Ash* was written, the fires of 2009 have eclipsed them all: 172 people in Victoria died as a direct result, and many more died in the preceding heatwave. In between the cataclysmic events of 1851 and 1898, there were other significant years of fires. Part of Angus McMillan's financial ruin was caused by fires sweeping through Bushy Park in 1861. More suffered the following year, with most of the country between the Mitchell and Avon rivers alight, and in 1870 Bairnsdale found itself encircled by flame.[81]

---

77   *GT*, 6 January 1898.
78   David Brown, Reminiscences of Brandy Creek, CGS, 696.
79   J Adams, *So tall the trees: A centenary history of the southern districts of the Narracan Shire*, Narracan Shire Council, Narracan, 1978, p. 54. See also *GT*, 21 February 1890, for a case where a Warragul farmer was fined for not giving notice. Also, there was the discussion in 1881 when the neighbours rushed over thinking there was a fire. He was backburning, and this drew serious criticism from the paper, referencing the story of Peter and the wolf.
80   Diary of Catherine Currie, 15 February 1881, CGS, 04381.
81   LH Campbell Coulson, 'Early history of Gippsland, pt 2', NLA, MS 8600, folder no. 13, pp. 32–3; *GT*, 22 February 1870.

Figure 7.3: The homestead saved, An incident of the Great Gippsland Fire of 1898, JA Turner, published 1908.
Source: Pictures Collection, State Library Victoria, Accession no. H33849.

Because bushfires were of extreme importance, local newspapers covered their extent and impact closely. The following quote from the *Morwell Advertiser* in 1888 illustrates both the extent of bushfires and conveys something of the fear and drama associated with them:

> The drought stricken condition of the country, and the long continuance of excessively hot weather rendered the occurrence of bush fires on a large scale inevitable. Fires are reported at Lakes Entrance, Toongabbie, Briagolong, Cowwar, Lindenow, Walhalla, Moondarra and different places south of the Sale Railway Line. During the past week a heavy pall of smoke has enshrouded the whole of north Gippsland, contributing not a little to intensify the discomfort caused by the excessive heat, while every evening the sun set like a globe of fire, and the sky was lurid with the glow of bushfires.[82]

---

82   *Morwell Advertiser*, 1 December 1888. Other examples of fear/anxiety comments in the papers include Dutson correspondent, *GT*, 8 April 1891, *Bairnsdale Liberal News*, 20 September 1879. The correspondents from Dargo, Carrajung, Stradbroke and Woodside in *GT*, 22 August 1886, all mention the welcome onset of rain and how it means a good season for crops.

Figure 7.4: Fire near Holey Plains State Park, close to Traralgon in the Latrobe Valley, July 2006.
Source: Author.

Diary evidence supports an interpretation of fear and anxiety. Owing to the thin smattering of population in 1851, few accounts of that massive fire's spread through Gippsland exist. Patrick Buckley, usually a terse, even taciturn, diarist, actually showed emotion. He said how afraid he was, especially of losing the whole farm, and wished himself and the staff safely on the beach.[83] George Auchterlonie felt 'suffocated' by them. Catherine Currie, a woman of anxious disposition, was frightened by the fires.[84]

---

83   Diary of Patrick Buckley, 6 February 1851. He expressed surprise that he lost no buildings and noted that native dogs and kangaroos had been lamed. D Parry-Okeden, 'The Parry Okeden family at Rosedale', *Gippsland Heritage Journal*, vol. 10, 1991, p. 51, for his remembrance of the Black Thursday fire. For the 1898 fires, see Diary of Charles Alfred Broome, 5 and 6 December 1898, SLV, MS 10774, Box 1542; Diary of Margaret McCann, 20 January and 31 December 1898.
84   A new area of research is the link between mental health and experience of natural disaster or environmental degradation, e.g. R Few, 'Health and climatic hazards: Framing social research on vulnerability, response and adaptation', *Global Environmental Change*, vol. 17, no. 2, 2007, pp. 281–95. doi.org/10.1016/j.gloenvcha.2006.11.001; P Speldewinde, A Cook, P Davies & P Weinstein, 'A relationship between environmental degradation and mental health in rural Western Australia', *Health and Place*, vol. 15, 2009, pp. 880–87. doi.org/10.1016/j.healthplace.2009.02.011. I suggest this here because Catherine Currie's breakdown may have been in part exacerbated by her feelings of fear caused by climatic extremes.

For her, the fire season could not pass soon enough, gratefully recording on 9 March 1881: 'No fear of fire now for this year. Some very heavy hail storms'. Even solid water was welcome after a taxing fire season.

The 1898 fires were not unexpected, but their scale and ferocity shocked everyone. In the spring of 1897, George Auchterlonie noted what 'an exceptionally dry spring it was'.[85] On 20 December 1897, the *Gippsland Times* Maffra correspondent wrote:

> The intense heat of Thursday and yesterday is causing land owners considerable anxiety as to the outbreak of bushfires. It is generally expected that fires will be more numerous this season unless more than ordinary precautions are taken in consequence of an abundance of grass. Already several narrow escapes have been experienced; the outbreaks being discovered before any good progress was made and at Ravenswood on Thursday about 100 acres was destroyed before the fire was mastered.

Mr J Gorman of Childers wrote to his sister: 'We are experiencing a phenomenally hot, dry season, usually we count on plenty of rain till about the middle of January, with occasional hot days; but we have no rain worth speaking of for two months'.[86] In such desiccated landscapes, ownership of a spring was literally a lifesaver. Copeland tells of a man and child who survived the fire by huddling under bags, which he periodically wetted in the nearby spring.[87] While his father and older siblings were out helping the neighbours save their property, Frank Savige, his mother and his younger siblings were able to save their own home because they could draw water from a spring.[88]

The combination of drought, heatwave and fire made everyone's worst nightmares come true, like the summer of 1881/82 where days of extreme heat, fire and dust storms prevailed. Because of the lack of official temperature measurement, it is not possible to determine how long and severe the heatwaves were, but newspaper descriptions suggest a summer best spent in deep shade by a river. In the same week the thermometer had reached 106°F in the shade, the Sale Council received a petition from 144 residents, including their own health officer and various doctors in town,

---

85   Quoted in *Coach News*, vol. 2, no. 2, December 1974, p. 5.
86   Quoted in *Coach News*, vol. 2, no. 2, December 1974, p. 8.
87   Copeland, *Path of progress*, p. 419.
88   Memories of Frank Savige, *Coach News*, vol. 5, no. 1, September 1977, p. 8.

begging that the bathing sheds be renovated.[89] The 'fearfully oppressive' weather included hot winds, a substantial dust storm and threatening fires.[90] Fears were expressed that should a westerly wind come, the fires that had been contained to the ranges above Toongabbie would escape to the flats. Between the smoke and dust, Maffra residents could 'scarcely see the opposite side of the street' in the week of 23 February. Around Tinamba the pastures were 'very much parched and burned up'.[91] At the beginning of March, the *Maffra Spectator* wrote: 'To give our readers an idea of the exceptionally dry season we are experiencing, and the great want of rain that exists, hereby is subjoined a return of the rainfall at Maffra for the month of February during the last three years:– 1882 0.6 inches, 1881 1.12 inches, 1880 4.49 inches'.

The difficult weather continued unabated. In the week before 20 February, there were fires at Merriman's Creek, Briagolong, Kilmany Park, Tanjil, Dutson and other localities. In a thunderstorm, lightning struck a swamp gum on Mr Cameron's property near Maffra, which spread flaming shards of timber through the tinder-like stubble. In mid-March, the heat was so fierce that animals at the various stock saleyards were dropping dead.[92]

In recent years, there has been research carried out on the psychological impacts of prolonged drought. Findings show a variety impacts, especially depression. Under conditions of climate change, which are likely to aggravate drought and fire weather, the Australian Psychological Association expects mental health impacts to increase.[93] Because colonial Gippslanders were pursuing the same farming enterprises, and shared common frames of reference about the desirability of agriculture, there are valid grounds to expect settlers to feel much the same as their modern counterparts. The newspaper and diary evidence shows that there is no doubt that colonial Gippslanders were threatened on most levels by the absence of water. Most obvious is the economic threat it posed: poor or nonexistent harvests, stock in poor condition or dying from thirst and starvation. Drought and depression were linked, while fire was linked to fear and anxiety. Their economic survival and their mental health were at stake.

---

89   *GT*, 20 January 1882.
90   The description 'fearfully oppressive' comes from the *Maffra Spectator*, 23 February 1882.
91   *Maffra Spectator*, 9 February 1882.
92   *Maffra Spectator*, 16 March 1882.
93   Australian Psychological Society, *Position statement on climate change*, www.psychology.org.au/community/topics/climate, accessed 4 April 2011.

Dealing with extremes

As the animals in Gippsland's stockyards found to their detriment in the appalling summer of 1881/82, dryness and heat is fatal. The water famine and the fire demon brought times of great emotional and financial distress for the settlers of Gippsland. They remained a constant threat, disrupting settlers' cherished notions of order, progress and prosperity, and put at risk the ideology of the yeoman farmer. They also put at risk the mythology of the abundant biblical garden, which they were trying to reproduce in the antipodes. These were the 'the two dominant social and economic philosophies of the time: the creation of a class of yeoman farmers and the attempt to make the deserts bloom'.[94] A number of writers have pointed out the religious imagery employed in the debate on irrigation in Victoria, which is perhaps best expressed in the cartoon of Deakin as Moses. Irrigation appeared as the most rational and logical solution to nature's threat to their achievement.[95]

Some of the most powerful metaphorical language was employed by the Royal Commission on Water in their descriptions of aridity. Deakin's report on the waterless American landscapes he saw was full of evocative imagery. He described the lands west of Kansas with terms like 'glaring barreness' and 'illimitable desert', which had 'gaping gulchures and fissures of unappeasable thirst'. In contrast, irrigation channels and ditches were places 'where one beholds industry and intelligence transmuting barren surfaces into orchards and fields of waving grain … It is here veritably the water of life'.[96] Francis Myers (Telemachus of the *Argus*) used equally strong language in the pamphlet he wrote for the Chaffey Brothers, architects of the irrigation scheme at Mildura, calling them apostles.[97]

---

94 E Harris, 'Development and damage: Water and landscape evolution in Victoria, Australia', *Landscape Research*, vol. 31, no. 2, 2006, p. 175. doi.org/10.1080/01426390600638687.
95 See HS Hawes, 'Commission to community: A History of salinity management in the Goulburn Valley, 1880–2007', MA thesis, March 2007, University of Melbourne, p. 28; M Bellanta, 'Irrigation millennium: Science, religion and the new garden of Eden', *Eras* (Monash University), no. 3, June 2002.
96 A Deakin, *Irrigation in western America, So far as it has relation to the circumstances of Victoria: A memorandum for the members of the Royal Commission on Water Supply*, Government Printer, Melbourne, 1885, p. 9.
97 F Myers, *Irrigation or the new Australia*, Chaffey Bros Ltd, Mildura Victoria, 1893, p. 2. Other biblical analogies included: 'They would dig, plant and water it; they would settle on it a population; they would transform a wilderness to a garden …', p. 5 and 'They saw sterility on either bank, and the river of life flowing between them', p. 6.

With this language, Deakin tapped into an enduring set of biblical metaphors about making deserts bloom. Locally, Gippsland residents bought into this imagery with their persistent portrayal of the catchment as 'the garden of Victoria'. At a public dinner in 1881, John King said:

> He came to Gippsland from the neighbourhood of the Goulbourn, after a period of terrible and prolonged drought, and when his eye first lighted on our fresh green grassy plains, with the rich juicy herbage reaching up to our saddle girths, his astonishment and delight could scarcely be described. It was like coming to a veritable paradise, and he could assure them he should never forget the days when Gippsland was without a fence, and (*sotto voce*) without a selector.[98]

Other gardenesque comparisons were used, for example JJ English, Sale's mayor, dubbed Gippsland as the Lombardy of Australia.[99]

Figure 7.5: Cowarr Butter Factory.
Source: Author.

---

98   Report of proceedings of a dinner held in the Duke of Manchester's honour, *GT*, 29 November 1880.
99   Discussion of English's article on artesian irrigation, *GT*, 4 March 1881. Other examples include *GT*, 3 April 1863, 16 July 1867, 5 April 1870 and 31 October 1874.

The emphasis on the productive, garden-like nature of the catchment was tied to an economic goal. When Gippslanders collaborated to send loads of produce to England in newly refrigerated containers in the 1890s, they were exporting the catchment's water in the guise of apples or butter. Chapter 6 illustrated how vital the concept of flow was to colonial Gippslanders. They wanted the products of their labour to flow out to the empire, and worked tirelessly to reshape Gippsland's hydrology in support of that vision. The water famine and the fire demon threatened their ability to participate in the imperial economy.

Yet, it is the nature of the hydrological cycle to periodically flux from wet to dry. The length of these periods of flux is highly variable across the globe, with Australia gaining the reputation for the most variable climate.[100] As Tuan and others have pointed out, European culture and science has always been disparaging about, if not actively hostile to, arid lands.[101] Falkenmark describes a blue water bias, which only values surface water and ignores water in other pathways of the hydrological cycle.[102] How might the history of hydrological science have been different if it had reflected the Australian context, where there is more water stored in soils than there has ever been in surface waters?[103] Perhaps soil moisture measurements might have been the first order of measurement, not rainfall. The apparent refusal to consider other meteorological measurements besides rainfall confirms the blue water bias held by European migrants. Had they been able to see beyond this, perhaps other ways of seeing the colonial landscape would have emerged over time.

---

100 Aplin, *Australians and their environment*, p. 49; Hanna et al., 'Climate change and rising heat', p. 15s.
101 For an example of the longstanding routine negativity, Prof RF Peel at his Presidential Address to the Institute of British Geographers said: 'Aridity, caused by the partial breakdown of the hydrological cycle, is a condition which has afflicted great tracts of the earth, throughout all of man's acquaintance with it and probably throughout geological existence'. RF Peel, 'The landscape in aridity, presidential address', *Transactions of the Institute of British Geographers*, no. 38, June 1966, pp. 1–23. doi.org/10.2307/621421. See also Y-F Tuan, *The hydrologic cycle and the wisdom of God: A theme in geoteleology*, University of Toronto Press, Toronto, 1968, p. vii; and J Linton, 'Is the hydrologic cycle sustainable? A historical-geographical critique of a modern concept', *Annals of the Association of American Geographers*, vol. 98, no. 3, 2008, p. 641. doi.org/10.1080/00045600802046619.
102 M Falkenmark, 'Freshwater as shared between society and ecosystems: From divided approaches to integrated challenges', *Philosophical Transactions: Biological Sciences*, vol. 358, no. 1440, 2003, p. 2040.
103 Aplin, *Australians and their environment*, p. 422.

What options did settlers have to limit the effects of the water famine and the fire demon? They had most capability to ameliorate drought, and this is where most turned their attention. Settlers had four options, the first of which was the spiritual option of praying for rain.

In doing so, Gippslanders were not departing from any cultural norm.[104] Historians of natural disasters have shown that in Europe prayer was a popular recourse, based upon a conception of the universe where God would punish the wicked for their sins through natural disasters.[105] They were also following Biblical instruction because, as the *Gippsland Times* noted:

> In Joel, chapter first, we are told that 'the seed is rotten under the clods, the harvest is perished. The herds of cattle are perplexed because they have no pasture; yea, the flocks of sheep are made desolate.' Under these circumstances, the people are directed to 'sanctify a feast, call a solemn assembly, gather the elders and all the inhabitants of the land into the house of God, and cry unto the Lord.'[106]

The earliest reference found in the *Gippsland Times* to the practice was in January 1866 and continues.[107] Examples include reporting in the *Gippsland Mercury* on 1 May 1877, after a harsh summer with major stock deaths and disease.[108] Usually, the intercession sought was rain, but there is also one instance of praying in relation to wind patterns. In 1882, the *Maffra Spectator* records settlers in Toongabbie praying for the wind not to change, because otherwise the fire would move down onto the flats.[109] In the report from May, readers were told that the government had set aside the previous day 'for humiliation and prayer for rain owing to the severe drought experienced in most parts of the colony'. Individually, even

---

104 T Fort, *Under the weather: Us and the elements*, Century, London, 2006, p. 10.
105 See, for example, F Mauelshagen, *Disaster and political culture in Germany since 1500: Re-historicising disaster*, Cambridge University Press, Cambridge, 2005; M Kempe, 'Noah's flood: The Genesis story and natural disasters in early modern times', *Environment and History*, vol. 9, no. 2, 2003, pp. 151–71. doi.org/10.3197/096734003129342809; A Nienhaus, 'Disaster coping and prevention in the Swiss Alps in the early 19th century', *IHDP Update* 02/2005, International Human Dimensions Programme on Global Environmental Change, Bonn, Germany.
106 *GT*, 3 January 1866.
107 *GT*, 3 January 1866. Also *GT*, 10 January 1865 and 30 March 1869; *GT*, 6 April 1869, noted that Roman Catholics had been performing a daily prayer for rain for the whole of the previous month; *GM*, 1 May 1877; *Morwell Advertiser*, 8 December 1888.
108 Advising readers of the times and locations of the services, *GT*, 3 May 1897 and 22 November 1897.
109 *Maffra Spectator*, 23 February 1882.

though surviving diaries rarely recorded such personal spiritual matters, there is indication of praying for rain, usually in reference to controlling fire. Catherine Currie's diary displays her anxiety openly, and she often prays for rain in February to protect her home and her family.[110] During the 1898 fires, people beseeched God for divine intervention, and some were converted when it came.[111]

However, praying for rain was a contested practice. In the summer of 1881/82, Bishop Moorhouse of Melbourne had been asked to authorise prayers for rain in services. He refused, saying that pleading for intervention in natural cycles was wrong, and further suggested that if the population had not conserved the water that it received, asking for intervention was downright insulting to God.[112] This message was repeated on 8 December 1888, when the *Morwell Advertiser* ran a lengthy article on the arguments for and against the practice of praying for rain, indicating a level of practice that was sufficiently widespread to warrant comment. The *Advertiser*'s journalist opposed the practice. 'Some clergymen', he wrote, 'are uttering a half hearted protest against praying for rain while facilities for irrigation surround us on every hand, facilities which no man attempts to make use of.' Why ask God to intervene when we are capable of doing it ourselves?

In his view, praying for rain demonstrated how few accepted or understood the fruits of scientific enquiry. He asserted that the world was governed by 'rigid unvarying' natural laws and, given this, no halfway intelligent ecclesiast could 'subscribe to the doctrine that the laws which govern meteorology are capable of being altered or modified in order to square with the little temporary exigencies of particular localities'. The writer's firm belief in a set of fixed laws meant that no amount of prayer could alter the course of natural events, which included drought. Belief in the efficacy of prayer meant that rain could be organised like the railway timetable, 'providing for the periodical recurrence, at convenient season, of the amount of rain or sunshine which the condition of any given district might seem to necessitate', which clearly was illogical.

Despite the jesting tone, the *Morwell Advertiser*'s journalist precisely summed up the desire to escape the variability of the hydrological cycle. If prayer worked, then it meant that men like Michael Landy didn't have

---

110 Diary of Catherine Currie, 19 February 1879 and 13–17 February 1895.
111 *Morwell Advertiser*, 4 February 1898.
112 *GT*, 1 March 1882, editorial discussing Bishop Moorhouse's refusal to pray for rain.

to testify to the royal commission that the rain didn't come till after the crops were ruined. Praying therefore was a good option and, more to the point, it was infinitely cheaper than all the others.

A second option to minimise exposure to drought was to chose one's location wisely. Gippsland was initially colonised as a response to the 1838 drought in New South Wales. John King's impressions have already been cited, and Angus McMillan was equally impressed with Gippsland's abundance. The selection era opened many of the squatting estates up to small agriculturalists. Most selections took place in the comparatively wetter early 1870s. Settlers could not have known that this was not the norm. Many would therefore have chosen an area whose appearance at the time of selection was uncharacteristic. However, because they selected in a wetter period, this established the benchmark for normal by which later years would be judged.

George Auchterlonie left the rainshadow area around Maffra and moved to Wilderness Creek south of Narracan, precisely because he wanted a more reliable rainfall. Gormandale was settled because it remained green during a drought, and so the farmers of Merriman's Creek decided to move there *en masse*.[113] While some settlers moved, the more common experience across the colonies was to demand assistance from the government. The fixed nature of farming drove the irrigation push, because they could not move to water as pastoralists could.[114]

The oft-told story of Goyder's line and the failure of farms beyond it is illustrative. The timing of selections in Gippsland, in those crucially wet early 1870s, is a mini version of the experience in South Australia. Connell reads the Goyder story as being a struggle between 'biophysical realities and human ambition'.[115] Colonial Gippslanders displayed a similar misperception, thinking that how it was when they first arrived was how the catchment was always going to be. The difference between arrival in Gippsland and perceptions of the climate is amply demonstrated in the evidence given to the Royal Commission on Water by William Palmer. 'Then your memory does not go back to the old times?' asked the commissioners. 'Not the old, moist times', said Palmer.[116]

---

113 K Huffer, *A history of Gormandale*, Gormandale Centenary Committee, Traralgon, 1982, p. 6.
114 Harris, 'Development and damage', p. 173.
115 D Connell, *Water politics in the Murray-Darling Basin*, Federation Press, Annandale, 2007, p. 8.
116 *Royal Commission on Water Supply*, p. 250.

Figure 7.6: Senescent pine windbreak, Churchill area, November 2011.
Source: Author.

Once a preferred location had been secured, the next option was property management and development. Settlers could augment supply, and lower evaporation rates by excavating dams and sinking wells, planting windbreaks, and in the home garden and orchard, using mulch. It is expected that trial-and-error testing of different species for their robustness also occurred, but that subject has not been investigated here.

Sinking wells presupposes the knowledge of groundwater, but wells were common in Europe and this was no great departure.[117] In 1877, Samuel Lacey, an engineer with a prosperous business in Sale and future joint venturer with August Niemann in artesian boring, exhibited his pumping windmill at the local agricultural show.[118] In later years, Lacey took to advertising their pumping windmills with this direct reference to lack of rainfall:

---

117 J Rattue, *The living stream: Holy wells in historical context*, Boydell Press, Woodbridge, Suffolk, 2001.
118 *GM*, 21 April 1877, report on the agricultural show. Mr Lacey's windmill is described as 'one of simple construction and which has been testified to by several cattle breeders as a most effective apparatus'.

> As Egypt does not on the clouds rely,
> But to the Nile owes more than the sky;
> So, what the Heaven's this thirsty land desires.
> Our tireless FRIEND, THE SAFETY MILL supplies.[119]

Windbreaks alleviated the desiccating effects of drying summer winds. It is striking to note, driving through Gippsland now, how many remnants of conifer windbreaks have survived. Howitt was adamant on the need for windbreaks in hops cultivation, for he found that the 'winds have a most injurious action in bruising the vines'.[120] For example, a report on an experiment in windbreaks was included in the *Gippsland Times* in 1897. Next to irrigation, windbreaks were praised as the most important thing a farmer could do to protect crops. A controlled experiment was performed, and the performance of the orchard inside the windbreak far outstripped the unprotected one.[121]

Almanacs and more specialist publications also give some insight into the ways in which settlers could attempt to manage hot and dry conditions. Much of this related to the home vegetable garden and orchard, which could be more intensively managed. Moreover, this information was known from as early as 1841 in *Kerr's Melbourne Almanac*. The January entry for the 'Farmer's Calendar' reads, 'in this colony from the sudden effects of the hot winds, no care can guard the farmer from occasional heavy losses by the falling off of the heads of the various cereal grasses, more particularly barley'. In early spring the advice was that 'the hot winds occur in the latter end of November, or beginning to end of December; if the wheat is not out of blossom before these winds come on, it is sure to be more or less blighted'.[122] To prevent evaporation from the soil surface, *Castner's Rural Australian* gave detailed instructions on how to mulch in November 1875.[123] In January 1875, it noted that with shading and mulching, gladioli made a brave addition to the summer flower garden.

---

119 *GT*, 18 November 1897, emphasis in original.
120 *GT*, 23 October 1882.
121 *GT*, 13 December 1897. See O Archibold, 'Living in a garden in a valley', *Australian Garden History*, vol. 11, no. 3, 1999, pp. 10–14, reference to planting pines as windbreaks in the Latrobe catchment. Also, for the Macalister catchment, see *GT*, 12 March 1883; *GT*, 23 November 1887, design of the Yarram Episcopalian church takes account of needing to break the east wind; *GT*, 16 June 1890, Michael Landy's farm on Freestone Creek 'occupies a well sheltered position, being protected on the north and west sides by the foothills which form an effective break for the hot winds of summer or the westerly gales of winter'; *GT*, 6 February 1891 and 10 February 1892, on the use of ti-tree as a windbreak at Prospect.
122 *Kerr's Melbourne Almanac*, 1841, pp. 208 and 213. See also *Port Phillip Patriot Almanac and Directory* for 1847, *Levy Brothers and Co. Victorian Almanac* for 1872, *S Mullen's Victorian Almanac* for 1882 all of which offer similar advice as to the problems of cultivation in January and February.
123 *Castner's Rural Australian*, November 1875. Also *Glass's Almanac*, 1862, February entry.

In February, the principal advice was to install water tanks. The advice from *Glass's Almanac* was to protect young trees from winds in February. The almanacs clearly show that the difficulties of the Australian summer were well understood from an early stage, and that there was plenty of pragmatic advice that individuals could act upon to prevent moisture loss from their crop, orchard, vegetable or ornamental garden.

Finally, the last option was irrigation. Colonial Victoria was the epicentre of irrigation development in Australia.[124] The scholarly attention given to the large, centralised schemes has obscured the many individual efforts made to practise irrigation at the domestic or farm scale. A report from 1869 suggests basic irrigation practices at the Aboriginal mission at Ramahyuck, while, in 1876, Blythe and Howitt were irrigating their large hops plantations on the Mitchell River.[125] The success of the artesian well at Sale prompted William Pearson to start boring at Kilmany Park to facilitate irrigation.[126] The Royal Commission on Water Supply visited Palmer's Charlcote estate at Clydebank and William Craig's Craigelee near Stratford, both of whom were practising irrigation on relatively large scales.[127] Other references have been found to individual farmers irrigating parts of their property.[128] And, as

---

124 J Powell, *Watering the garden state: Water land and community in Victoria 1834–1988*, Allen and Unwin, Sydney, 1989.
125 *GT*, 10 August 1869 and 29 December 1876.
126 *GT*, 3 March 1881.
127 *Royal Commission on Water Supply*, Evidence from 26 June 1885.
128 Examples include *GT*, 21 January 1884; Park House, 28 February 1882, 'The machinery consists of a Wood's string reaper and binder, threshing machine, reaper, mower, double and single furrow ploughs, chaff cutter &c., all by the best makers, and a McComas water lift. Although the latter may be a very suitable pump for drawing water within a few feet of the surface, I do not think it adapted for lifting from 15 to 20 feet below the surface, the power required being too great, and also the speed at which it is necessary to drive it, are drawbacks to its more common use. One thing in its favour, its non-liability to get out of repair'. Hales, *Gippsland estates*, p. 7; 'From the early 1880s onwards trials were carried out on irrigation methods. In 1884, a Californian pump was imported that used one horse in a horse works. That horse, walking around in a circle, pumped the water up from the creek or river to irrigate the flats. Later, on the river flat, Tom Kendal and Henry Drew purchased a stationary steam engine to drive a pump that was shared between the two farms. In the same area, John Higgins used a steam pump in conjunction with a giant flume to carry water to a series of irrigation ditches he had dug across the paddocks.' L Barraclough & M Higgins, *A valley of glens: The people and places of the Upper Macalister River*, Kapana Press, Bairnsdale 1986, p. 40; 'The water race has been used to irrigate the river flats were he grew crops of oats, peas and potatoes which he sold to the nearby farming community': 'Return to Swingler's Flat', *Coach News*, vol. 14, no. 4, June 1987, p. 15; Extracts from report of an excursion to Gormandale: Cr Hiam of Rosedale Shire (after who Hiamvale was named) at his property in the 1880s (by then a pine forest) had built an irrigation system from a spring some distance away. *Traralgon DHS Bi Monthly Bulletin*, vol. 12, no. 2, May 1981, p. 10; Notes of a talk given by Jean Galbraith: 'Pipes were made from cylindrical bark taken from young saplings to water the plants'. *Traralgon DHS Bi Monthly Bulletin*, vol. 3, no. 3, July 1972, p. 4; Bushy Park: 'The benefits of irrigation however attracted attention, and Mr Mackintosh is now seriously considering the advisability of erecting a large pumping plant and going in extensively in irrigation. The house is pleasantly situated on the banks of the Avon and is surrounded by a fine old orchard containing a great variety of trees … Water is raised from the river by means of a Tangye engine and pump, and laid onto the house and garden'. *GT*, 29 April 1887.

Blackburn notes, the role and importance of Chinese market gardeners and their irrigation and land management practices in making irrigation seem feasible should not be understated. Blythe, as mentioned above, is noted as using Chinese labour to construct his irrigation system.[129]

The experience of drought had already made the presence or absence of water a significant issue by the time irrigation's most celebrated promoter, Alfred Deakin, became a minister. At the end of 1884, he was appointed chair of a royal commission to investigate the applicability of irrigation to Victoria. He travelled in California between December 1884 and May 1895, and his report recommending irrigation was presented in June 1885. The Irrigation Act was introduced into parliament in June 1886.[130] As Powell demonstrates in *Watering the Garden State*, most of the emphasis in this era was in northern Victoria, especially areas abutting the Murray River. The comparative dryness of the climate in the inland north makes the fervour for irrigation more understandable. In the GLC, the interest was less pronounced, but the drier eastern sections of the GLC followed the broad trend of interest in irrigation. There was widespread support for irrigation as a concept, and most in Gippsland believed in its progressiveness.[131]

One of the strongest public advocates of irrigation was the Bishop of Melbourne. In December 1883, he arrived in Sale to preside at the foundation stone ceremony of St Paul's Church. His address used irrigation as his metaphor. He considered irrigation to be an alignment of humans with the forces of nature, in much the same way that dedication to Christ was aligning oneself with the true nature of prosperity.[132] The bishop told of the experiences of settlers on the northern Victorian plains, who he described as belonging to 'a grand, an imperial race – one that never knows when it is beaten, that cannot say die'. He described their ruin when rains 'failed', and then pointed out that he, the wise man of God, had advocated for irrigation five years previously. Using catchment imagery and language, there was rain on the mountains, and rivers that

---

129 G Blackburn, *Pioneering irrigation in Australia to 1929*, Australian Scholarly Publishing, Melbourne 2004, pp. 43 and 63–5.
130 JA La Nauze, *Alfred Deakin: A biography*, Melbourne University Press, Melbourne, 1965, pp. 84–6.
131 *GT*, 5 September 1884, is a good example. Councillor Davis of Maffra Shire gives a public lecture exhorting the young men of the district to take up the cause of irrigation because 'it would increase the wealth of the district to a great extent, and in that they were all interested'.
132 *GT*, 19 December 1883.

would transport the rain past them, if only they would go to the trouble and expense of getting it. This, in the bishop's view, was 'humouring Nature', or 'bringing their energies into line with the natural energies around them'.[133] In this view, irrigation seems less like an intervention in the hydrological cycle than an accommodation with it.

Bairnsdale established the first irrigation trust in 1891, but there had been concerted lobbying in the decade prior, especially by Michael Landy who was a significant farmer in Briagolong. He was a consistent advocate for irrigation from as early as 1879 when he gave evidence to the inquiry into land selection.[134]

The increase in newspaper reporting about irrigation from approximately 1884 is the combination of the establishment of the Royal Commission on Water Supply, and the cumulative experience of desiccating winds, variable rainfall, crop failures and a run of bad fires. The slowly worsening dryness prompted collective action by several of the shires in 1885. They extended an invitation to the royal commission to inspect the region. Hoping to gain grants for schemes of irrigation, the visit was intended to establish in the minds of the commissioners an image of Gippsland as a place of hydrological deficiency. An extension of their previous *modus operandi*, Gippslanders were very experienced at lobbying government for support for capital works. Between the railway, de-snagging the rivers and the creation of the entrance, they had already appropriated millions of pounds of colonial revenue in the endeavour to overcome Gippsland's hydrological barriers to progress.

The commission visited in June 1885, inspected parts of the Thomson, the Macalister and the Mitchell rivers and took evidence from 11 men. Charles Geoghan, Secretary of the Borough of Sale; Jan August Niemann, water borer; Thomas Lloyd Flegge, farm manager; Michael Landy and William Palmer, graziers; and George Jones, the engineer at Maffra Shire, were examined in Sale. In Bairnsdale, the commission examined Peter Bredt, Bairnsdale Shire Secretary; Joseph Taylor, hops and seeds grower; William Ross and Andrew Macarthur, both shire councillors; and James Bankin, a mixed farmer.

---

133 *GT*, 19 December 1883.
134 JM Powell (ed.), *Yeomen and bureaucrats: The Victorian Crown Lands Commission 1878–79*, Oxford University Press, Melbourne, 1973, p. 353.

The evidence proffered is striking in a number of ways. Only two of the 11, Niemann and Flegge, depart from the script of enthusiasm and boosterism. Niemann was more interested in protecting his water-boring business, so his evidence must be read in this light. 'I wish to say that bringing water to Sale from the River is like bringing coals to Newcastle.'[135] He argued for an artesian supply, which would be more reliable that dam water distributed through irrigation channels. Flegge described his experience of irrigating from an artesian bore drilled by Niemann. While agreeing that irrigation could be good, he found it too expensive and ill suited to his very flat lands. He pointed out that very expensive infrastructure would be needed to retain the constant flow of the ground water, along with distribution channels and thought the expense not worthy of the result. He also pointed out the porosity of the soil as a problem.[136]

The commissioners probed both the costs of works installed by individual farmers and the anticipated costs of the grand schemes presented for their consideration. They also tried to gain an understanding of the expected increase in production as a result of irrigation, by understanding the losses from drought. Only Landy, Palmer, Taylor and Bankin could give even approximate figures of the costs of their works and returns. Palmer said that his carrying capacity went from one sheep to 10.[137] Landy based his calculations on the difference between wet and dry years and estimated his losses from drought over the past three years at £20 an acre. Most reckoned that pastures could hold three head to an acre under irrigation, compared to one, dry.

What is most striking in the evidence is the near complete lack of reliable figures from anyone, compared to their ability to make sweeping claims for the desirability and utility of irrigation. Landy claimed that dams in the Alps could water 'the whole of Victoria'. George Jones presented a plan to dam the Macallister and irrigate as far south as Rosedale without having made an estimate of the average flow rate. The slightly less bold Peter Bredt told the commissioners that 'unlimited supply' of water in the Mitchell River catchment could supply 'half of Victoria', because of the vastness of the Mitchell's catchment and the snow which fed it. He could not answer questions about how the water might be brought from the Alps, but suggested that 'it is a mere matter of engineering to bring the water

---

135 *Royal Commission on Water Supply*, Niemann, p. 244.
136 *Royal Commission on Water Supply*, Flegge, p. 247.
137 *Royal Commission on Water Supply*, account of site visit on 26 June 1885.

from a higher level to lower'. Bredt at least had an estimate of flow rates, a 6 mile an hour current, although he did not disclose his method of calculation to the commission.[138] By this stage, the commissioners were finding such boosterism tedious. They grilled Bredt, who appeared foolish when forced to admit that he did not know of any catchment that could supply water to catchments beyond it. The *Bairnsdale Advertiser* did not report this aspect of proceedings.

The best data the witnesses had to give the commissioners was rainfall data, and even then, they fudged. Bredt submitted returns saying: 'Formerly our rainfall went as high as 5 and 6 ft per annum. Last year we had 21.17 inches, the year prior to that 25.85'. However, he had to admit that that his rubbery estimate of rainfall in feet was in a flood year, and before rain gauges were installed.[139]

Only Bairnsdale farmer Joseph Taylor explicitly identified evaporation as a problem. He told the commission that much of the water he applied in summer evaporated before it could do any good. He preferred winter irrigation because it would ensure enough soil moisture would be present during the growing season. He also thought that irrigation was better for non-alluvial lands, believing that evaporation from the river condensed on alluvial lands and gave it an advantage that non-alluvial lands didn't have.[140]

There was no one who thought irrigation an inherently bad thing, even if their experience of it had not been trouble free. They had all suffered in the recent drought, or when rain came at the 'wrong' times. They all had a clear conception of irrigation's benefits, which Landy summarised for all. 'We all have the idea that, if this colony is to go on progressing, there must be some general scheme of irrigation, to supplement natural rainfall and equalise the seasons, and to give some certainty to production.'[141] In this single sentence, Landy stated the argument of this research. Colonial Gippslanders believed in growth as much as they believed in God, and both were good. The greatest amount of growth would occur with reliable and predictable water supply. Permanently flowing rivers were the most desirable, but Gippslanders were also showing themselves to be willing experimenters with ground water. They were less fussy about where the

---

138 *Royal Commission on Water Supply*, Landy, p. 248, Jones, p. 251, Bredt, p. 252.
139 *Royal Commission on Water Supply*, Bredt, p. 254.
140 *Royal Commission on Water Supply*, Taylor, p. 255.
141 *Royal Commission on Water Supply*, Landy, p. 249.

water came from, as long as it was there and was fit for the intended purpose. Irrigation was seen as a technical and rational solution to even out the vagaries of nature, just as the efforts to open the permanent entrance were seen.

Interest in irrigation continued, even in the wetter parts of the GLC. In September 1887, the *Morwell Advertiser* remarked that the waters from approximately 20 feet below the Macallister floodplain were suitable for irrigation because they did not injure plant life.[142] Further calls for a widespread scheme of irrigation came from the editor of the *Morwell Advertiser*, writing in December 1888:

> the voice of the country should be unanimous in demanding that the government of Victoria should see to it that steps should at once be taken with a view to the mitigation of the horrors of future drought, if indeed it is not possible to prevent the occurrence of droughts, or at least of an insufficient supply altogether, by means of a system of irrigation on a grand and comprehensive national scale.[143]

Irrigation was given a further encouraging push when Alfred Deakin was invited by the Sale Commercial Association to lecture in Sale in January 1890. However, the Bairnsdale Council had already announced its intention to form an irrigation trust.[144] Michael Landy's previous attempts to encourage irrigation had failed due to fears of the cost, not because of any objection to the idea of irrigation. Bairnsdale district residents were about to enact the same debate.

Deakin's lecture was extensively reported by the *Gippsland Times*. He emphasised the predictability and permanency of water:

> In a country such as this, it was impossible to farm on scientific lines without a regular and adequate supply of moisture. It was unjustifiable that a man prepare his land and put in expensive crops, with a certainty staring him in the face that unless the clouds sent sufficient rain his labour would be useless. Farming on these lines meant 'betting on the clouds', but give the farmer the certainty of moisture and he has the keystone of success.[145]

---

142 *Morwell Advertiser*, 17 September 1887. This could be taken to indicate an implicit understanding of the impact of salt on plants, and possibly the presence of a saline aquifer already present within the catchment. Salinity has become a problem in the flatter and lower drained portions of the GLC in the late twentieth century requiring management plans.
143 *Morwell Advertiser*, 3 December 1888.
144 *GT*, 17 January 1890, Maffra correspondent.
145 *GT*, 27 January 1890.

Figure 7.7: Carl Walter's image of a prospector on Freestone Creek, 1867.
Source: National Library of Australia, Accession no. 4587417.

Immediately after Deakin's speech, a group of interested landowners got together to discuss opportunities for extending irrigation in Gippsland. The men included Michael Landy and George Jones. Jones produced a map of the County of Tanjil, in which he had marked out places suitable for damming and the likely places that could be developed as irrigation areas. Dam sites included Alick's Downfall, on the Avon River and Freestone Creek, which would irrigate the areas around Maffra, Sale and the Heart, Boisdale to Marlay Point, and Briagolong to Stratford incorporating Bushy Park. They agreed to organise a series of public meetings, through the respective local governments, to gauge support for a scheme. History was repeating itself, as George Jones had not varied the plan he presented to the royal commission in 1885.

7. BETWEEN 'THE WATER FAMINE AND THE FIRE DEMON'

After the lecture in Sale, Deakin proceeded to Bairnsdale, where he attended a dinner given in honour of local members Allan McLean and Henry Foster. The dinner was also the public announcement of the 'new and great enterprise': the irrigation trust. In his speech, Deakin linked concepts of flow and irrigation. He encouraged Bairnsdale residents to take irrigation seriously, because the amount of irrigated produce would be greater and the railways would derive more revenue. This would lend further weight to any claims from Bairnsdale for new lines, stock or infrastructure when it came to debating the proposed new Bill.[146] Deakin thus linked very explicitly the flow of water with the flow of goods and revenue.

The Bairnsdale farmers accepted the encouragement. The trust revived a plan originally floated in 1885 to construct a weir at Glenaladale.[147] It received a loan of £106,662 to carry out the works, which is where the trouble started.[148] The royal commission evidence showed that very few settlers had a good grasp of economics or of hydraulics. These twin failings would eventually destroy the Bairnsdale Trust. The loan proved to be woefully inadequate for the job and, as the works proceeded, more and more money was required. A faction of farmers formed against the trust, protesting the escalating costs. Throughout the fighting, construction work on the weir at Glenaladale continued. However, the collapse of the trust in the face of the opposition, combined with a flood that damaged the wall, sank the hopes of the irrigationists.[149] The available flow of money could not support the wish for flowing water out of season. The expansion of irrigation in the Maffra district would have to wait for another two decades.[150]

The failure of the first irrigation scheme in Gippsland shows that there were nuances in the devotion to permanent flowing water. The divisions were predicated upon the need of some farmers to avoid increasing their personal level of debt. A water rate would be enduring and, no doubt, escalating, and a farmer could quite reasonably decide that the proposed

---

146 *GT*, 27 January 1890.
147 *Bairnsdale Advertiser*, 5 May 1885.
148 Adams, *Path among the years: History of the Shire of Bairnsdale*, Bairnsdale Shire Council, Bairnsdale, 1987, p. 116.
149 Adams, *Path among the years*, pp. 116–18.
150 This has been told by Meredith Fletcher in both *The small farm ideal* and *Avon to the Alps*. I have not included it here because the time lag is too great.

water rates might be more productively spent elsewhere on the farm. Had the personal costs to farmers been more reasonable, there is no doubt that the trust would have succeeded.

Had the irrigationists been as scientific and rational as they proclaimed, and based the trust's figures on well-thought-out planning and financing, there would have been no need for opposition. The irrigationists fell because of their own boosterism, not because anyone objected to irrigation itself. The ideology of the yeoman farmer and the garden mythology permeated the social fabric so strongly that anyone opposing irrigation could not be taken seriously.[151] Irrigation was progressive, and as both Michael Landy and Alfred Deakin proclaimed, it provided certainty in an uncertain world. Cloud betting was not the way forward for a brave, pioneering white race.

---

151 Reasons for this have been explored by K Proust, 'Learning from the past for sustainability: Towards an integrated approach', PhD thesis, The Australian National University, 2004; Hawes, 'Commission to community'; M Sexton, *Silent flood: Australia's salinity crisis*, ABC Books, Sydney, 2003.

CHAPTER 8

Mirror, mirror? The reflective catchment

Water is my eye, most faithful mirror ...

Newton Faulkner[1]

Be praised, My Lord, through Sister Water;
She is very useful, and humble, and precious, and pure.

St Francis of Assissi[2]

Water is the earth's natural mirror. Peering into the waters of the Gippsland Lakes Catchment (GLC), we see a story on many levels. At one level, it is a variation on humanity's love affair with and dependency on water. At another, it is about what people see and how they see it. More fundamentally, it is about learning from the ways in which that love affair and those perceptions shaped practices in the past, and analyses the impacts of those on the present.

For most of history, humans have lived within the controls set by nature. Hours of activity were curtailed by the amount of daylight, and warfare was a summertime activity. As long as the changes in environmental conditions were reasonably regular and predictable, societies could prepare and survive. It was the unpredictable and unforeseeable changes in the environment that threatened survival. The amazing thing about the Victorian era was that for the first time it looked as if humans could be free from those ecological limits. Imagine the optimism such a thought created – the conviction that 'nature' was there to be mastered, engineered and trained, and not just in its local patterns but globally!

---

1 From the song 'Teardrop' on Newton Faulkner's album, *Handbuilt By Robots*.
2 St Francis of Assissi, *Canticle of the sun*, c. 1225, webapp.stthomas.edu/recyclingquotes/catagory.html?catagory=15, accessed 4 September 2007.

Recently, the concept of the Anthropocene has been formulated to account for the extent to which human activities associated with the Industrial Revolution began to significantly affect world environmental systems. This book – in dealing with an outpost of nineteenth-century imperial aspirations, capacities and experiences – has sought to understand the intersection of that faith in transformation and that accelerated impact. Through the figure of the hydrological cycle, it has sought to comprehend the allure and comprehension of water as a resource, the interventions in water management as a part of patterns of colonial settlement, and the enduring impact of such an amalgam of factors on a catchment system. And, as an integrated study, it has sought not to simply separate 'cause' and 'effect' in terms of people's mishandling of an ecology, but to account for the ways in which that ecology was made comprehensible in the first place.

From our vantage point in the twenty-first century, living the effects of those colonists' optimism, we know they were wrong, just as we appreciate how much we have – in the short term – benefited from their faith in progress. We have the improved life expectancy and universal education, as well as the extinctions and the pollution, to prove it. But in 1838, the world was their oyster. This was the partly the gift that science and the Industrial Revolution gave our colonial forebears, but – as this book also shows – that science and industry were also meshed within rich cultural fabrics of faith, custom, labour and trade that are less easily separated out as 'causes' to be indicted. Rather, like the hydrological cycle itself, that optimism worked within a system that was hard to grasp in its entirety, and in which effects ran in wide and enduring channels. The combination of narcissism and undreamt of power the settlers of Gippsland carried with them transformed not only the Lakes catchment, but also the world systems in which they understood themselves.

What do people see? As Michael Cathcart writes:

> The notion of 'projection' can imply that the world is merely a blank field, waiting to be imprinted with our landscapes of desire: That there is only language. The reality is that the world answers back. Project whatever you like. If you fail to find water, you will die.[3]

---

3   M Cathcart, *The water dreamers: The remarkable history of our dry continent*, Text Publishing, Melbourne, 2009, p. 3.

Yet people also see the world through filters, and those of the Victorian settlers included their belief in being the favoured parts of God's creation and their belief in science and progress. Perception is as much a cultural act as it is a biophysical one. Sewall, in *Sight and Sensibility*, takes a tour through definitions of seeing across Hindu, Buddhist, ancient Greek, medieval Christian, eighteenth-century Enlightenment and twentieth-century neurophysiological versions of seeing. She demonstrates that perception is a cultural act. James Hillman has said that 'we see what our ideas let us see'.[4] This book is an extended excursion into the cultural perception of the hydrological cycle held by nineteenth-century white settler immigrants to Gippsland, seeking equally to engage with what shaped those perceptions – from the transitional mix of versions of the cycle accessible to them (as discussed in Chapter 2) – to the justifications given for profound interventions in that cycle in the name of commerce, comfort and certainty. And, in building on a reconstruction of that perception, this research has also traced actions that had an impact on the hydrological cycle of the catchment, with ramifications that speak to contemporary concerns about the integrity of ecological processes, of sustainability and ecological change.

What emerges from reading the sources is the settlers' dominant narrative of lack and deficiency in the hydrology around them. The rivers were too sluggish, or too shallow. The banks were too steep. The bogs were too boggy or the sands too sandy. There was too much rain, or not enough. All of these were impediments to the flow of goods and people in the imperial economy. Millions of pounds of public and private money were spent to rectify perceived hydrological barriers in the pursuit of a connected imperial economy based on agricultural trade with Britain and her colonies. The evidence marshalled here for the multiple influences of water on colonial Gippslanders confirms the importance of water to every aspect of their lives – that there was no simple lack of 'care' in their actions. Gippslanders were as complex and dynamic as the hydrological cycle itself, as shown by the interrelationships between their economic, material, social, intellectual and spiritual lives.

As McNeill notes in *Something New Under the Sun*, 'the health, wealth and security of any and all societies depended upon getting sufficient supplies of sufficiently clean water to the right place at the right time, without doing too much damage in the process'.[5] But what is significant

---

4   L Sewall, *Sight and sensibility: The ecopsychology of perception*, Jeremy P Tarcher/Putnam Books, New York, 1999, p. 54.
5   JR McNeill, *Something new under the sun: An environmental history of the twentieth-century world*, WW Norton and Co., New York, 2000, paperback edition in 2001, p. 5.

and worth noting, both in the particularities of a case study as well as the generalities of an environmental process, is the mix of available knowledge, technology and will that transforms a system to meet those ends, and leaves enduring impacts.

The Kurnai had permanent access to good-quality fresh water, in the same catchment, and their changes were minimal. Warner describes the contrast between hunter-gatherer societies and modern ones as moving 'from low, slow impacts associated with small populations with low levels of technology to large, high consumer populations, with a technology where almost anything is possible'.[6] All the problems facing the catchment now, such as eutrophication, salinisation and erosion, can be traced back to changes set in motion within the first 70 years of settlement. The near collapse of the entire Lakes ecosystem in closing decades of the twentieth century is attributable to a significant amount of organic pollution generated by various physical alterations such as drainage and agricultural development, combined with the stratification of the water column created by the permanent entrance.

Before the arrival of Angus McMillan, the Lakes were a fresh to brackish water system fed by five rivers that flowed through a largely vegetated catchment. At the turn of the millennium, there are still five rivers, but some bear little resemblance to their former selves. The Latrobe in particular is not the Latrobe of 1838, having been dredged, channelled and dammed along its course. In addition there are miles and miles of drains, delivering stormwater and its toxicants into the rivers and lakes. Whole ecological communities have been destroyed or remain critically endangered.[7] The entire assemblage of fringing vegetation species around the lakes has been changed, from fresh/brackish species to saline species.

The full scale of change apparent in the health of the catchment cannot be fully laid at the feet of nineteenth-century colonisers. There was a 100-year time lag in the system for some of the changes, which, incidentally, makes the GLC a handy analogy for the current debate on the need for climate change mitigation and adaptation. But everything was well underway, clearing, draining, dredging and snagging. The twentieth century merely intensified the effort. Seemingly small and positive changes in human

---

6   RF Warner, 'Do we really understand our rivers? Or rivers in the pooh-semper in excreta', *Australian Geographical Studies*, vol. 34, no. 1, 1996, p. 11. doi.org/10.1111/j.1467-8470.1996.tb00099.x.

7   I Lunt, 'The distribution and environmental relationships of native grasslands on the lowland Gippsland Plain, Victoria: An historical study', *Australian Geographical Studies*, vol. 35, no. 2, July 1997, pp. 140–52. doi.org/10.1111/1467-8470.00015.

lifestyles combined synergistically, and over the long term, to bring about a near collapse of the lakes aquatic ecosystem. Driving those changes was a conflict between the catchment as it was and the catchment as the colonists thought it should be.

Modern Australian society is as much in thrall to the glamour of progress and development as colonial Gippslanders.[8] Because we now have laws, policies and programs to protect water and biodiversity, and a science that more readily understands and accommodates systemic effects, we like to think we have moved on. Yet by almost any indicator or framework for assessing impacts on ecological processes, the situation, both domestically and globally, is getting worse.[9] The *World's Water Report* for 2008–09 suggests that many catchments around the planet have reached their 'peak ecological demand', defined as the point at which the benefits delivered to humans through abstraction are outweighed by the cost of the damage caused by the abstraction of water.[10] At current rates of consumption (unevenly distributed between developed and developing countries) of natural resources, by 2030 we will need two planets.[11] Yet, as the Biosphere II project showed so unflinchingly, our capacity to replicate ecological processes is inadequate. There is no second planet, and we can't do as good a job as Nature does.

---

8   For example, see H MacKay, *What makes us tick? The ten desires that drive us*, Hatchette, Sydney, 2010, ch. 8.
9   The cumulative impacts of settlement lead to the introduction of the Victorian River Health Strategy in 2001. The 2009 audit demonstrates the comprehensive level of rehabilitation required, e.g. 'Seventy five percent of the lower Nicholson River is now excluded from stock, with 60 per cent fully revegetated. Works are continuing to establish a stock free riparian zone connecting the mountain reaches with the Lakes by 2013. Two other systems are nearing completion: the Wonnangatta River above the Wongungarra River confluence (Mitchell River catchment), and the Thurra River'. Victoria, Department of Sustainability and Environment, *Securing our rivers for future generations: Victorian river health program report card 2002–2009*, the Department, Melbourne, 2009, p. 18. The 2008 Victorian State of the Environment report showed that 'only one fifth of major rivers and tributaries in Victoria were in good or excellent condition', that '21 fish species, 11 frog species and 29 species of waterbirds are threatened, and only 14% of riverside vegetation along major rivers and streams in Victoria was found to be in good condition'. An assessment in 1994 found that one-third of Victorian wetlands had been destroyed, with most of that loss, fully 90 per cent, occurring on private land and largely drained for agriculture. Commissioner for Sustainability and Environment, *State of the Environment Victoria 2008, Summary: Living well within our environment – Are we? Can we?*, the Commissioner, Melbourne, 2008, p. 9. Globally, biodiversity loss rates have not slowed, protection for key habitat is insufficient, more species are being driven towards extinction, 13 per cent of the world's population doesn't have access to clean drinking water and fully half of the population of developing countries do not have sanitation. United Nations, *Millenium development goals report*, United Nations, New York, 2010, ch. 7.
10   PH Gleick, H Cooley, MJ Cohen, M Morikawa, J Morrison & M Palaniappan, *The world's water 2008–2009: The biennial report of freshwater resources*, Island Press, Washington DC, 2009.
11   *The footprint network*, footprintnetwork/org/index.php/GFN/page/world_footprint, accessed 10 December 2010.

How does an environmental history of a regional catchment in the nineteenth century help us when faced with such a situation? It helps by providing that proverbial mirror for ourselves. The mirror shows that there is only a sliver of difference between then and now when it comes to thinking about growth, progress, development and all those other superficially good things about Western civilisation. So, what is that hair breadth? It could be summed up in a single word. Connection. We live in a connected world, and there are profound differences between which connections different cultures value.

Colonial Gippslanders did want to live in a connected world. They appreciated that they were in an increasingly sophisticated and complex web of global economic and political connections (sound familiar?). They were, after all, one of the more far-flung outposts of British imperialism, although none the less patriotic for their distance from Queen Victoria. They were indefatigable about becoming better connected with that world, primarily though transport infrastructure and communication technology. Again, sound familiar? To build such infrastructure networks frequently involved major engagement with the hydrological cycle. To sell their apples and butter in a London winter involved the hydrological cycle at every stage, whether as an input or impediment. That kind of input/output thinking is characteristic of their quantitative and volumetric vision of hydrology. They couldn't see the hundreds of litres of water that went into the butter, and the cow that produced the milk. And they definitely couldn't see the rising groundwater caused by the clearing of forests to make the pastures, or lament the homeless birds and incinerated marsupials in their clearing fires.

It does matter that colonial Gippslanders replumbed the catchment. Nor were they the only ones. Just about every catchment in Australia is degraded somehow, whether that be through low flows, over-clearing or weed infestation. To rectify the damage, national, state and local policies on integrated catchment management require millions and millions of taxpayer funds.[12] Ironically, many of the actions are the absolute reverse

---

12   In 2008/09 spending on the Avon and Perry rivers and Freestone and Valencia creeks (draining into Lake Wellington) was $537,500, plus commencement of a $1.2 m project to protect farmland along the Avon from flood-induced erosion. For the Latrobe catchment, the 2008/09 spend was $2,509,500; $803,500 was spent in 2008/09 on the Thomson catchment. All the above from West Gippsland Catchment Management Authority website. See also, for a broader view of investment into remediation, LTH Newnham & JJ Drewry, *Modelling catchment scale nutrient generation*, Technical Report 28/05, National River Contaminants Program of Land and Water Australia, CSIRO Land and Water, Canberra, 2006.

of what colonial settlers thought of as progress – for example, the Victorian River Health progress report released in 2009 shows large woody debris being placed back into rivers, to rebuild instream habitat.

Our colonial forebears realised how essential water was for their own lives and livelihoods. They failed to see that every other species had the same needs. The majority of Gippslanders were incapable of perceiving the breadth of ecological connections that supported them. Not that they are to be judged harshly for this. Their personal and collective history was about maintaining human life in the face of the exigencies of nature. Ours is opposite. We need to maintain nature in the face of the exigencies of human life. In the 119 years that have passed since the arrival of the twentieth century, the inheritors of the Western rationalist tradition are trying to learn what the Kurnai already knew. Humans are only one species among many, sharing a living world where life is only made possible by the presence of water. 'However slight,' suggests Stephanie Dowrick, 'any shift in perception to a more inclusive and respectful vision is always significant – and never for our own sakes only.'[13]

Figure 8.1: Swans feeding at dusk, Lakes Entrance, November 2011.
Source: Author.

13　S Dowrick, *Seeking the sacred: Transforming our view of ourselves and one another*, Allen and Unwin, Crows Nest, 2010, p. 69.

Bibliography

Victorian Government documents

Agricultural and livestock statistics of Victoria for the year ending 31 March 1865 with preliminary statistical notes, Votes and Proceedings of the Legislative Assembly and Papers, First Session, 1866.

Alphabetical List of the Rivers, Rivulets, Creeks, Channels, Aqueducts, Lakes, Reservoirs, Swamps, Inlets, Loughs, and Straits Referred To in the Order in Council, dated 23rd May 1881 (Gazette, 27 May 1881, p. 1389) permanently reserving as sites for public purposes, the Crown Lands forming the Bed, or such part of the Bed as Indicated, of each of certain rivers rivulets etc and the Crown Lands within the limits specified.

Commissioner for Sustainability and Environment, *State of the environment Victoria 2008, Summary: Living well within our environment – Are we? Can we?*, the Commissioner, Melbourne, 2008.

Commissioner of Public Works, *Estimates of expenditure for 1866*, Votes and Proceedings of the Legislative Assembly and Papers, First Session, 1866.

Conditions of a lease of the Koo Wee Rup Swamp under the Land Act 1865, Votes and Proceedings of the Legislative Assembly and Papers, Second session, p. 317, 1866.

Deakin, A, *Irrigation in Western America, so far as it has relation to the circumstances of Victoria*, A memorandum for the members of the Royal Commission On Water Supply, Government Printer, Melbourne, 1885.

Deaths from Miasmatic disease to 1864; Extracted from *Statistics of Victoria*, 1864, vital Statistics etc, in Acts and Proceedings of the Legislative Assembly and Papers, First Session, 1866.

Department of Sustainability and Environment, *Gippsland Lakes Ramsar site: Strategic management plan*, East Melbourne, July 2003.

Harris, G, G Batley, I Webster, R Molloy & D Fox, *Gippsland Lakes environmental audit: Review of water quality and status of aquatic ecosystems of the Gippsland Lakes*, prepared for the Gippsland Coastal Board by CSIRO Environmental Projects Office, Melbourne, October 1998.

Record of results of observations in meteorology and terrestrial magnetism made at the Melbourne Observatory, and at other localities in the Colony of Victoria, Australia, prepared by Pietro Barrachi, Govt Astronomer, Govt Printer, Melbourne, 1919.

Registrar General's Office, *Statistics for the Colony of Victoria for the year 1864*, Votes and Proceedings of the Legislative Assembly and Papers, First Session, 1866.

Report of the Professional Board on Reclamation of Swamps, in Votes and Proceedings of the Legislative Assembly and Papers, First Session, 1866.

Report on the physical character and resources of Gippsland, by the Surveyor General and the Secretary for Mines, with a Map and Geological Section, John Ferres, Govt Printer, 1874.

Victoria, Department of Sustainability and Environment, *Securing our rivers for future generations: Victorian river health program report card, 2002–2009*, the Department, Melbourne, 2009.

Victoria, Royal Commission on Water Supply, *Further Progress Report with appendices thereto, Extracts from Minutes of Committee: Together with Minutes of Evidence etc (Aug 31, 1885)*, VPP, vol. 3, Government Printer, Melbourne, 1885.

Newspapers and almanacs

Argus (Melbourne)

Bairnsdale Advertiser

Bairnsdale Liberal News and North Gippsland District Advertiser

Calvert's Illustrated Almanac for Victoria, 1859

Castner's Rural Australian

Clarson, Massina and Co's Weather Almanac and General Guide for Handbook for Victoria for 1877

Dr LL Smith's Medical Almanac

Gippsland Farmer's Journal

Gippsland Mercury

Gippsland Times

Glass's Almanac, 1862

Kerr's Melbourne Almanac, 1841

Illustrated Australian News for Home Readers

Levy Brothers and Co. Victorian Almanac, 1872

Maffra Spectator

Moe Register

Morwell Advertiser

Port Phillip Patriot Almanac and Directory

Port Phillip Separation Merchants and Settlers Almanac, diary and Directory for Melbourne and the District of Port Phillip, 1846

S Mullen's Victorian Almanac

Sands & McDougall's Commerical and general Melbourne Directory, 1862

Stevens and Bartholomew's Sandhurst, Castlemaine, Echuca, Maldon, Dunolly, Maryborough, Back Creek, and Avoca district directory for 1867.

Traralgon Record

Victorian Almanc, 1871

Victorian Temperance Pioneer, or Monthly Magazine

Walhalla Chronicle

Watmuff and Smith's Australian Almanac, 1870

Watmuff's Australian Almanac, 1877

Primary sources

Anon., *Guide for Excursionists from Melbourne 1868*, H Thomas, Melbourne, 1868.

Barfus, Francis, A Visit to the Mission Station Ramahyuck, SLV, MS 12645, Box 3486/3.

Brodribb, WA, *Recollections of an Australian Squatter 1838–1883*, John Ferguson Sydney in association with the Royal Australian Historical Society, Sydney, 1978. First published in 1883 by John Woods and Co., 13 Bridge St, Sydney.

Brown, D, Reminiscences of Brandy Creek, CGS, 696.

Butler's Woods Point and Gippsland General Directory, 1866. Compiled by Henry Young and John Dixon. Originally published by Butler and Brooke, Melbourne, this facsimile edition published by Kapana Press, Bairnsdale, 1985. Np 994.56 B987.

Campbell Coulson, LH, 'Early History of Gippsland pt 2', NLA MS 8600.

Diary of AM Caughey, SLV, MS 8735, MSB 434.

Diary of Angus McMillan during the 1864 Alpine expedition, SLV, MS 9776, box 268/2.

Diary of Annie Prout, SLV, MS 12306, box 3054/5.

Diary of Catherine Currie, CGS, 04381.

Diary of Charles Alfred Broome, SLV, MS 10774, box 1542.

Diary of Duncan Johnston, CGS, 00317.

Diary of George Glen Auchterlonie, CGS, 4060.

Diary of Margaret McCann, SLV, MS 9632, MSB 480.

Diary of (Margaret) Maggie Lamb, 'Everyday Life on a Moondarra Farm', 1910, CGS, 05317.

Diary of Mark Daniel, Daniel Family Papers, SLV, MS 10222.

Diary of Patrick Coady Buckley, CGS, 2806.

Diary of William Hastings, SLV, MS 8919, MSB 445.

Howitt, AW, 'Geology and Gold Mines of Swifts Creek', originally published as *The Diorites and Granites of Swift's Creek, and Their Contact Zones, with Notes on Auriferous Deposits* (1880) with an introductory essay by PD Gardner and an index by Margaret Gardner. First published in the *Transactions and Proceedings of the Royal Society of Victoria*, vol. 16, April 1880. Facsimile reproduction by Ngarak press, Ensay, 1996.

Huffer, K, 'Dedicated to the pioneers of Gormandale: To those who succeeded and those who failed', Typescript, NLA Npf994.56 H889.

Hunter Family Papers, SLV, PA 99/64.

Journal of the Rev. Francis Hales (edited by Michael H Wilson), SLV, MS 12950, box 1716/14.

King Station Day books, SLV, MS 11396, MSB 404.

King, Anna Josepha, 'A little bit of early Gippsland and sea travelling in the sixties', SLV, MS 9699, MSB 78.

Letterbooks of FA Hagenauer, 2 vols, NLA MS 3343.

Letters of E Franklin Tregaskis, SLV, MS 11955, 2494/19.

McMillan, A, 'Diary of the Alpine Track', *Gippsland Heritage Journal*, vol. 6, 1989, pp. 51–6.

Memoirs of John Joseph O'Connor, SLV, MS 10409, MSB 208.

Memoirs of William Pearson, SLV, MS 9583/74, MSB 475.

Meyrick, FJ, *Life in the Bush 1840–1847: A Memoir of Henry Howard Meyrick*, Thomas Nelson and Sons, London, 1939.

Middleton and Manning's Directory, 1884.

Papers of CJ Tyers, SLV, MSM 157–8.

Papers of George Campbell Curlewis, NLA, MS 1007.

Peppercorne, FS, *On Australian Meteorology and Hydrography*, Coupland Harding, Hastings St, Napier, New Zealand, 1879.

Rawlinson, TE, *Papers and Reports Read Before the Philosophical Institute and the Royal Society of Victoria, and Other Documents*, Wilson and MacKinnon Printers, Melbourne, 1865.

Reminiscences of Caleb Burchett, SLV, MS 8814, MSB 436, unpaginated.

Spurrell, W, 'History of Nicholson, 1876–1976', SLV, MS 10774, Box 1542.

Strzelecki, EP, *Physical description of New South Wales and Van Diemen's Land*, Longman, Brown, Green and Longmans, London, 1845.

Thomson Family Papers, NLA, MS 8600.

Secondary sources

Abernethy, B & ID Rutherford, 'Where along a river's length will vegetation most effectively stabilise stream banks?', *Geomorphology*, vol. 23, 1998, pp. 55–75. doi.org/10.1016/S0169-555X(97)00089-5.

Abernethy, B & I Rutherford, 'Riverbank reinforcement by riparian roots', *Second Annual Stream Management Conference*, Adelaide, Sth Australia, 8–11 February 1999.

Ackroyd, P, *Thames: Sacred river*, Chatto and Windus, London, 2007.

Adams, J, *So tall the trees: A centenary history of the southern districts of the Narracan Shire*, Narracan Shire Council, Narracan, 1978.

Adams, J, *Path among the years: History of the Shire of Bairnsdale*, Bairnsdale Shire Council, Bairnsdale, 1987.

Adams, J, *From these beginnings: History of the Shire of Alberton*, Alberton Shire Council, Yarram, 1990.

Allaby, M (ed.), *A dictionary of ecology*, Oxford University Press, 2006. *Oxford Reference Online*. Oxford University Press, www.oxfordreference.com/views/ENTRY.html?subview=Main&entry=t14.e5956, accessed 16 July 2008.

Allison, A, H Barrows, C Blake, Carr Arthur, A Eastman & H English (eds), *Norton anthology of poetry*, 3rd edn, WW Norton and Co., New York, 1970.

Allon, F & Z Sofoulis, 'Everyday water, cultures in transition', *Australian Geographer*, vol. 37, no. 1, 2006, pp. 45–55. doi.org/10.1080/00049180500511962.

Anderson, SC & BH Tabb, *Water, leisure and culture: European historical perspectives*, Berg, Oxford, 2002.

Andrews, M, *Landscape and Western art*, Oxford History of Art, Oxford University Press, Oxford, 1999.

Andrews, T, *Legends of the earth, sea and sky: An encyclopedia of nature myths*, ABC Clio, Santa Barbara, 1998.

Aplin, G, *Australians and their environment: An introduction to environmental studies*, Oxford University Press, Melbourne, 1988.

Archibold, O, 'Living in a garden in a valley', *Australian Garden History*, vol. 11, no. 3, Nov/Dec 1999, pp. 10–14.

Armstrong, K, *A history of god: From Abraham to the present, the 4000 year quest for god*, Vintage, London, 1999.

Arthington, AH & JB Pusey, 'Flow restoration and protection in Australian rivers', *River Research Application*, no. 19, 2003, pp. 377–95.

Australian Academy of Technological Sciences and Engineering, *Technology in Australia 1788–1988: A condensed history of Australian technological history and adaptation during the first two hundred years*, University of Melbourne, Melbourne, 2000.

Australian Psychological Society, *Position statement on climate change*, www.psychology.org.au/community/topics/climate, accessed 4 April 2011.

Barraclough, L & M Higgins, *A valley of glens: The people and places of the Upper Macalister River*, Kapana Press, Bairnsdale, 1986.

Barrett, C(K), *A busy life: The reminiscences of Charlie Bartlett born at Providence Ponds, near Stratford, Victoria, in 1890*, Stratford Historical Society, Stratford, 1980.

Beattie, J, 'Exploring trans-Tasman environmental connections 1850s to 1900 through the imperial careering of Alfred Sharpe', *Environment and Nature in New Zealand*, vol. 4, no. 1, April 2009, pp. 37–57.

Bellanta, M, 'Irrigation millennium: Science, religion and the new Garden of Eden', *Eras* (Monash University), no. 3, June 2002. Available from www. arts.monash.edu.au/cgi-bin/inc/print?page=/eras/edition_3/bellanta.htm (page discontinued).

Berry, T, *The dream of the Earth*, Sierra Club Books, San Francisco, 1998.

Bijker, WE, 'Dikes and dams, thick with politics', *Isis*, vol. 98, 2007, pp. 109–23. doi.org/10.1086/512835.

Bird, ECF, *A Geomorphological study of the Gippsland Lakes*, Research School of Pacific Studies, Department of Geography Publication G/1, The Australian National University, Canberra, 1965.

Bird, ECF & J Lennon, *Making an entrance: The story of the artificial entrance to the Gippsland Lakes*, James Yeates & Sons, Bairnsdale, 1989.

Biswas, Asit, K, *History of hydrology*, North Holland Publishing Company, Amsterdam, 1970.

Black, B, 'Oil Creek as industrial apparatus: Re-creating the industrial process through the landscape of Pennsylvania's oil boom', *Environmental History*, vol. 3, no. 2, April 1998, pp. 210–28. doi.org/10.2307/3985380.

Blackbourn, D, *The conquest of nature: Water, landscape and the making of modern Germany*, WW Norton and Co., New York, 2005.

Blackburn, G, *Pioneering irrigation in Australia to 1929*, Australian Scholarly Publishing, Melbourne, 2004.

Bonyhady, T, *The colonial earth*, Melbourne University Press, Melbourne, 2002.

Bras, RL, 'A brief history of hydrology', *Bulletin of the American Meteorological Society*, vol. 80, no. 6, June 1999, pp. 1151–65. doi.org/10.1175/1520-0477-80.6.1151.

Breen, SD, 'Ecocentrism, weighted interests and property theory', in M Humphrey (ed.), *Political theory and the environment: A reassessment*, F Cass, London, 2001, pp. 36–51.

Bride, TF, *Letters from Victorian pioneers: A series of papers on the early occupation of the colony, the aborigines, etc. / addressed by Victorian pioneers to Charles Joseph La Trobe*, CE Sayers (ed.), Lloyd O'Neill for Currey O'Neill, Sth Yarra, Melbourne, 1983.

Brightling, P, 'A sketch on the Avon; Tuesday October 3, 1843', *Gippsland Heritage Journal*, no. 7, December 1989, p. 38.

Brimblecombe, P, *The big smoke: A history of air pollution in London since medieval times*, Methuen, London, 1987.

Brinsmead, GSJ, 'A geographical study of the dairy manufacturing industry in Gippsland 1840–1910', MSc thesis, University of Melbourne, December 1977.

Broome, R, *The Victorians: Arriving*, Fairfax, Syme and Weldon Associates, McMahons Point, NSW, 1984.

Brunckhorst, D, 'Understanding design for planning alternative landscape futures to adapt to climate change: Learning from temporal inconsistencies in vulnerability and adaptation studies', Paper delivered at the *1st International Conference on Adapting to Climate Change: Preparing for the Unavoidable Impacts of Climate Change*, Gold Coast, Australia, 29 June to 1 July 2010.

Brutsaert, W, *Evaporation into the atmosphere*, D Reidel Publishing, Dordrecht, 1982. doi.org/10.1007/978-94-017-1497-6.

Bullock, A & M Acreman, 'The role of wetlands in the hydrological cycle', *Hydrology and Earth System Sciences*, vol. 7, no. 3, 2003, pp. 358–89. doi.org/10.5194/hess-7-358-2003.

Bunyan, J, *The pilgrim's progress*, JM Dent and Son for Everyman's Library, London, 1978.

Burningham, N, *Messing about in earnest*, Fremantle Arts Centre Press, Fremantle, 2003.

Cafiso, JL, *A Nieman family history*, the author and EH Nieman, Morwell, n.d.

Camuffo, D, C Bertolin, PD Jones, R Cornes & E Garnier, 'The earliest daily barometric pressure readings in Italy: Pisa AD 1657–1658 and Modena AD 1694, and the weather over Europe', *The Holocene*, vol. 20, no. 3, 2010, pp. 337–49. doi.org/10.1177/0959683609351900.

Cannon, M, *Australia in the Victorian age: Life in the cities*, 3rd edn, Penguin Books, Ringwood, Vic., 1988.

Cathcart, M, *The water dreamers: The remarkable history of our dry continent*, Text Publishing, Melbourne, 2009.

Chappelle, FH, *Wellsprings: A natural history of bottled spring waters*, Rutgers University Press, New Brunswick, New Jersey, 2005.

Chodron, P, *The places that scare you: A guide to fearlessness*, Element, Hammersmith, London, 2001.

Christie, RW, *Victoria's forgotten goldfield: A history of the Dargo and Crooked River goldfield*, High Country Publishing, Dargo, 1996.

Cioc, M, *The Rhine: An eco-biography*, University of Washington Press, Seattle, 2002.

Ciriacono, S, *Building on water: Venice, Holland and the construction of the European landscape in early modern times*, Berghahn Books, New York, 2006.

Clark, AE, *The church of our fathers: Being the history of the Church of England in Gippsland 1847–1947*, Diocese of Gippsland, Sale, April 1947.

Clarke, R & J King, *The water atlas: A unique visual analysis of the world's most critical resource*, The New Press, New York, 2004.

Coach News, Journal of the Moe and District Historical Society.

Coles, B & J Coles, *People of the wetlands: Bogs, bodies and lake dwellers; a world survey*, Thames and Hudson, London, 1989.

Collett, B, *Wednesdays closest to the full moon: A history of South Gippsland*, University of Melbourne Press, Melbourne, 1994.

Committee of the South Gippsland Pioneer's Association, *The land of the lyre bird: A story of early settlement in the great forest of South Gippsland*, Shire of Korumburra for the South Gippsland Development League, Clayton, Vic., 1966 (original published by Gordon and Gotch, Melbourne, 1920).

Connell, D, *Water politics in the Murray-Darling Basin*, Federation Press, Annandale, 2007.

Cooper Marcus, C, *House as a mirror of self: Exploring the deeper meaning of home*, Nicolas-Hays Inc, Lake Worth, Florida, 2006.

Copeland, H, *The path of progress: From the forests of yesterday to the homes of today*, Shire of Warragul, Warragul, 1934.

Cosgrove, D & G Petts (eds), *Water, engineering and landscape*, Belhaven Press, London, 1990.

Court, J, *From squatters hut to city, Traralgon, 1840–1976*, City of Traralgon, Traralgon, 1976.

Cowan, IR, 'Water use in higher plants', in *Water: Planets, plants and people*, Australian Academy of Science, Canberra, 1978, pp. 71–107.

Cox, N, 'Residents of Gipps' Land 1851', *Gippsland Heritage Journal*, vol. 8, 1995, p. 48.

Croutier, AL, *Taking the waters: Spirit, art, sensuality*, Abbeville Press, New York, 1992.

Daley, C, *The story of Gippsland*, Whitcombe & Tombs for the Gippsland Municipalities Association, Melbourne, 1960.

Daley, C, 'How the Hunters came to Gippsland', *Gippsland Heritage Journal*, vol. 3, no. 2, 1988, pp. 3–10.

Dante, *Inferno*, trans. Henry Wadsworth Longfellow (1909) www.online-literature.com/dante/Inferno, accessed 6 May 2008.

Darby, HC, *The draining of the Fens*, Cambridge University Press, Cambridge, 1956.

Darby, HC, *The changing fenland*, Cambridge University Press, Cambridge, 1983.

Davie, T, *Fundamentals of hydrology*, Routledge Fundamentals of Physical Geography, Routledge, London, 2003.

Davison, G, *Narrating the nation in Australia*, The Menzies Lecture, 2009, Menzies Centre for Australian Studies, Kings College, London, University of London, 2009.

Day, D, 'Beyond the biodome', 17 January 2005, www.thespacereview.com/article/305/1, accessed 3 January 2011.

de Angeles, L, E Restall Orr & T van Dooren (eds), *Pagan visions for a sustainable future*, Llewelleyn Publications, Woodbury, Minnesota, 2005.

Dingle, T, *The Victorians: Settling*, Fairfax, Syme and Weldon Associates, McMahons Point, NSW, 1984.

Dispenza, J, *God on your own: Finding a spiritual path outside religion*, Josey Bass, San Francisco, 2006.

Dobson, M, '"Marsh fever" – The geography of malaria in England', *Journal of Historical Geography*, vol. 6, no. 4, 1980, pp. 357–89. doi.org/10.1016/0305-7488(80)90145-0.

Douglas, K, *Under such sunny skies: Understanding weather in colonial Australia, 1860–1901*, Bureau of Meteorology (Metarch Series), Canberra, 2007.

Dovers, S (ed.), *Australian environmental history: Essays and cases*, Oxford University Press, Melbourne, 1994.

Dovers, S (ed.), *Environmental history and policy: Still settling Australia*, Oxford University Press, Melbourne, 2000.

Dow, C, 'Tantungalung Country: An environmental history of the Gippsland Lakes', PhD thesis, Monash University, 2004.

Dowrick, S, *Seeking the sacred: Transforming our view of ourselves and one another*, Allen and Unwin, Crows Nest, NSW, 2010.

Drysdale, M, 'Survival of culture: Alice Thorpe and her basket', *Gippsland Heritage Journal*, no. 15, December 1993, pp. 50–2.

DuNann Winter, D, *Ecological psychology: Healing the split between planet and self*, Harper Collins College Publishers, New York, 1996.

Dundes, A (ed.), *The flood myth*, University of California Press, Berkeley, 1998.

Dunstan, D, *Governing the metropolis: Melbourne 1851 to 1891*, Melbourne University Press, Melbourne, 1984.

East Gippsland Historical Society Newsletter, East Gippsland Historical Society.

European Union Water Time Project, www.watertime.net.

Fairweather, KMcD, *Time to remember: The history of gold mining on the Tambo and its tributaries*, the author, Doctors Flat, 1975.

Falkenmark, M, 'Freshwater as shared between society and ecosystems: From divided approaches to integrated challenges', *Philosophical Transactions: Biological Sciences*, vol. 358, no. 1440, Freshwater and Welfare Fragility: Syndromes, Vulnerabilities and challenges, 29 December 2003, pp. 2037–49.

Falkenmark, M & C Folke, 'The ethics of socio-ecohydrological catchment management: Towards hydrosolidarity', *Hydrology and Earth System Sciences*, vol. 6, no. 1, 2002, pp. 1–9. doi.org/10.5194/hess-6-1-2002.

Feliks, Y, *Nature and man in the Bible: Chapters in biblical ecology*, Soncino Press, London, 1981.

Few, R, 'Health and climatic hazards: Framing social research on vulnerability, response and adaptation', *Global Environmental Change*, vol. 17, no. 2, 2007, pp. 281–95. doi.org/10.1016/j.gloenvcha.2006.11.001.

Fletcher, M, 'The small farm ideal: Closer settlement in the Maffra District 1911–1938', MA thesis, Monash University, 1987.

Fletcher, M, *Avon to the Alps: A history of the Shire of Avon*, Shire of Avon, Stratford, 1988.

Fletcher, M, 'The Cunninghame letters', *Gippsland Heritage Journal*, vol. 16, June 1994, pp. 42–9.

Fletcher, M & L Kennett, *Changing landscapes: A history of settlement and land use at Driffield*, International Power, Hazelwood, 2003.

Fletcher, M & L Kennett, *Wellington landscapes: History and heritage in a Gippsland shire*, Maffra and District Historical Society, Maffra, 2005.

Fort, T, *Under the weather: Us and the elements*, Century, London, 2006.

Frawley, K, 'Evolving visions: Environmental management and nature conservation in Australia', in S Dovers (ed.), *Australian environmental history: Essays and cases*, Oxford University Press, Melbourne, 1994.

Frost, W & S Harvey, 'Forest industries or dairy pastures? Ferdinand von Mueller and the 1885–93 Royal Commission on Vegetable Products', *Historical Records of Australian Science*, vol. 11, no. 3, June 1997, pp. 431–7. doi.org/10.1071/HR9971130431.

Fumagalli, V, *Landscapes of Fear*, Polity Press, Cambridge, 1994.

Fussell, GE, 'The dawn of high farming in England: Land reclamation in early Victorian days', *Agricultural History*, vol. 22, April 1948, pp. 83–95.

Garcier, R, 'The placing of matter: Industrial water pollution and the construction of social order in nineteenth-century France', *Journal of Historical Geography*, vol. 36, no. 2, 2010, pp. 132–42. doi.org/10.1016/j.jhg.2009.09.003.

Garden, D, 'Catalyst or cataclysm? Gold mining and the environment', *Victorian Historical Journal*, vol. 72, no. 1, 2001, pp. 28–44.

Garden, D, *Droughts, floods* and *cyclones, El Ninos that shaped our colonial past*, Australian Scholarly Publishing, North Melbourne, 2009.

Gardner, PD, *Through foreign eyes: European perceptions of the Kurnai tribes of Gippsland*, 2nd edn, Ngarak Press, Ensay, 2004.

Gibbs, LM, '"A beautiful soaking rain": Environmental value and water beyond Eurocentrism', *Environment and Planning D: Society and Space*, vol. 28, 2010, pp. 363–78. doi.org/10.1068/d9207.

Giblett, R, *Postmodern wetlands: Culture, history, ecology*, Edinburgh University Press, Edinburgh, 1996.

Gillbank, L, 'Ferdinand Mueller in Gippsland', *Gippsland Heritage Journal*, no. 10, June, 1991, pp. 3–11.

Gippsland Illustrated 1904, first facsimile edition by Kapana Press, Bairnsdale, 1987 (first published by the Periodicals Publishing Co. in Melbourne, 1904).

Glacken, C, *Traces on the Rhodian shore: Nature and culture in Western thought from ancient times to the end of the eighteenth century*, University of California Press, Berkeley, 1967.

Glantz, M, *Drought follows the plow*, Cambridge University Press, Cambridge, 1994.

Gleick, PH, H Cooley, MJ Cohen, M Morikawa, J Morrison & M Palaniappan, *The world's water 2008–2009: The biennial report of freshwater resources*, Island Press, Washington DC, 2009.

Glendinning, C, *My name is Chellis and I'm in recovery from Western civilization*, Shambala Press, Boston, 1994.

Glick, T, *Irrigation and society in medieval Valencia*, Belknap Press of Harvard University Press, Cambridge, MA, 1970. doi.org/10.4159/harvard.9780674281806.

Goodison, L, *Moving heaven and earth: Sexuality, spirituality and social change*, Pandora Press, Hammersmith, London, 1992.

Goodman, D, *Gold seeking: Victoria and California in the 1850s*, Allen and Unwin, St Leonards, NSW, 1994.

Goulding, M, 'Placing the past: Aboriginal historical places in East Gippsland, Victoria', *Historic Environment*, vol. 13, no. 2, 1992, pp. 40–51.

Grayson, RB, C Kenyon, BL Finlayson & CJ Gippel, 'Bathymetric and core analysis of the Latrobe River delta to assist in catchment management', *Journal of Environmental Management*, vol. 52, 1998, pp. 361–72. doi.org/10.1006/jema.1998.0181.

Griffiths, T & L Robin (eds), *Ecology and empire: Environmental history and settler societies*, University of Washington Press, Seattle, 1997.

Griffiths, T, *Forests of ash: An environmental history*, Cambridge University Press, Cambridge, 2001.

Guillerme, AE, *The age of water: The urban environment in the north of France, AD 300–1800*, Texas A&M University Press, Texas, 1988.

Hales, J & J Little (eds), *Gippsland estates 1882: A series of articles which appeared in the Gippsland Mercury from January to August 1882*, Maffra and District Historical Society Bulletin, Supplementary Issue no. 4, Maffra.

Halliday, S, *Water: A turbulent history*, Sutton Publishing, Phoenix Mill, Gloucestershire, 2004.

Hanna, E, T Kjellstrom, C Bennett & K Dear, 'Climate change and rising heat: Population health implications for working people in Australia', *Asia Pacific Journal of Public Health*, Supp to vol. 23, no. 2, Special Issue on Climate Change, March 2011, pp. 14s–26s.

Harris, E, 'Development and damage: Water and landscape evolution in Victoria, Australia', *Landscape Research*, vol. 31, no. 2, 2006, pp. 169–81. doi.org/10.1080/01426390600638687.

Harris, G, 'Inland waters', theme commentary prepared for the 2006 Australian State of the Environment Committee, Department of the Environment and Heritage, Canberra, 2006, www.environment.gov.au/soe/2006/publications/commentaries/water/index.html, accessed 20 August 2008.

Harvey, G, *Contemporary paganism: Listening people, speaking earth*, New York University Press, New York, 1997.

Haslam, S, *The historic river: Rivers and culture down the ages*, Cobden of Cambridge Press, Cambridge, 1991.

Hatfield, G, *Memory, wisdom and healing: The history of domestic plant medicine*, Sutton Publishing, Gloucestershire, 2005, 2nd edn.

Hawes, HS, 'Commission to community: A history of salinity management in the Goulburn Valley, 1880–2007', MA thesis, University of Melbourne, March 2007.

Heffeman, MJ, 'Bringing the desert to bloom: French ambitions in the Sahara Desert during the late nineteenth century – the strange case of "La Mer interieure"', in D Cosgrove & G Petts (eds), *Water, Engineering and Landscape*, Belhaven Press, London, 1990.

Hibben, J, 'The Disher family in the nineteenth century', BA Hons thesis, University of Melbourne, 1978.

Hillel, D, *The natural history of the Bible: An environmental exploration of the Hebrew scriptures*, Columbia University Press, New York, 2006.

Hobday, R, *The light revolution: Health, architecture and the sun*, Findhorn Press, UK, 2006.

Hobsbawm, E, *On history*, Abacus Books, London, 2002.

Horton, T, *Turning the tide: Saving the Chesapeake Bay*, Island Press, Washington DC, 1991.

Hoskins, WG, *The making of the English landscape*, Pelican, Middlesex, 1978.

Hubbard, P, R Kitchin & G Valentin, *Key thinkers in space and place*, Sage Publications, London, 2004.

Hubbart, J with J Kundell, 'History of hydrology', in CJ Cleveland (ed.), *Encyclopedia of Earth*, Environmental Information Coalition, National Council for Science and Environment, Washington DC (first published in the Encyclopedia of Earth, 20 May 2007), last revised 11 February 2008, retrieved 25 February 2009, editors.eol.org/eoearth/wiki/History_of_hydrology.

Huffer, K, *A history of Gormandale*, Gormandale Centenary Committee, Traralgon, 1982.

Hughes, JD, *What is environmental history?*, Polity Press, Cambridge, 2006.

Hulme, M, S Dessai, I Lorenzoni & DR Nelson, 'Unstable climates: Exploring the statistical and social constructions of 'normal' climate', *Geoforum*, vol. 40, 2009, pp. 197–206. doi.org/10.1016/j.geoforum.2008.09.010.

Hurd, B, *Stirring the mud: On swamps, bogs and human imagination*, Beacon Press, Boston, 2001.

Hutton, D & L Connors, *A History of the Australian environment movement*, Cambridge University Press, Cambridge, 1999.

Jankovic, V, *Reading the skies: A cultural history of English weather, 1650–1820*, Manchester University Press, Manchester, 2000.

Johnson, K, 'Creating place and landscape', in S Dovers (ed.), *Australian environmental history: Essays and cases*, Oxford University Press, Melbourne, 1994, pp. 43–4.

Jones, R, 'An ideal holiday experience: Guesthouses of the Gippsland Lakes', *Gippsland Heritage Journal*, no. 17, 1994, pp. 25–30.

Kaijser, A, 'System building from below: Institutional change in Dutch water control systems', *Technology and Culture*, vol. 43, no. 3, July 2002, pp. 521–48. doi.org/10.1353/tech.2002.0120.

Karterakis, SM, BW Karney, B Singh & A Guergachi, 'The hydrologic cycle: A complex concept with continuing pedagogical implications', *Water Science and Technology: Water Supply*, vol. 7, no. 1, 2007, pp. 23–31. doi.org/10.2166/ws.2007.003.

Keating, J, *'And the drought walked through': A history of water shortage in Victoria*, Dept of Water Resources, Melbourne, 1992.

Keen, I, 'The anthropologist as geologist: Howitt in colonial Gippsland', *The Australian Journal of Anthropology*, vol. 11, no. 1, 2000, pp. 78–90. doi.org/10.1111/j.1835-9310.2000.tb00264.x.

Kemp, D, *Maffra: The history of the shire to 1975*, Shire of Maffra, Bairnsdale, 1975.

Kempe, M, 'Noah's flood: The Genesis story and natural disasters in early modern times', *Environment and History*, vol. 9, no. 2, 2003, pp. 151–71. doi.org/10.3197/096734003129342809.

Kennett, L, *Index to the Gippsland Times 1861–1900*, Centre for Gippsland Studies, Monash University, Churchill, 1995.

La Nauze, JA, *Alfred Deakin: A biography*, Melbourne University Press, Melbourne, 1965.

Lacey, SL, *The Laceys of Gippsland: The history of a pioneer firm 1870–1970*, the author, Sale, n.d.

Larkcom, J, 'The floating gardens of Amiens', *Journal of the Royal Horticultural Society*, vol. 121, no. 9, September 1996, pp. 540–43.

LaTrobe Leopold, D, 'My first and I trust my last New Year's Day at Omeo', from the 1863 Diary of WH Foster, Police Magistrate and Goldfields Warden, *Gippsland Heritage Journal*, no. 8, 1990, pp. 23–5.

Lee, RL & AB Fraser, *The rainbow bridge: Rainbows in art, myth and science*, Pennsylvania State University Press, Pennsylvania, 2001.

Legg, SM, 'Arcadia or abandonment; the evolution of the rural landscape in South Gippsland, 1870–1947', MA thesis, Dept of Geography, Monash University, 1984.

Legg, SM, 'Farm abandonment in South Gippsland's Strzelecki Ranges, 1870–1925: Challenge or tragedy?', *Gippsland Heritage Journal*, vol. 1, no. 1, 1986, pp. 14–21.

Leslie, JW & HC Cowie, *The wind still blows: Extracts from the Diaries of Rev WS Login, Mrs H Harrison and Mrs W Montgomery*, the authors, Sale, 1973.

Linn, R, *Battling the land: Two hundred years of rural Australia*, Allen and Unwin, St Leonards, NSW, 1999.

Linton, J, 'Is the hydrologic cycle sustainable? A historical-geographical critique of a modern concept', *Annals of the Association of American Geographers*, vol. 98, no. 3, 2008, pp. 630–49. doi.org/10.1080/00045600802046619.

Liverman, DM, 'Conventions of climate change: Constructions of danger and the dispossession of the atmosphere', *Journal of Historical Geography*, vol. 35, 2009, pp. 279–96. doi.org/10.1016/j.jhg.2008.08.008.

Longhurst, R & W Douglas, *The Brisbane River: A pictorial history*, WD Incorporated, Brisbane, 1997.

Lunt, I, 'Snakes Ridge views', *Gippsland Heritage Journal*, no. 13, 1993, pp. 37–8.

Lunt, I, 'The distribution and environmental relationships of native grasslands on the lowland Gippsland Plain, Victoria: An historical study', *Australian Geographical Studies*, vol. 35, no. 2, July 1997, pp. 140–52. doi.org/10.1111/1467-8470.00015.

MacKay, H, *What makes us tick? The ten desires that drive us*, Hatchette, Sydney, 2010.

Macleod, P, *From Bernisdale to Bairnsdale: The story of Archibald and Colina Macleod and their descendants in Australia, 1821–1994*, the author, Nar Nar Goon North, 1994.

MacLeod, R, *The Commonwealth of science: ANZAAS and the scientific enterprise in Australasia, 1888–1988*, Oxford University Press, Melbourne, 1988.

Macy, J & M Young Brown, *Coming back to life: Practices to reconnect our lives, our world*, New Society Publishers, Gabriola Island, Canada, 1998.

Maddern, IT, *Light and life: A history of the Anglican Church in Gippsland*, Enterprise Press, Sale, n.d.

Main, G, 'Industrial earth: An ecology of rural place', PhD thesis, Centre for Resources, Environment and Society, The Australian National University, 2004.

Mauelshagen, F, *Disaster and political culture in Germany since 1500: Re-historicising disaster*, Cambridge University Press, Cambridge, 2005.

McDonald, N, 'On epigraphic records: A valuable tool in reassessing flood risk and long-term climate vulnerability', *Environmental History*, vol. 12, no. 1, January 2007, pp. 136–40. doi.org/10.1093/envhis/12.1.136.

McLeary, A & T Dingle, *Catherine: On Catherine Currie's diary*, Melbourne University Press, Melbourne, 1998.

McLoughlin, L, *The Middle Lane Cove River: A history and a future*, Centre for Environmental and Urban Studies, Macquarie University, North Ryde, 1985.

McNeill, JR, *Something new under the sun: An environmental history of the twentieth-century world*, WW Norton and Co., New York, 2001.

Meindl, CF, 'Past perceptions of America's great wetland, Florida's Everglades in the early twentieth century', *Environmental History*, vol. 5, no. 3, July 2000, pp. 378–395. doi.org/10.2307/3985482.

Merchant, C, *The death of nature: Women, ecology and the scientific revolution*, Harper Collins, San Francisco, 1980.

Middleton, WEK, *A history of the theories of rain and other forms of precipitation*, Oldbourne, London, 1965.

Millennium Ecosystem Assessment, *Ecosystems and human wellbeing wetlands and water synthesis*, World Resources Institute, Washington DC, 2005.

Mingay, GE, *Parliamentary enclosure in England: An introduction to its causes, incidence and impact 1750–1850*, Addison Wesley and Longman, London, 1997.

Moe and District Historical Society, *A pictorial history of Moe and District*, Moe City Council, Moe, 1988.

Morgan, P, 'The Campbell incident', *Victorian Historical Journal*, vol. 68, no. 1, April 1997, pp. 20–8.

Morgan, P, *The literature of Gippsland: The social and historical context of early writings with bibliography*, Centre for Gippsland Studies, Churchill, 1983.

Morgan, P, *The settling of Gippsland: A regional history*, Gippsland Municipalities Association, Leongatha, 1997.

Morrison, L, 'The newspapers of Gippsland, 1855–1890', *Gippsland Heritage Journal*, vol. 6, 1989, pp. 3–10.

Morton, J, 'Tiddalik's travels: The making and remaking of an Aboriginal flood myth', *Advances in Ecological Research*, vol. 39, 2006, pp. 139–58. doi.org/10.1016/S0065-2504(06)39008-3.

Moyal, A, *A bright and savage land*, Penguin, Ringwood, Vic., 1993.

Moyal, A, *Platypus*, Allen and Unwin, Crows Nest, NSW, 2002.

Myers, F, *Irrigation or the new Australia*, Chaffey Bros Ltd, Mildura, Vic., 1893.

Nace, R, *General evolution of the concept of the hydrological cycle*, Three Centuries of Scientific Hydrology: Key papers submitted on the occasion of the celebration of the Tercentenary of Scientific Hydrology, 9–12 September 1974, UNESCO, Paris, 1974.

Newham, LTH & JJ Drewry, *Modelling catchment-scale nutrient generation*, Technical Report 28/05, National River Contaminants Program of Land and Water Australia, CSIRO Land and Water, Canberra, 2006.

Newson, M, *Hydrology and the river environment*, Clarendon Press, Oxford, 1994.

Newton Verrier, N, *Coming home to self: The adopted child grows up*, Gateway Press, Baltimore, 2003.

Nienhaus, A, 'Disaster coping and prevention in the Swiss Alps in the early 19th century', *IHDP Update* 02/2005, International Human Dimensions Programme on Global Environmental Change, Bonn, Germany, 2005.

Nunn, HW, 'Hodgkinson, Clement (1818–1893)', *Australian Dictionary of Biography*, vol. 4, Melbourne University Press, Melbourne, 1972, pp. 403–4.

Oelschlaeger, M, *The idea of wilderness: From prehistory to the age of ecology*, Yale University Press, New Haven, 1991.

Ostrom, E, *Governing the commons: The evolution of institutions for collective action*, Cambridge University Press, Cambridge, 1990. doi.org/10.1017/ CBO9780511807763.

Owen, S, C Pooley, A Folkard, G Clark & N Watson, *Rivers and the British landscape*, Carnegie Publishing, Lancaster, 2005.

Parry, M, 'Copenhagen number crunch', Nature Reports Climate Change, Published online 14 January 2010. doi.org/10.1038/climate.2010.01.

Parry-Okeden, D, 'The Parry Okeden family at Rosedale', *Gippsland Heritage Journal*, vol. 10, 1991, pp. 46–51.

Paul, AM, 'How the first nine months shapes the rest of your life: The new science of fetal origins', *Time*, 4 October 2010, pp. 28–34.

Peel, RF, 'The landscape in aridity, presidential address', *Transactions of the Institute of British Geographers*, no. 38, June 1966, pp. 1–23. doi.org/10.2307/621421.

Perkins, M, *Visions of the future: Almanacs, time and cultural change 1775–1870*, Clarendon Press, Oxford, 1996. doi.org/10.1093/acprof:oso/9780198121787 .001.0001.

Perry, TM, 'Climate, caterpillars and terrain: A study of grazing expansion in New South Wales 1813–1826', *Australian Geographer*, vol. 7, no. 1, May 1957, pp. 3–14. doi.org/10.1080/00049185708702319.

Phillip, JR, 'Water on the earth', in AK McIntyre (ed.), *Water: Planets, plants and people*, Australian Academy of Science, Canberra, 1978, pp. 35–59.

Pick, D, 'Roma or Morte: Garibaldi, nationalism and the problem of psycho-biography', *History Workshop Journal*, vol. 57, no. 1, 2004, pp. 1–33.

Pigram, JJ, *Issues in the management of Australia's water resources*, Longman Cheshire, Melbourne, 1986.

Pisani, DJ, 'Beyond the hundredth meridian: Nationalising the history of water in United States', *Environmental History*, vol. 5, no. 4, October 2000, pp. 466–82. doi.org/10.2307/3985582.

Ponting, C, *A new green history of the world: The environment and the collapse of great civilisations*, Vintage Originals, London, 2007.

Porter, H, *Bairnsdale: Portrait of a country town*, John Ferguson, Sydney, 1977.

Powell, JM, *The public lands of Australia Felix: Settlement and land appraisal in Victoria 1834–91 with special reference to the Western Plains*, Oxford University Press, Melbourne, 1970.

Powell, JM (ed.), *Yeomen and bureaucrats: The Victorian Crown Lands Commission, 1878–9*, Oxford University Press, Melbourne, 1973.

Powell, JM, 'Snakes and cannons: Water management and the geographical imagination in Australia', in S Dovers (ed.), *Environmental history and policy: Still settling Australia*, Oxford University Press, Melbourne, 2000, pp. 47–71.

Powell, J, 'Environment and institutions: Three episodes in Australian water management, 1880–2000', *Journal of Historical Geography*, vol. 28, no. 1, 2002, pp. 100–14. doi.org/10.1006/jhge.2001.0376.

Powell, J, *Environmental management in Australia, 1788–1914*, Oxford University Press, Melbourne, 1976.

Powell, J, *Watering the Garden State: Water land and community in Victoria, 1834–1988*, Allen and Unwin, Sydney, 1989.

Power, J, 'Squatters and selectors: An historical geography of the central Gippsland Plains 1840–1880', BA Hons thesis, University of Melbourne, 1979.

'Primary sources: McMillan's letter to the Colonist', *Gippsland Heritage Journal*, vol. 3, no. 1, 1988, pp. 39–41.

Prince, H, 'A marshland chronicle, 1830–1960: From artificial drainage to outdoor recreation in Wisconsin', *Journal of Historical Geography*, vol. 21, 1995, pp. 3–22. doi.org/10.1016/0305-7488(95)90003-9.

Prince, H, *Wetlands of the American Midwest: A historical geography of changing attitudes*, University of Chicago Press, Chicago, 1997. doi.org/10.7208/chicago/9780226682808.001.0001.

Proust, K, 'Learning from the past for sustainability: Towards an integrated approach', PhD thesis, The Australian National University, 2004.

Pyne, S, *Burning bush: A fire history of Australia*, Allen and Unwin, North Sydney, 1992.

Radin, D, *The noetic universe: The scientific evidence for psychic phenomena*, Corgi Books, London, 2009.

Rattue, J, *The living stream: Holy wells in historical context*, Boydell Press, Woodbridge, Suffolk, 2001.

Ravensdale, JR, *Liable to floods: Village landscape on the edge of the Fens AD 450 to 1850*, Cambridge University Press, Cambridge, 1974.

Reisner, M, *Cadillac desert: The American West and its disappearing water*, Viking, New York, 1986.

Rezende, L, *Chronology of science*, Checkmark Books, New York, 2006.

Richard, E, 'Australian Federation, religion and James Bryce's nightmare', in *Intellect and emotion: Perspectives on Australian history; essays in honour of Michael Roe*, jointly published by Centre for Australian Studies, Deakin University, and Centre for Tasmanian Historical Studies, University of Tasmania, Geelong, Vic., 1998, pp. 151–69.

Roberts, J & G Sainty, *Listening to the Lachlan*, Sainty and Associates, Murray-Darling Basin Commission, Potts Point, NSW, 1996.

Roberts, L, 'An Arcadian apparatus: The introduction of the steam engine into the Dutch landscape', *Technology and Culture*, vol.54, no. 2, April 2004, pp. 251–76. doi.org/10.1353/tech.2004.0089.

Rocco, F, *The miraculous fever tree: Malaria and the quest for a cure that changed the world*, Harper Collins, Great Britain, 2003.

Roe, M, *Nine Australian progressives: Vitalism in bourgeois social thought*, University of Queensland Press, St Lucia, 1984.

Rolls, E, 'More a new planet than a new continent', in S Dovers (ed.), *Australian environmental history: Essays and cases*, Oxford University Press, Melbourne, 1994.

Rose, D, 'Fresh water rights and biophilia: Indigenous Australian perspectives', *Dialogue*, vol. 23, no. 3, 2004, pp. 35–43.

Roszak, T, *The voice of the earth*, Simon and Schuster, New York, 1992.

Roszak, T, ME Gomes & AD Kanner (eds), *Ecopsychology: Restoring the earth, healing the mind*, Sierra Club Books, San Francisco, 1995.

Royal Historical Society of Victoria, *River of history: Images of the Yarra since 1835*, Royal Historical Society of Victoria, Melbourne, 1998.

Ruess, M, 'Learning from the Dutch: Technology, management, and water resources development', *Technology and Culture*, vol. 43, no. 3, July 2002, pp. 465–72. doi.org/10.1353/tech.2002.0135.

Rupp, R, *Four elements: Water, air, fire, earth*, Profile Books, London, 2005.

Sadleir, J, *Recollections of a Victorian police officer*, George Robertson and Co., Melbourne, c. 1913.

Savige, RM, *History of the Savige family*, the author, Frankston, 1966.

Schmemann, A, *Of water and the spirit: A liturgical study of baptism*, St Vladimir's Seminary Press, New York, 1974.

Schmitt, DJ, 'Provincial pestilence: A study of the impact and response to typhoid, tuberculosis and diphtheria in the Gippsland Hospital catchment, 1866–1931', Masters thesis, Monash University, March 1993.

Scott, A, *Tuggerah Lakes: Way back when*, Sainty and Associates, Sydney, 2002.

Searle, S, *The rise and demise of the black wattle bark industry in Australia*, CSIRO Division of Forestry, Canberra, 1991.

Seddon, G, *Landprints: Reflections on place and landscape*, Cambridge University Press, Cambridge, 1997.

Seddon, G, *Searching for the Snowy*, Allen and Unwin, St Leonards, NSW, 1994.

Seed, J, J Macy, P Fleming & A Naess, *Thinking like a mountain: Towards a council of all beings*, New Society Publishers, Gabriola Island, Canada, 1998.

Seibold-Bultmann, U, 'Monster soup: The microscope and Victorian fantasy', *Interdisciplinary Science Reviews*, vol. 25, no. 3, June 2000, pp. 211–19. doi.org/10.1179/030801800679242.

Sellye, J, *Beautiful machine: Rivers and the republican plan*, Oxford University Press, New York, 1991.

Servan-Schreiber, D, *Anti-cancer: A way of life*, Scribe Publications, Carlton, 2008.

Sewall, L, *Sight and sensibility: The ecopsychology of perception*, Jeremy P Tarcher/Putnam Books, New York, 1999.

Sewall, L, 'The skill of ecological perception', in T Roszak, ME Gomes & AD Kanner, *Ecopsychology: Restoring the earth, healing the mind*, Sierra Club Books, San Francisco, 1995, pp. 201–15.

Sexton, M, *Silent Flood: Australia's Salinity Crisis*, ABC Books, Sydney, 2003.

Shaw, AGL, *Gipps–LaTrobe Correspondence: 1839–1846*, Melbourne University Press at the Miegunyah Press, Melbourne, 1989.

Sheldrake, R, *The rebirth of nature: The greening of science and God*, Century, London, 1990.

Sher, K, 'Catchment management and littoral consequences: A case study approach from Victoria, Australia', BA Hons (Geography) thesis, Monash University, 1986.

Sherratt, T, L Robin & T Griffiths, *A change in the weather: Climate and culture in Australia*, National Museum of Australia Press, Canberra, 2005.

Sheriff, C, *The artificial river: The Erie Canal and the paradox of progress, 1817–1862*, Hill and Wang, New York, 1996.

Simmons, IG, *An environmental history of Great Britain: From 10,000 years ago to the present*, Edinburgh University Press, Edinburgh, 2001.

Sinclair, P, 'Making the deserts bloom: Attitudes towards water and nature in the Victorian irrigation debate, 1880–1890', MA thesis, University of Melbourne, 1994.

Sinclair, P, *The Murray: A river and its people*, Melbourne University Press, Melbourne, 2001.

Smith, DC, 'Salt marshes as a factor in the agriculture of north eastern America', *Agricultural History*, vol. 63, no. 2, Spring 1989, pp. 270–94.

Smith, V, *Clean: A history of personal hygiene and purity*, Oxford University Press, Oxford, 2007.

Solecki, WD, J Long & CC Harwell, 'Human-environment interactions in south's Florida's Everglades region: Systems of ecological degradation and restoration', *Urban Ecosystems*, vol. 3, nos 3–4, 1999, pp. 305–43. doi.org/10.1023/A:1009560702266.

Speich, D, 'Draining the marshlands, disciplining the masses: The Linth Valley hydro engineering scheme (1807–1823) and the genesis of Swiss national unity', *Environment and History*, vol. 8, no. 4, 2002, pp. 429–47. doi.org/10.3197/096734002129342729.

Speldewinde, P, A Cook, P Davies & P Weinstein, 'A relationship between environmental degradation and mental health in rural Western Australia', *Health and Place*, vol. 15, 2009, pp. 880–7. doi.org/10.1016/j.healthplace.2009.02.011.

Steenhuis, L, *Donnelly's Creek: From rush to ruin of a Gippsland mountain goldfield*, Paoletti's Maps and Videos PL, Lagwarrin, Vic., 2001.

Strang, V, *The meaning of water*, Berg Publishing, Oxford, 2004.

Synan, P, *Precious water: 100 years of reticulated supply in Sale, 1888–1988*, City of Sale, Sale, 1988.

Synan, P, *Gippsland's lucky city: A history of Sale*, City of Sale, Sale, 1994.

Synan, P, *Highways of water: How shipping on the Lakes shaped Gippsland*, Landmark Press, Drouin, 1989.

Tacey, D, *Re-enchantment: The new Australian spirituality*, Harper Collins, Sydney, 2000.

Taffs, KH, 'The role of surface water drainage in environmental change: A case example of the Upper South East of South Australia; an historical review', *Australian Geographical Studies*, vol. 39, no. 3, 2001, pp. 279–301. doi.org/10.1111/1467-8470.00147.

TeBrake, W, 'Taming the waterwolf: Hydraulic engineerings and water management in The Netherlands during the middle ages', *Technology and Culture*, vol. 43, no. 3, July 2002, pp. 475–99. doi.org/10.1353/tech.2002.0141.

Te Chow, V, 'Hydrology in Asian civilization', *Water International*, vol. 1, no. 2, 1976.

Tibby, J, 'Explaining lake and catchment change using sediment derived and written histories: An Australian perspective', *Science of the Total Environment*, vol. 310, nos 1–3, 1 July 2003, pp. 61–71. doi.org/10.1016/S0048-9697(02)00623-X.

Tindley, A, 'The Iron Duke: Land reclamation and public relations in Sutherland 1868–95', *Historical Research*, vol. 82, no. 216, May 2009, pp. 303–19. doi.org/10.1111/j.1468-2281.2007.00441.x.

Tracey, MM, 'No water no gold: Applied hydrology in nineteenth century gold mining', *Australasian Mining History Conference*, University of Melbourne, Melbourne, 1996, pp. 76–84.

Traralgon and District Historical Society Bi Monthly Bulletin, Newsletter of the Traralgon and District Historical Society.

Tolley, C, *Anglican Parish of Bairnsdale, 120th Anniversary*, Church of St John the Baptist, Barinsdale, 1987.

Tuan, Y-F, *The hydrologic cycle and the wisdom of God: A theme in geoteleology*, University of Toronto Press, Toronto, 1968.

Tvedt, T & T Oestigaard (eds), 'A history of the ideas of water: Deconstructing nature and constructing society', *History of Water, Series 2, The ideas of water from antiquity to modern times*, IB Tauris, London, 2010.

United Nations, *Millenium development goals report*, United Nations, New York, 2010.

United Nations World Commission on Environment and Development, *Our common future: The World Commission on Environment and Development*, Oxford University Press, Oxford, 1987.

van Dam, PJEM, 'Ecological challenges, technological innovations: The modernisation of sluice building in Holland 1300–1600', *Technology and Culture*, vol. 43, no. 3, July 2002, pp. 500–20. doi.org/10.1353/tech.2002.0144.

van Dam, P, 'Sinking peat bogs: Environmental change in Holland 1350–1550', *Environmental History*, vol. 6, no. 1, 2001, pp. 32–46.

Vilesis, A, *Discovering the unknown landscape: A history of America's wetlands*, Island Press, Washington DC, 1997.

Walker, G, *An ocean of air: A natural history of the atmosphere*, Bloomsbury, London, 2007.

Wardy, V, *Beneath blue hills: A history of Mewburn Park, Tinamba and Riverslea*, Kapana Press, Bairnsdale, 1994.

Warner, RF, 'Do we really understand our rivers? Or rivers in the pooh-semper in excreta', *Australian Geographical Studies*, vol. 34, no. 1, 1996, pp. 3–17. doi.org/10.1111/j.1467-8470.1996.tb00099.x.

Watson, D, *Caledonia Australis: Scottish highlanders on the frontier of Australia*, Random House, Milsons Point, NSW, 1997.

Watson, L, *Supernature: A natural history of the supernatural*, 3rd edn, Sceptre Books, Great Britain, 1986.

Watson, M, 'William Odell Raymond', *Gippsland Heritage Journal*, vol. 2, no. 2, 1987, p. 35.

Weir, J, 'Connectivity', *Australian Humanities Review*, no. 45, 2008, pp. 153–64.

Weir, J, *Murray River Country: An ecological dialogue with Traditional Owners*, Aboriginal Studies Press, Canberra, 2009.

White, R, *Inventing Australia*, Allen and Unwin, North Sydney, 1981.

White, R, *The organic machine: The remaking of the Columbia River*, Hill and Wang, New York, 1995.

Wikander, O (ed.), *Handbook of ancient water technology*, Technology and Change in History, vol. 2, Brill, Leiden, 2000.

Williams, M, *The draining of the Somerset Levels*, Cambridge University Press, Cambridge, 1970.

Williams, M, *Deforesting the Earth, from prehistory to global crisis*, University of Chicago Press, Chicago, 2003.

Willmoth, F, 'Dugdale's History of Imbanking and Drayning: A royalist antiquarian in the 1630s', *Historical Research*, vol. 71, no. 176, 1998, pp. 281–302.

Withers, CWJ, 'On georgics and geology: James Hutton's "Elements of Agriculture" and agricultural science in eighteenth-century Scotland', *Agricultural History Review*, vol. 42, no. 1, 1994, pp. 38–48.

Wittfogel, K, *Oriental despotism*, Yale University Press, New Haven, 1957.

Worster, D, *Rivers of empire: Water, aridity, and the growth of the American West*, Pantheon Books, New York, 1985.

Worster, D, *Nature's economy: A history of ecological ideas*, Cambridge University Press, Cambridge, 1994.

Wright, L, *Clean and decent: The fascinating history of the bathroom and the water closet*, Classic Penguin, England, 2000.

Index

Page numbers in *italics* indicate illustrations. Footnote numbers are indicated by 'n' following a page number.

Aboriginal people *see* Indigenous Australians
Acacia mearnsii, 141–2
Adams, J, 149, 150
Agenda 21 conference, Rio, 1992, 36
agriculture, 25, 36, 66–75
    agricultural shows, 104, 118, 204, 225, 244
    dairy industry, 119–20
    desirability of, 16–17, 66, 67, 72–5, 128, 237
    and dryness (amount and seasonality), 221, 223–5, 226
    exports, 16
    foundation of civilisation, 17
    impact on ecology, 226
    impact on landscape, 67, 68, 70, 71–2
    land clearing, 73–4, 142
    lot size, 74–5
    vs pastoralism, 66, 68–9, 73, 75
    rain, 72, 116–20, 219–20, 221
        see also rainfall
    water use, 16–17, 23, 41, 72
    wetland drainage, 80, 173, 193, 195, 196–7, 199, 202–8, 210, 259n9
    see also pastoralism
ague *see* malaria
alchemy, 22, 24, 27, 88
alcohol consumption, 65, 134
    see also temperance movement
algal blooms, 1, 4, 178
Allison, Sergeant, 52
almanacs
    astronomical phases in, 97, 99
    drainage advice, 204–5
    hot weather advice (prevention of moisture loss), 245–6
    importance to settlers, 98–9
    significant dates in, 66
    weather advice, 89, 95, 97–102
Anaxagoras of Clazomene, 23, 31, 215–16
Anaximander of Miletos, 215n9
ancient civilisations
    concepts of the hydrological cycle, 22–6
    drainage and irrigation, 25, 196
    evaporation, 215
    floods, 24–5
    see also Greek civilisation
Anthropocene, 256
Aplin, G, 35–6
aridity, 35, 238–40
    see also deserts; dry weather
Aristotle, 23, 27, 92, 215n9, 216
artesian water, 29, 64, 137, 139–40, 179, 244, 246, 249
    see also groundwater
artistic pursuits, 10, 50–6, 182, 186
Auchterlonie, G, 45, 96, 101, 107, 116, 119, 122, 140, 171, 193, 227–8, 232, 235, 236, 243

Augustine, Saint, 27
Aurora Australis, 100
Avon River, 2, 55, 112, 142, *145*, 147, 148, 152, 183, 233
    dam proposed, 252
    erosion, 156, *157*, 158
    wetlands, 195

backwater
    definition of, 171
    drownings in, 192
    drought refuge, 193
    flooding of, 194
    see also wetlands
Bairnsdale, 2, 92
    agricultural industries, 223
    common, 174
    establishment, 149
    fire, 233
    floods, 107, 108, 113
        see also floods
    irrigation, 248–51, 253
    port, 154
    rainfall, 91, 219, 228, 230
    rivalry with Sale, 2, 158
    road draining, 209
    tannery, 142, 143
    temperance meetings, 135
    temperature, 219
    water supply, 64–5, 137
Barfus, F, 188
bark hut, 52, *53*, 110, 116
Barton, MA, 205
bathing, 86n4, 102
    18th- and 19th-century practices, 65
    Roman practices, 25
Bean, Rev., 147
Benambra (Strathdownie), 90
Bible
    creation story: world designed for human benefit, 42–3
    description of God, 188
    on deserts, 218
    Ecclesiastes verse on rivers, 26–7
    imagery about precipitation, 120–2
    importance of, 27, 42
    praying for rain, 241–3
    on water, 26–7, 48–50, 120–1, 123, 133
    see also Christianity
biodiversity
    impact of dredging, 156
    introduced species, 104
    Lakes, 163–4, 178, 258, 261
    loss of, 4, 164, 259
    protection of, 163, 168, 259
    reduction in, 68, 78, 156, 164
    species threatened, 259n9
    understanding of, 10–11, 67
    vegetation loss, 68, 141
    wetlands, 168, 203
    see also ecological processes; vegetation; waterbirds
Biosphere II project, 1, 259
Blackbourn, D, 15
Blind Joe's Creek, 108
Boggy Creek, *109*
bogs *see* wetlands
boiling down process, 71–2
Bolden, L, 205
Bonyhady, T, 13
Bramah, J, 65
Bras, R, 34, 37
Bredt, P, 248–50
Breen, S, 194
Briagolong, 228, 234, 237, 248, 252
bridges, 31, 121, 130, 146–53, 197, 202
    economic value, 150, 151–2
    flood damage, 104, 110–2, 152
    Swing Bridge, Sale, xxiv, 152, *153*
Brinsmead, GSJ, 120
Brodribb, WA, 62, 81
Broome, CA (Alf), 175, 207
Broome, R, 39, 58
Brown, D, 97

Brundtland Report, 36
Buckley, PC, 96, 146, 173–4, 179, 180, 189, 206, 223, 224, 235
Bundalaguah, 150, 174, 194
Bunyan, J, *Pilgrim's Progress*, 50, 191–2
Burchett, C, 45, 224
bushfires *see* fire
Bushy Park, *71*, 148, 233, 246n128, 252
Butterfield, T, 114

Campbell, J, 152n67, 180
Campbell, M, 161–2
Campbell Pratt and Co, 127
canals, 14, 41, 57–8, 154, 161, 190
Cannon, M, 43
Cansick, P, 116, 142
Castelli, B, 29, 87n6
catchment degradation, 36, 137, 158, 161, 164, 178, 256, 259–61
    see also Gippsland Lakes catchment
Cathcart, M, 92, 256
Caughey, Miss, 45, 107, 149–50
Chambers, Rev. CJ, 188
Chappelle, FH, 64
Chinese market gardeners, 72, 110, 194, 205, 247
Chodron, P, 14–15
cholera, 40, 60, 64, 65–6, 186
Christianity, 26–30
    attitudes to Indigenous people, 81
    churches, 43–5
    hierarchical nature of, 27–8, 47
    humans separate from nature concept, 11–12, 15, 79
    importance to settlers, 43–50
    punishment and reward, 47–50
    supporting temperance, 133
    water symbolism in, 41–50
    see also Bible; divine design
Cicero, 15
Cioc, M, 34, 59, 110

clean water, health benefits of, 64–5, 71, 130, 186–7
    see also drinking water; water quality; water supply
cleanliness, 65, 191
clergy
    views on alcohol, 134–5
    views on irrigation, 242, 247
Clifton's Morass, 170, 179
climate
    climate change, 237, 240, 258
    El Nino/La Nina, 227
    and health, 60–1, 90
    perceptions of, 90–2, 160, 220, 226, 243, 257
    unpredictability of, 17, 35–6, 173–4, 240, 255
    see also drought; rainfall
clouds, 48, 88, 89, 100n50, 120–2, 251
Clydebank, 99, 110, 219, 229, 246
colonisation, moral integrity of, 43, 73–5, 83
    see also progress, ideology of
'common water', 40, 83
commons, 150, 174, 177, 193–4
connected model of the hydrological cycle, 22, 34–7, 38
connectivity, 1, 20, 34–7, 38, 260
Coode, Sir John, 154, 163
Cooper's Creek, *74*
Copeland, H, 52, 66, 190, 236
Cowarr
    backwater, 193
    Butter Factory, *239*
Crapper, T, 65
'Cringle, Tom' (William Walker), 163
Crooked River, 148, 151
crops *see* agriculture
Crossley, Ada, 54
Cumberland Plain (Sydney), 69
Cunninghame, M, 107, 160
Currie, C, 95, 116, 193, 233, 235, 242

293

dairy industry, 119–20, 224, 229–30, *239*, 260
Dalton, J, 30–1, 216
Daniel, J, 148
Dante, 186
Darcy, H, 31
Dargo, 2, 109, 148
Dargo River, 112, 117, *145*
Darwinian theory, 43
Davie, T, 31–2
Davison, G, 43
Dawson, WT, 45, 78, 103, 151, 161
De La Methiere, Jean-Claude, 30
Deakin, A, 238–9, 247, 251–2, 254
Democritus, 216
Dendy, H, 116
Descartes, 216
deserts, 48–50, 220
   Biblical references, 218
   'deserts blooming' metaphor, 218–19, 238–9
   see also aridity
destructive capacity of water, 24–5
   see also floods
diaries
   as historical sources, 19
   weather observations recorded, 95–6, 97
Dickens, C, 55, 186n66
Dingle, T, 82, 95
Dispenza, J, 47, 49
divine design
   hydrological cycle, 22, 26–30, 38, 42–3, 82, 90
   of world for human benefit, 42–3
Dow, C, 128, 175, 178, 180, 198
Dowd's Morass, 170
drainage
   for agriculture, 73, 80, 173, 195, 196–7, 199, 202–8, 210, 259n9
   in antiquity, 25, 154
   council role in, 139, 207–10

ecological impacts from, 169, 179, 258–9
failures, 203–4
to improve travel, 199, 202
practices in Europe, 33, 41, 59–60, 197
urban areas, 65, 139–40, 179, 207–10, 231
viewed as progress, 80, 208
dredging, 15–16, 156–8, 163, 258
drinking water, 30, 64–5, 71, 137, 230–1, 259n9
see also clean water; water supply
drought, 20, 69n90, 70n72
   and bushfires, 232–7
   definitions, 227
   as divine punishment, 50
   driving colonial exploration, 69–71
   duration, 215, 220
   experience of, 164, 228–9
   feelings about, 225, 227, 231–2
   impact on agriculture, 227–9, 249
   impact on settlers, 119, 231–2, 237
   as part of climate, 17
   pastoral frontier extended by, 70
   refuges from, 176, 177, 193, 198
   settlers' amelioration efforts, 241–54
   vulnerability to, 226
   water carting, 230
   see also evaporation; irrigation
dry weather, 225–37
   advantages, 220–5
   see also drought
dualism, 12, 48
   in Christianity, 47–51
   embodied/external, 6
   flowing/still waters, 48–9, 168
   in Greek medical theory, 60–1
   health/sickness, 60–1
   high/low, 54, 64, 185–6
   humanity/nature, 15

light/dark, 54, 80
movement/stillness, 48, 57, 82, 132
objective/subjective, 6–7
open/closed, 183–4
progress/stagnation, 77, 82
wet/dry, 48–50, 57
wilderness/cultivation, 77, 82
Duffy, Sir Charles Gavan, 170
DuNann Winter, D, 12, 38
Dunstan, D, 65–6
Dutch hydraulic engineering, 26n15, 34, 207

Eagle Point, 134
Eagleson, PS, 34
ecological perception, 9–12, 86, 111, 217
ecological processes, 1, 2, 4, 5–9, 20, 38, 168, 176, 189, 261
   damage to, 36, 57, 68, 70, 72, 78, 156, 161, 163–4, 258–61
   definition, 5–6, 78
   dependence on, 1, 7–8, 14
   difficulty teaching, 6–7
   impact of agriculture, 203, 226
   invisibility of, 6
   nature of, 14–15
   perception of, 9–11, 13, 78–9, 257
   time lags, 11, 25–6
   writing about, 8–9
   see also biodiversity; hydrological cycle
ecopsychology
   addiction theory, 14
   definitions of, 12
   gestalt approach, 9, 18, 20, 40
   human separation from nature (assumption), 11–12, 15, 79
   worldview, 11–12, 58, 128, 130, 136, 242, 256, 261
   see also ecological perception
Eliot, TS, 21

Ely, R, 42
Empedocles, 60–1
English, JJ, 67, 239
Erica, 86
erosion, 6, 68, 72, 156, 163–4, 168, 203, 258
   Avon River, 156, *157*, 158
   loss of riparian vegetation, 156, *157*, 258, 259n9
   Mitchell River, 158
   shoreline, 4, 163–4
estuaries, 131, 154, 160–1, 163
European Union Water Time Project, 33
evaporation, 8
   ancient beliefs, 215–16
   in Bible, 48
   defined, 214
   dual exhalation theory, 215n9, 216
   impact on agriculture, 227–30, 250
   impact on water quality, 230–1
   measurement, 217, 219
   rates of, 214, 227, 230
   scientific understanding of, 29, 30–1, *33*, 216–17
   settlers' understanding of, 217–20
   see also drought

Falkenmark, M, 35, 240
Feliks, Y, 24, 49
fen, definition of, 170
   see also wetlands
Fens (England), 62, 172–3
fire
   1851 (Black Thursday), 233, 235
   1881/82 summer, 236–7, 242
   1898, 177, 181, 233, *234*, 236, 242
   1939, 233
   2006, *236*
   2009, 233
   arson, 181
   bushfires, 17, 232–7

deaths from, 233
fear of, 233, 235
land clearing technique, 224, 233
in theory of humours, 60
weather conditions for, 232, 234, 236–7
Flegge, TL, 248, 249
Flooding Creek, 72, 111, 137, 149
see also Sale
floods, 103–14
1863, 108, 110
1870, 103–4, 107–10, 121
Brisbane, 20
damage to infrastructure, 104, 110, 112–13, 151, 152
deaths from, 110
as divine punishment, 24, 49–50, 121, 133
effect of topography, 24–5
impact on settler life, 104, 107, 112–13
measurement of, 108–11
Murray Darling, 20
rainfall and, 103–11
and sandbar at entrance to lakes, 113–14, 160–1
snow melt, 112
unpredictability, 108
warnings, 110, 112
worsened by wetland drainage works, 204
flow
experiments, 18, 30
of goods and revenue, 128, 253, 256
meanings, 131
metaphorical, 129–36, 253
positive meaning of, 85, 131–6
regularising and improving water flow, 56–60, 128–9, 144, 162–4
see also lakes, permanent entrance
runoff, 30

symbolism in religion, 133
variability, 128
see also rivers; streamflow
flowing water
benefits of, 66, 137
in defined channels, as ideal, 41
flow-dependent industries, 140–3
metaphor used by temperance movement, 133–6
motif in religion, 48–9, 133
see also rivers; streamflow; water supply
Foster, H, 253
Francis of Assisi, Saint, 27, 255
Fraser, AB, 93
Frawley, K, 11, 13
Freestone Creek, 252
fruit-growing industry, 223

Gaffney's Creek A1 Mine, *76*
Gellion, J, 127, 146, 148
Giblett, R, 50, 55
Gipps, Governor, 73, 101
Gippsland
compared to Europe, 51, 58, 257
descriptions of, 2, 90
exploration of area, 70–1, 81
maps, *xv–xxiv*
mild, dry weather benefits, 220–5
portrayed as 'garden', 239–40, 243, 254
settlement, 69, 73–4, 77, 82–3, 220, 243–4
topography, *xviii*, 73
Gippsland Lakes catchment, 2, 41
ecological condition, 1, 4, 258–61
maps, *xv–xvii*
signage, *3*
Gippsland Lakes Environmental Audit 1998, 4
Glacken, G, 61
Glenaladale, 253
Glencoe Station, 180
Glengarry, 120

gold industry, 137, 154n69, 156, 162
gold rushes, 68, 69, 75
Gorman, Mr J, 236
Gormandale, 119, 205, 243
Goyder's line, 243
grasslands, 68, 71, 74, 90, 184
Gray, F, 155
Great Chain of Being, 27
Greek civilisation
    hydrological cycle concepts, 22–5, 28–9, 38, 215–16
    medical theories, 60–1
    weather theories, 92
Gregory's Cottage, Cooper's Creek, Walhalla Road, *74*
Gregson, Flora, 10
Griffiths, T, 14, 233
Groom, AC, 202–3
groundwater, 32, 35–6, 250, 260
    artesian water, 29, 64, 139–40, 179, 244, 246, 249
    research, 30, 31
    for town water supply, 80, 139–40
    see also irrigation; soil moisture
growth *see* progress, ideology of
Gunn, N, 115
Guthridge, N, 74, 175n30

Hagenauer, FA, 81, 186–7, 189
Hales, Rev. F, 183, 188
Hales, S, 216
Halley, E, 29, 30, 216
Hanneman, Herr, 96–7
Harris, E, 173
Haslam, S, 58, 65, 129–30, 137
Hatfield, G, 93
health benefits
    of clean water, 64–5, 71, 130, 186–7, 257
        see also water supply
    of high ground, 61–2, 64, 185
    see also cholera; illness; miasma; typhoid fever; water supply; water-borne disease

The Heart (pastoral run), 72n96, 74, 75, 197, 252
The Heart Morass, 206
heatwaves, 225–6, 233, 236–7
Heberden, W, 88
Herakleios of Ephesos, 215n9
Heyfield, 108, 142
Hillman, J, 257
histories
    early Australian environmental, 13
    environmental, 5–8, 12, 260
    of hydrology, 29, 38, 240
    of rivers, 15–16, 34, 57
Hobsbawm, Eric, 7
Hobson's Crossing, 148
Holland *see* the Netherlands
Holland, G, 114
hop kilns, 117, *222*, 223
hops, 223, 228, 245, 246
Horsley brothers, 205
Horton, R, 32
Horton, T, 154
Hoskins, WG, 57
Howard, L, 88
Howitt, A, 222, 245, 246
humidity, 223
humours, theory of, 60–1
Hunter, Alick, 148
Hurd, Barbara, 1
hydraulic engineering, 26n15, 59
hydraulics, 25, 26n15, 32–4
    consequences of ignorance of, 253–4
    Roman mastery of, 25
hydrological cycle, 6–9, 13, 256
    agricultural impacts on, 36
    ancient theories of, 22–6
    beliefs of settlers about, 22, 41, 82–3, 86, 124–5, 128
    catchments, 2, 4
    elements as tool of divine punishment, 24, 49–50
    Euro-centric perceptions, 35
    global water cycle, *33*

297

interconnectivity, 34–7, 38, 260
manufacturing processes impact
    on, 16
phases of, 8
research, 30–1
'stop-go' nature, in Australia,
    35–6
vegetation role, 34
water balance equation, 31–3
wind role, 27
see also ecological processes;
    evaporation; precipitation;
    rainfall; rivers
hydrological cycle, concepts of, 21–2,
    38
ancient Greek, 22–5, 28–9, 38
Roman, 25–6
Middle Ages, 26–8
Renaissance, 28–9
18th century, 30–1, 38
19th century, 30–1, 38, 256
20th century, 29, 31–7
21st century, 31–7
hydrological cycle, models of
bias to flow, 35
connected model, 22, 34–7, 38
divine design model, 22, 26–30,
    38, 82, 90
prejudice against aridity, 35
rainfall-driven, quantitative
    model, 22, 23–4, 30–4
underground purification model,
    22, 23–4, 27, 29
vertical model, 22, 23–4, 26,
    27–8, 38
hydrology, 22, 88
see also hydrological cycle
hygiene *see* sanitation
hyper-separation, 48

illness
effects of cold and wet, 114–15
water-borne disease, 60–6, 136,
    186

Indigenous Australians
employment, 206
experience of drought, 226–7
land management practices, 67
missions, 189–90
relationship with environment,
    14, 23, 67, 97, 226–7, 258
settler attitudes to, 14, 81, 189–90
settler violence towards, 81, 180–1
surface waters, use of 128
use of wetlands, 128, 181, 182
water dreaming, 37
weather lore, 94
see also massacres
Industrial Revolution, 15, 31, 79, 256
industrialisation, 33–4
infrastructure *see* water infrastructure
irrigation
in antiquity, 25, 196
Chinese expertise, 247
failures, 253–4
linked to flow of goods and
    revenue, 253
lobbying for, 243, 247–53
market gardens, 72, 247
response to drought, 238, 246–54
Royal Commission, 238, 246,
    247, 248–51
and salinity, 25
trusts, 248, 251, 253
Isidore, Bishop of Seville, 121

Jankovic, V, 87
Johnston, D, 101, 114, 115, 118,
    140, 148
Jones, G, 249, 252
Judaeo-Christian traditions, 15, 42
see also Christianity

Karterakis, SM, 6
King, Dr, 90
King, J, 80, 179, 181, 189, 239, 243
knowledge, 25, 39, 40, 87, 93
cultural, 217

embodied, 6, 7, 217
formal, 217
objective and subjective, 6–7, 8
Koo-Wee-Rup Swamp, 199, 202, 204
Kurnai people *see* Indigenous Australians

La Trobe, Governor, 101
Lacey, S, 139, 244–5
lakes
definition of, 169
permanent entrance, 154, 160, 161, 162, 163–4, 195
sandbar, 108, 113–14, 158, *159*, 160–1
see also wetlands
lakes (specific lakes)
Guthridge, 169
King, 2, 169, 170, 171, 175
Victoria, 169, 175
Wellington, 2, 169, 175, 195, 205
see also Gippsland Lakes catchment
Lakes Entrance, 2, *163*, *261*
wildflowers, *10*
land clearing, 73–4, 142, 184–5, 224, 233
land selection, 73, 74, 81, 82, 95, 119, 137, 162, 184, 194
landscape
Australian/British contrasts, 56–8, 75, 77, 82
concept of, 5, 41, 51
medical geography concepts, 61–4
perceptions of, 13–14, 257
settler preferences, *44*, 56–7, 60, 73, 82, 183–5, 240
vegetation loss, *xxii–xxiii*, 68, 73–4, 78, 156
'war' on, by settlers, 66–75
wilderness, 75, 77, 81
see also deserts; lakes; mountains; rivers; vegetation; wetlands
Landy, M, 213, 228, 242, 248, 249, 250, 251, 252, 254

Latrobe River, 2, *105*, 144, 149, 151, 152, *153*, 156, 195, 199, 204, 258
leather *see* tanning industry
Lee, RL, 93
Legg, S, 73, 119
Leonardo da Vinci, 28–9
Leukipos, 216
libraries, 51, 79n121, 124, 132
Licola, 92
Linn, R, 66
Linton, J, 33, 35
literature, 50–6, 124, 131
Livin, 144
Livingstone Creek, 113, *144*
Login, Rev. WS, 124
London, sewerage, 65
Longford, 134n19, 142, 150n61, 174n26
Loy Yang, 151
Lunt, I, 39, 68

Macalister River, 2, 148, 248, 249
Macalister Swamp, *177*
McArdell, P, 161
McCann, M, 45, 122, 207, 223, 225n43
McDonald, Dr, 138
McIntosh, A, 149
MacKillop, G, 90
McLaren, C, 181
McLean, A, 253
McLeary, A, 82, 95
MacLeod, D, 171
MacLeod, I, 95, 171
MacLeod's Morass, 170, *183*
McMillan, Aleck, 207
McMillan, Angus, 70–1, 90, 117, 147–8, 181, 183, 233, 243
McNeill, JR, 257
Maffra, *69*, 92, 119, 150, 208, 228, 230, 237, 243, 253
Main, G, 67
malaria, 60–4

Manchester (Duke of), 67
marsh gas, 179n37
marshes *see* wetlands
massacres, 180–1
Mechanics Institutes, 181
    role of, 51, 79
    Sale, 79, 124, 132
medicinal plants, 58
Melaleuca ericifolia, 164
Melbourne
    Gippsland's isolation from, 2
    public health problems, 65–6
    rainfall (1870), 107
mental health impacts of drought, 231–2, 237
mere, definition of, 170
    see also wetlands
Merriman's Creek, 119, 243
meteorology, 87–9, 125
    astronomical theory of, 97–100
    development in Australia, 89
    equipment, 87, 89, 97, 101
        see also rain gauges
    forecasting *see* weather forecasts
    observation stations, 89–90
    rainfall data *see* rain gauges; rainfall
    record keeping, 87, 101, 219
    traditional weather lore, 92–4, 96–7
Meyrick, H, 114
miasma, 60–1, 64, 170, 186
Middle Ages
    Christian ideology, 26–7
    hydrological cycle concepts, 26–8
Middleton, WEK, 88, 100
Mingay, GE, 173
mining, *76*
    flood impact on, 109, 113
    settlements, 2
    tailings, 113, 137, 156
    threat to agrarian ideals, 75
    see also gold rushes
Mirboo, cleared land, *184*
mist, 100n50

Mitchell River, 2, 71, 127, 133, 142, 149, 156, 158, 246, 248, 249
Moe, 114, 149, 210
Moe Flat, *203*
Moe River, 199
Moe Swamp, 177, 181, 188, 199–204
Montgomery, E, 148
moon's influence on weather, 97–101
Moorhouse, Bishop, 79, 242, 247–8
morasses, 170, 179, *183*, 206
    see also wetlands
Morwell, 151, 208, 230
Mosquito Creek, *63*
mosquitoes, 61, 62
mountains
    healthy air of, 62, 64, 185
    summer retreats, 61, 62
mud, 127, 147, 167, 170, 180, 188, 191–2, 208, 210, 213
music, 19, 50–6
Myers, F, 238

Nace, R, 23, 28
Narracan, 73, 108, 114, 119
nature
    connectivity in, 7, 34–7, 38, 260
    control of, 14–17, 75–82, 162, 238, 255–6, 258
    fear of, 24
    hostile, 77
    human separation from nature, 11–12, 15, 79
    learning about, 39–40
    as a resource, 74
        see also rainfall-driven, quantitative model; water
    technology to transcend, 15–16
    variability, 89
    see also ecological processes; landscape; romanticism; wilderness
Neerim, bark hut, *53*
the Netherlands, 26n15, 34, 59
Nicholson River, 2, 175

INDEX

Niemann, JA, 139, 248–9
Nightingale, F, 65
Nuntin Creek, 150

O'Connor, J, 82
O'Connor, M, 114
Oelschlaeger, M, 77
Omeo, 2

pagan allusions, 86
Palmer, W, 243, 249
pastoral runs
    Bushy Park, *71*, 148, 233, 246n128, 252
    The Heart, 72n96, 74, 75, 197, 206, 252
pastoralism, 66–75
    vs agriculture, 66, 68–9, 73, 75
    ecological impacts of, 68–72
    impact of drought, 230, 249
    over-exploitation of land, 68–70
    squatting runs, *xix–xx*, 73
    water use, 71–2
    water-quality impacts, 71–2
    wetland use, 173–4, 180, 198–9
    see also agriculture
Pearson, W, 199–200, 246
Peppercorne, FS, 89
Perkins, M, 95, 97
Perrault, P, 30, 31, 216
    On the Origin of Springs, 29, 88
planetary observances, 97–8
plants
    domestic and medicinal uses, 58
    time-telling by, 93
    weather forecasting by, 96–7
    see also vegetation
Plato, 23, 27
Pliny, 23
poetry, 50–6, 123
    by settlers, 159–60
Poowong, 45
Port Albert, 81, 189
    canal proposed, 161

potable water *see* drinking water
Powell, JM, 13
praying for rain, 241–3
precipitation, 87–9
    experience of, 6, 7, 11
    imagery and affect, 120–4
    measurement, 87
        see also rain gauges
    and streamflow, 30, 31
    see also rainfall; snow
process, definitions of, 5
processes, ecological *see* ecological processes
progress, ideology of, 13, 14, 38, 75–82, 256–60
    18th-century development of idea, 77–8
    definitions of progress, 77–81
    dependence on water for, 4, 71–2, 74, 250
    dualism, 77, 82
    measures of, 79–80
    newspaper coverage of progress, 79–81
    signs of, 78–82
property ownership and wetland changeability, 193–4
Prout, A, 175
psychology, traditional vs ecopsychology, 12
    see also ecopsychology
Punt Lane, Sale, *xxiv*, 104, 111, 181, 192
Pyne, S, 224

railways, 58, 132, 159, 253
    impact in Britain, 58
    reserves, 68
    sign of progress, 14, 131–2
'rain follows the plough' debate, 219–20
rain gauges, 29, 87, 88, 107
rainbows, 93
rainfall, 30
    and agriculture, 72, 116–20, 221
        see also agriculture

301

data, 91, 101, 250
effects of rain (beneficial), 102, 113–14, 118–20, 221
effects of rain (negative), 102–17, 221, 223
expectations of settlers, 228
experience of, 6, 7, 11
imagery and affect, 120–4
measurement, 87–9, 107, 219
praying for, 241–3
protection from cold and wet, 114–17
research, 88–9
source of rivers, 25, 88
timing, 118–19, 122
variability, 72, 85–6, 91–2, 96, 124–5, 128, 228–9, 240
see also floods
rainfall-driven, quantitative model of the hydrological cycle, 22, 23–4, 30–4
rainshadow, 92, 228, 243
Ramahyuck Mission, 189–90, 246
Ramsar, 163, 168
Rawlinson, TE, 144, 156
Ray, J, *The Wisdom of God*, 30
Raymond, WO, 55
Reeves, J, 161
Renaissance concepts of the hydrological cycle, 28–9
Rhine River, 15, 59, 110, 195
riparian vegetation, 144, 164
    domestic and medicinal uses, 58
    loss of, 156, *157*, 258, 259n9
rivers
    Bible references, 26–7, 48
    characteristics of Gippsland rivers, 144, 146–7
    crossing points and bridges, 146–53
        see also bridges
    dredging and snagging, 15–16, 156
    flow metaphors in religion, 48–9, 132
    health audit, 259

historic uses in Europe, 58–9, 130, 154–5
histories of, 15–16, 34, 57, 58–9
riparian vegetation, 58, 144, 156, *157*, 164, 259n9
sandbars at river mouths, 158
settlement choices and patterns, 58–9
theories on origin of, 22–9, 88
trained and transformed, 16, 56–60, 128–9, 154–5, 156, 164–5, 258
transport role, 59, 67, 152, 154–6
see also floods; flowing water; streamflow; and specific rivers: Avon; Crooked; Dargo; Latrobe; Macalister; Mitchell; Moe; Nicholson; Rhine; Tambo; Thames; Thomson; Tiber; Wonnangatta
roads
    flood damage, 104
    poor standard of, 58, 67, 104, *106*, 107, 159, 188, 190–1, 209–10
roadmaking, 150–1, 209
Robinson, GA, 220
Robinson, Kate, 192
Rodgers, J, 159–60
Rolls, E, 11
Roman concepts of the hydrological cycle, 25–6
romanticism, 51
Rome, malaria infections and death rate, 61
Rose, D, 37
Rosedale, 120, 142
Ross, Councillor, 46
Ross, Sir Ronald, 61
Roth, L, 51, 124
Royal Commission on Water Supply, 238, 243, 246, 247, 248–51
Royal Societies, 77, 88
runoff, 30, 32, 161, 231, 245
Rupp, R, 24

Sadleir, J, 151
Sale, *xxi*, 2
  buildings, 52, *76*
  climate, 90
  establishment, 149
  floods, 104, 107, 110–11, 112, 151
  grazing common, 174, 177, 193–4
  market gardens, 72
  Mechanics institute, 79
  port, 154
  public health problems, 66
  Punt Lane, *xxiv*, 104, 111, 181, 192
  rainfall, 91
  rivalry with Bairnsdale, 2, 158
  road draining, 209
  Swing Bridge, xxiv, 152, *153*
  tannery products trade, 142
  water supply, 33, 66, 80, 137, 138–40
  see also floods
salinity, 25, 163, 258
sanitation, 34, 65–6
Savige, Frank, 236
science
  Darwinian theory, 43
  experiments, 29, 216
  exploration, 11
  scientific instruments, 29, 87
  scientific method, 29
  tool for progress, 77, 79–80, 197, 256–9
  see also hydrology; meteorology; thermodynamics
Scotland
  progress societies, 77
  settlers from, 54, 170
sea water as source of rivers (theory), 30, 88, 215n9
sedimentation, 4, 156, 158
Sewall, Laura, 9–10
sewerage *see* water infrastructure

Shakespeare, W, 55, 124
sheep industry
  boiling down, 71–2
  see also pastoralism
Sheepwash Creek, 108
ship groundings, 158, *159*
shipping, 159, 161–2
sickness *see* water-borne disease
silt jetties, 2
Simmons, IG, 56
'slough of despond' metaphor, 50, 191–2
Small, Davy, 52
Smith, Adam, 77
snow, 103, 109, 123
  snow melt, 112, 249
Snow, J, 64
soil moisture, 111–12, 175, 221, 240, 250
  see also groundwater
squatting industry *see* pastoralism
St Patrick's Creek, 148
stagnant water, 62, 63, 171
stillness (water), valued in European history, 187–8
  see also wetlands
Stradbroke, 207
Strang, V, 7, 40
Stratford, 2, 55, 91, 92, 142, 246
Strathdownie (Benambra), 90
streamflow, 30, 31, 35
Stringer's Creek, 138
Strzelecki, EP, 70, 213, 220
sun/sunshine, 48, 54, 122, 123
surface waters, European norm (blue water bias), 35, 240
  see also lakes; rivers; wetlands
swamps *see* wetlands
Swan Reach, 114
Swift's Creek, 92
Swing Bridge, Sale, *xxiv*, 152, *153*
Synan, P, 51, 78

tallow, 71n96
Tambo River, 2, 115, 148, 156, *229*
Tambo Valley, *44*

Tanjil Bren, 91
tanning industry, 71–2, 140–3, 223
Tarrago Creek, *3*
Taylor, J, 250
TeBrake, W, 26n15, 34
temperance movement, 133–6
temperature
  heatwaves, 225–6, 233, 236–7
  measurement, 219
Tennyson, A (Alfred, Lord Tennyson), 55, 131
Thales of Miletos, 23, 215n9
Thames River, 65, 155n70
thermodynamics, advances in, 88–9
Thomson, R, 99
Thomson River, 2, 138, 144, 248
Tiber River, 61
Tiddalik the frog (myth), 41–2, 227
time lags in ecological processes, 11, 25–6
Toongabbie, 92, 210, 237, 241
Townley, R, 87
trade, within empire, 67, 240, 257
Traralgon, 46n24, 55, 80, 92, 204, 208
travel and transport, 158–9, 260
  impediments to, 188–9, 199, 202
  by river, 59, 67, 152, 154, 156
  see also canals; railways; roads; shipping
Tuan, Yi-Fu, 27, 29–30, 240
Tyers, 206
Tyers, C, 159, 199, 202
typhoid fever, 64, 66, 186, 187

underground purification model of the hydrological cycle, 22, 23–4, 27, 29
United Nations
  Agenda 21, 36
  Brundtland Report, 36
  World Commission on Environment and Development, 36

Vallisnieri, A, 88
vegetation
  adaptation to drought, 35
  changes to, 258
  in hydrology studies, 34
  loss of, xxii–xxiii, 68, 73–4, 78, 164
  structural difference, 183
  see also grasslands; plants
vertical model of the hydrological cycle, 22, 23–4, 26–8, 38
Vileisis, A, 193
violence, 81, 180–2
Vitruvius (Marcus Vitruvius Pollio), 25
von Guerard, Eugene, 71
Von Mueller, Ferdinand, 115

Walhalla, 2, *8*, *56*, 72, 138, 210, 231
Walker, William (Tom Cringle), 163
Warner, RF, 258
Warragul, 92, 188, *209*, 228, 230
waste disposal, 71–2, 130, 137–8, 140, 179, 231
  see also water pollution
water, 7
  beliefs of settlers about, 22, 41, 83
  Bible references, 26–7, 48–50, 133
  carting of, 230
  cleansing properties of, 40, 64, 133, 135, 137
  descriptions and metaphors, 121, 124
  destructive capacity, 24–5
  drinking water, 30, 64–5, 71, 137, 230–1, 259n9
  embodied, 16, 17, 214, 227
  essential to life, 7, 41, 214, 227
  explorers' perceptions, 70
  flow-dependent industries, 140–3
  interactions with, 40–1
  in literature and music, 50–6, 186
  qualities of, 36–7, 139
  in religious traditions, 41–50, 133, 135

as a resource, 33, 67–8, 70–2, 74, 75, 256, 257, 260
stagnant, 62, 63, 168, 171
  see also wetlands
stillness, valued, 187–8
symbolism in Christianity, 41–50
transport role, 57–60, 67, 152, 154, 156
ubiquity, 9, 40
utility of, 34, 40
water balance equation, 31–3
water infrastructure, 14, 15, 33–4
  in ancient societies, 25
  dams proposed, 249, 252
  for sanitation, 34, 65–6, 179
  waterwheels, 154–5
water policy, 20, 32, 169, 259
water pollution, 6, 113, 137–8, 258
  from tanning industry, 71–2
  Thames River, 65
  see also cholera
water quality, 33, 64–5, 71–2, 176, 203, 230–1
water supply, 64–6, 80, 136–40
  Bairnsdale, 64–5, 137
  Sale, 33, 66, 80, 137, 138–40
Water Time Project, 33
water vapour, 22, 30, 31, *33*, 88–9
waterbirds, 113, 168, 175, *176*, 178, 261
water-borne disease, 60–6, 136, 186
waterhole, definition of, 170–1
  see also wetlands
Waterhole Creek, 151
waterwheels, 154–5
Watson, Rev. Canon, 121–2
wattle bark, 141–2, 189
Watts, Rev. J, 45
weather
  advice in almanacs, 97–102
  observation by settlers, 95–6
  traditional weather lore, 92–4, 96–7
  see also floods; rainfall
weather forecasts, 94, 96
  planetary observances, 97–100
  from plant behaviour, 96–7
weather observation stations, 89–90
  see also meteorology
Weir, J, 48
wells *see* artesian water
West Gippsland, *xvii*
wet weather, 223
  protection from cold and wet, 114–17
  too wet, 103–14
  welcome rain, 118–20
  see also floods; rainfall
wetlands, *115*, 164, *177*, *183*
  Aboriginal use of, 128, 181, 182
  algal blooms, 1, 4, 178
  biodiversity, 168, 203
  changeability, 193–4
  danger, 179–80, 192–3
  definitions, 168–71
  draining of, 59–60, 80, 169, 172–3, 195, 199–204, 210–11, 259n9
  drought refuges, 176, *177*, 193, 198
  ecological functions, 176, 203, 210–11
  enclosure of, 173
  and illness, 60–6
  impediment to travel, 188–9
  Iron age use, 172
  language about, 169–72
  metaphor for spiritual wilderness, 47, 48, 50, 191–2
  odour, 178–9, 186
  perceptions of, 169, 170, 171–2, 178–9, 182–96, 198
  places of refuge and shelter, 176, 178, 179–82
  and property systems, 193–4
  Ramsar, 163, 168
  uses of, 172–82
  violence in, 180–2
  and water quality, 176, 203
  waterbirds, 113, 168, 175, *176*, 178, 259n9

305

White, G, 87
White, T, 114
Wikander, O, *Handbook of Ancient Water Technology*, 25
wilderness, 75, 77, 81
    taming of *see* progress, ideology of
Wilson, D, 79
wind, 27, 100n50
windbreaks, *244*, 245
Wombat Creek, *146*
women, 45, 52
Wonnangatta River, 148
Wordsworth, W, 124
Wren, C, 87
Wurruk, 142, 200

Xenophanes of Colophon, 23

Yan Yean, 66
Yarragon, 188, 199, *200–1*
yeoman farmer ideology, 73, 74–5, 238, 254
Yinnar church, *44*

www.ingramcontent.com/pod-product-compliance
Lightning Source LLC
Chambersburg PA
CBHW061254230426
43665CB00027B/2934